Handbook of Life Cycle Assessment (LCA) of Textiles and Clothing

The Textile Institute and Woodhead Publishing

The Textile Institute is a unique organisation in textiles, clothing and footwear. Incorporated in England by a Royal Charter granted in 1925, the Institute has individual and corporate members in over 90 countries. The aim of the Institute is to facilitate learning, recognise achievement, reward excellence and disseminate information within the global textiles, clothing and footwear industries.

Historically, The Textile Institute has published books of interest to its members and the textile industry. To maintain this policy, the Institute has entered into partnership with Woodhead Publishing Limited to ensure that Institute members and the textile industry continue to have access to high calibre titles on textile science and technology.

Most Woodhead titles on textiles are now published in collaboration with The Textile Institute. Through this arrangement, the Institute provides an Editorial Board which advises Woodhead on appropriate titles for future publication and suggests possible editors and authors for these books. Each book published under this arrangement carries the Institute's logo.

Woodhead books published in collaboration with The Textile Institute are offered to Textile Institute members at a substantial discount. These books, together with those published by The Textile Institute that are still in print, are offered on the Elsevier website at http://store.elsevier.com/. Textile Institute books still in print are also available directly from the Institute's website at www.textileinstitutebooks.com.

A list of Woodhead books on textile science and technology, most of which have been published in collaboration with The Textile Institute, can be found towards the end of the contents pages.

Related titles

Assessing the Environmental Impact of Textiles and the Clothing Supply Chain
(ISBN 978-1-78242-104-7)

The Global Textile and Clothing Industry: Technological advances and Future Challenges
(ISBN 978-1-84569-939-0)

Handbook of Sustainable Textile Production
(ISBN 978-0-85709-136-9)

Woodhead Publishing Series in Textiles:
Number 172

Handbook of Life Cycle Assessment (LCA) of Textiles and Clothing

Edited by

Dr Subramanian Senthilkannan Muthu

The Textile Institute

ELSEVIER

AMSTERDAM • BOSTON • CAMBRIDGE • HEIDELBERG
LONDON • NEW YORK • OXFORD • PARIS • SAN DIEGO
SAN FRANCISCO • SINGAPORE • SYDNEY • TOKYO
Woodhead Publishing is an imprint of Elsevier

WP
WOODHEAD PUBLISHING

Published by Woodhead Publishing in association with the Textile Institute
Woodhead Publishing is an imprint of Elsevier
80 High Street, Sawston, Cambridge, CB22 3HJ, UK
225 Wyman Street, Waltham, MA 02451, USA
Langford Lane, Kidlington, OX5 1GB, UK

Notices

Knowledge and best practice in this field are constantly changing. As new research and experience broaden our understanding, changes in research methods, professional practices, or medical treatment may become necessary.

Practitioners and researchers must always rely on their own experience and knowledge in evaluating and using any information, methods, compounds, or experiments described herein. In using such information or methods they should be mindful of their own safety and the safety of others, including parties for whom they have a professional responsibility.

To the fullest extent of the law, neither the Publisher nor the authors, contributors, or editors, assume any liability for any injury and/or damage to persons or property as a matter of products liability, negligence or otherwise, or from any use or operation of any methods, products, instructions, or ideas contained in the material herein.

ISBN: 978-0-08-100169-1 (print)
ISBN: 978-0-08-100187-5 (online)

British Library Cataloguing-in-Publication Data

A catalogue record for this book is available from the British Library

Library of Congress Control Number: 2015939622

For information on all Woodhead Publishing publications visit our website at http://store.elsevier.com/

Contents

List of contributors **xi**
Woodhead Publishing Series in Textiles **xiii**
Introduction **xxi**

Part One Key indicators and processes in LCA **1**

1 Carbon footprints in the textile industry **3**
G. Peters, M. Svanström, S. Roos, G. Sandin, B. Zamani
1.1 Introduction **3**
1.2 Key issues in calculating carbon footprints **5**
1.3 Carbon footprint standards **9**
1.4 Calculating carbon footprints of textile manufacturing processes **12**
1.5 Calculating carbon footprints of textile products **16**
1.6 Carbon footprint labelling of textile products **21**
1.7 Future trends **22**
1.8 Conclusions **23**
1.9 Further information and advice **24**
Acknowledgement **25**
References **26**

2 Energy footprints in the textile industry **31**
S. Palamutcu
2.1 Introduction: key issues in calculating energy footprints **31**
2.2 Methodologies and standards **41**
2.3 Calculating energy footprints of textile products (applications and examples) **44**
2.4 How to reduce energy footprint in textile and clothing products **47**
2.5 How to reduce energy footprint in textile and clothing supply chain **52**
2.6 Energy analysis in the textile industry **53**
2.7 Energy analysis from the view of environmental approaches **54**
2.8 Future trends **55**
2.9 Conclusions **56**
2.10 Additional readings about energy in the textile industry **57**
References **57**

3 Ecological footprints in the textile industry **63**
E. Roca, D.M. Herva
3.1 Introduction **63**
3.2 Methodologies and standards **65**
3.3 Calculating ecological footprint of a textile manufacturing plant **67**
3.4 Calculating ecological footprints of textile products: evaluation of footwear as case study **74**
3.5 Future trends and challenges in calculating the EF of textile products **78**
3.6 Conclusions **79**
Acknowledgments **80**
References **80**

4 Measuring the reusability of textile products **83**
S.S. Muthu
4.1 Introduction **83**
4.2 Importance of reusability in LCA **84**
4.3 Implications of reusability in textiles and clothing products **84**
4.4 Importance of reusability in LCA studies **85**
4.5 Quantification of reusability of textiles and clothing products **87**
4.6 Future trends and remarks **90**
4.7 Conclusions **90**
References **91**
Further reading **92**

5 Environmental impacts of the use phase of the clothing life cycle **93**
S.S. Muthu
5.1 Introduction: the use phase in the life cycle of clothing **93**
5.2 Quantifying the use phase of the clothing life cycle **93**
5.3 Consumer use phase in LCA studies of textiles and clothing products – implications for LCA **98**
5.4 Challenges and conclusions **100**
References **100**
Further reading **101**

6 Open- and closed-loop recycling of textile and apparel products **103**
A. Payne
6.1 Introduction **103**
6.2 Recycling textiles and apparel **103**
6.3 Open-loop recycling **106**
6.4 Closed-loop recycling **111**
6.5 Barriers to effective OLR and CLR **117**
6.6 Future trends in the apparel industry **118**
6.7 Conclusions **120**
List of abbreviations **121**
References **121**

7 Life cycle assessment method for environmental impact evaluation and certification systems for textiles and clothing **125**
S.H. Eryuruk
7.1 Introduction **125**
7.2 Life cycle assessment and environmental impacts of textiles and clothing **126**
7.3 Standards and associations for environmental management systems **133**
7.4 Eco-labelling and certification of textiles and clothing **136**
7.5 Conclusions **144**
References **144**

8 Environmental impact assessment methods for textiles and clothing **149**
B. Resta, S. Dotti
8.1 Introduction **149**
8.2 Review of EIA process and methods **151**
8.3 An EIA tool kit for the textile and clothing industry: the SustainTex system **159**
8.4 Illustrative example **173**
8.5 Conclusions and future trends **186**
References **188**

Part Two Applications of LCA in the textile industry **193**

9 Life cycle assessment of cotton textiles and clothing **195**
S. Rana, S. Karunamoorthy, S. Parveen, R. Fangueiro
9.1 Introduction **195**
9.2 LCA of cotton textiles **201**
9.3 LCA of technical cotton textiles **207**
9.4 Discussion of LCA results **211**
9.5 Factors influencing sustainability of cotton textiles **212**
9.6 Conclusions **214**
References **215**

10 LCA of wool textiles and clothing **217**
B.K. Henry, S.J. Russell, S.F. Ledgard, S. Gollnow, S.G. Wiedemann, B. Nebel, D. Maslen, P. Swan
10.1 Introduction **217**
10.2 Wool LCA methodological and data challenges **224**
10.3 LCA case study for wool apparel **235**
10.4 Summary and future developments in wool LCA studies **249**
References **250**

11 Life cycle assessment of silk production – a case study from India **255**
M.F. Astudillo, G. Thalwitz, F. Vollrath
11.1 Introduction **255**
11.2 LCI of Indian silk **259**
11.3 Results and discussion **264**
11.4 Conclusions **270**
References **270**

12 Comparative life cycle assessment of natural and man-made textiles **275**
S.S. Muthu
12.1 Introduction **275**
12.2 Natural and man-made textiles **276**
12.3 LCA studies **276**
12.4 Concluding remarks **280**
References **280**
Further reading **281**

13 LCA of cotton shopping bags **283**
S.S. Muthu
13.1 Introduction **283**
13.2 Shopping bags and their environmental impacts **284**
13.3 Cotton shopping bags **285**
13.4 LCA studies on cotton shopping bags **287**
13.5 Conclusions **297**
References **298**

14 LCA of fibre-reinforced composites **301**
A.D. La Rosa, G. Cicala
14.1 Introduction **301**
14.2 Types of fibre-reinforced composites **303**
14.3 LCA method and models **306**
14.4 LCA of composites made out of different materials: examples and case studies **311**
14.5 Conclusions **318**
Acknowledgment **318**
References **318**

15 Life cycle assessment and the environmental and social labels in the textile and clothing industry **325**
M. Koszewska
15.1 Environmental and social labels: definitions **325**
15.2 Classifications of social and environmental labels **327**
15.3 Life cycle assessment and greenwashing in textiles and clothing **331**

15.4 A comparison of the selected environmental and social labels in terms of LCA **333**
15.5 The utility of environmental and social labels for consumers – barriers and challenges **335**
15.6 Future trends and concluding remarks **340**
References **342**
Sources for further information **344**

16 Clothing disposal habits and consequences for life cycle assessment (LCA) **345**
K. Laitala, I.G. Klepp
16.1 Introduction **345**
16.2 Methods **350**
16.3 Results **354**
16.4 Discussion **359**
16.5 Conclusions **361**
Acknowledgments **362**
References **362**

Index **367**

List of contributors

M.F. Astudillo University of Oxford, Oxford, UK

G. Cicala University of Catania, Catania, Italy

S. Dotti University of Bergamo, Dalmine, Bergamo, Italy

S.H. Eryuruk Istanbul Technical University, Istanbul, Turkey

R. Fangueiro Fibrous Materials Research Group, School of Engineering, University of Minho, Guimarães, Portugal

S. Gollnow PE International Australasia, Wellington, New Zealand

B.K. Henry Queensland University of Technology, Brisbane, Queensland, Australia

D.M. Herva Torres Quevedo Program at Cetus Group, Cerceda, A Coruña, Spain

S. Karunamoorthy PSG Institute of Advanced Studies, Coimbatore, India

I.G. Klepp National Institute for Consumer Research (SIFO), Oslo, Norway

M. Koszewska Lodz University of Technology, Lodz, Poland

K. Laitala National Institute for Consumer Research (SIFO), Oslo, Norway

A.D. La Rosa University of Catania, Catania, Italy

S.F. Ledgard AgResearch Ruakura Research Centre, Hamilton, New Zealand

D. Maslen The New Zealand Merino Company Limited, Christchurch, New Zealand

S.S. Muthu Global Sustainability Services, SGS Hong Kong Limited, Hong Kong

B. Nebel PE International Australasia, Wellington, New Zealand

S. Palamutcu Pamukkale University, Denizli, Turkey

S. Parveen Fibrous Materials Research Group, School of Engineering, University of Minho, Guimarães, Portugal

A. Payne Queensland University of Technology, Brisbane, Queensland, Australia

G. Peters Chalmers University of Technology, Gothenburg, Sweden

S. Rana Fibrous Materials Research Group, School of Engineering, University of Minho, Guimarães, Portugal

B. Resta University of Bergamo, Dalmine, Bergamo, Italy

E. Roca Sustainable Processes and Products Engineering Group (PRODES), School of Engineering, University of Santiago de Compostela, Santiago de Compostela, Spain

S. Roos Chalmers University of Technology, Gothenburg, Sweden; Swerea IVF, Mölndal, Sweden

S.J. Russell University of Leeds, UK

G. Sandin Chalmers University of Technology, Gothenburg, Sweden; SP Technical Research Institute of Sweden, Borås, Sweden

M. Svanström Chalmers University of Technology, Gothenburg, Sweden

P. Swan Australian Wool Innovation Limited, Sydney, Australia

G. Thalwitz University of Oxford, Oxford, UK

F. Vollrath University of Oxford, Oxford, UK

S.G. Wiedemann FSA Consulting, Toowoomba, Queensland, Australia

B. Zamani Chalmers University of Technology, Gothenburg, Sweden

Woodhead Publishing Series in Textiles

1 **Watson's textile design and colour Seventh edition**
Edited by Z. Grosicki
2 **Watson's advanced textile design**
Edited by Z. Grosicki
3 **Weaving Second edition**
P. R. Lord and M. H. Mohamed
4 **Handbook of textile fibres Volume 1: Natural fibres**
J. Gordon Cook
5 **Handbook of textile fibres Volume 2: Man-made fibres**
J. Gordon Cook
6 **Recycling textile and plastic waste**
Edited by A. R. Horrocks
7 **New fibers Second edition**
T. Hongu and G. O. Phillips
8 **Atlas of fibre fracture and damage to textiles Second edition**
J. W. S. Hearle, B. Lomas and W. D. Cooke
9 **Ecotextile '98**
Edited by A. R. Horrocks
10 **Physical testing of textiles**
B. P. Saville
11 **Geometric symmetry in patterns and tilings**
C. E. Horne
12 **Handbook of technical textiles**
Edited by A. R. Horrocks and S. C. Anand
13 **Textiles in automotive engineering**
W. Fung and J. M. Hardcastle
14 **Handbook of textile design**
J. Wilson
15 **High-performance fibres**
Edited by J. W. S. Hearle
16 **Knitting technology Third edition**
D. J. Spencer
17 **Medical textiles**
Edited by S. C. Anand
18 **Regenerated cellulose fibres**
Edited by C. Woodings
19 **Silk, mohair, cashmere and other luxury fibres**
Edited by R. R. Franck

20 **Smart fibres, fabrics and clothing**
Edited by X. M. Tao
21 **Yarn texturing technology**
J. W. S. Hearle, L. Hollick and D. K. Wilson
22 **Encyclopedia of textile finishing**
H-K. Rouette
23 **Coated and laminated textiles**
W. Fung
24 **Fancy yarns**
R. H. Gong and R. M. Wright
25 **Wool: Science and technology**
Edited by W. S. Simpson and G. Crawshaw
26 **Dictionary of textile finishing**
H-K. Rouette
27 **Environmental impact of textiles**
K. Slater
28 **Handbook of yarn production**
P. R. Lord
29 **Textile processing with enzymes**
Edited by A. Cavaco-Paulo and G. Gübitz
30 **The China and Hong Kong denim industry**
Y. Li, L. Yao and K. W. Yeung
31 **The World Trade Organization and international denim trading**
Y. Li, Y. Shen, L. Yao and E. Newton
32 **Chemical finishing of textiles**
W. D. Schindler and P. J. Hauser
33 **Clothing appearance and fit**
J. Fan, W. Yu and L. Hunter
34 **Handbook of fibre rope technology**
H. A. McKenna, J. W. S. Hearle and N. O'Hear
35 **Structure and mechanics of woven fabrics**
J. Hu
36 **Synthetic fibres: Nylon, polyester, acrylic, polyolefin**
Edited by J. E. McIntyre
37 **Woollen and worsted woven fabric design**
E. G. Gilligan
38 **Analytical electrochemistry in textiles**
P. Westbroek, G. Priniotakis and P. Kiekens
39 **Bast and other plant fibres**
R. R. Franck
40 **Chemical testing of textiles**
Edited by Q. Fan
41 **Design and manufacture of textile composites**
Edited by A. C. Long
42 **Effect of mechanical and physical properties on fabric hand**
Edited by H. M. Behery
43 **New millennium fibers**
T. Hongu, M. Takigami and G. O. Phillips

44 **Textiles for protection**
Edited by R. A. Scott
45 **Textiles in sport**
Edited by R. Shishoo
46 **Wearable electronics and photonics**
Edited by X. M. Tao
47 **Biodegradable and sustainable fibres**
Edited by R. S. Blackburn
48 **Medical textiles and biomaterials for healthcare**
Edited by S. C. Anand, M. Miraftab, S. Rajendran and J. F. Kennedy
49 **Total colour management in textiles**
Edited by J. Xin
50 **Recycling in textiles**
Edited by Y. Wang
51 **Clothing biosensory engineering**
Y. Li and A. S. W. Wong
52 **Biomechanical engineering of textiles and clothing**
Edited by Y. Li and D. X-Q. Dai
53 **Digital printing of textiles**
Edited by H. Ujiie
54 **Intelligent textiles and clothing**
Edited by H. R. Mattila
55 **Innovation and technology of women's intimate apparel**
W. Yu, J. Fan, S. C. Harlock and S. P. Ng
56 **Thermal and moisture transport in fibrous materials**
Edited by N. Pan and P. Gibson
57 **Geosynthetics in civil engineering**
Edited by R. W. Sarsby
58 **Handbook of nonwovens**
Edited by S. Russell
59 **Cotton: Science and technology**
Edited by S. Gordon and Y-L. Hsieh
60 **Ecotextiles**
Edited by M. Miraftab and A. R. Horrocks
61 **Composite forming technologies**
Edited by A. C. Long
62 **Plasma technology for textiles**
Edited by R. Shishoo
63 **Smart textiles for medicine and healthcare**
Edited by L. Van Langenhove
64 **Sizing in clothing**
Edited by S. Ashdown
65 **Shape memory polymers and textiles**
J. Hu
66 **Environmental aspects of textile dyeing**
Edited by R. Christie
67 **Nanofibers and nanotechnology in textiles**
Edited by P. Brown and K. Stevens

68 **Physical properties of textile fibres Fourth edition**
W. E. Morton and J. W. S. Hearle
69 **Advances in apparel production**
Edited by C. Fairhurst
70 **Advances in fire retardant materials**
Edited by A. R. Horrocks and D. Price
71 **Polyesters and polyamides**
Edited by B. L. Deopura, R. Alagirusamy, M. Joshi and B. S. Gupta
72 **Advances in wool technology**
Edited by N. A. G. Johnson and I. Russell
73 **Military textiles**
Edited by E. Wilusz
74 **3D fibrous assemblies: Properties, applications and modelling of three-dimensional textile structures**
J. Hu
75 **Medical and healthcare textiles**
Edited by S. C. Anand, J. F. Kennedy, M. Miraftab and S. Rajendran
76 **Fabric testing**
Edited by J. Hu
77 **Biologically inspired textiles**
Edited by A. Abbott and M. Ellison
78 **Friction in textile materials**
Edited by B. S. Gupta
79 **Textile advances in the automotive industry**
Edited by R. Shishoo
80 **Structure and mechanics of textile fibre assemblies**
Edited by P. Schwartz
81 **Engineering textiles: Integrating the design and manufacture of textile products**
Edited by Y. E. El-Mogahzy
82 **Polyolefin fibres: Industrial and medical applications**
Edited by S. C. O. Ugbolue
83 **Smart clothes and wearable technology**
Edited by J. McCann and D. Bryson
84 **Identification of textile fibres**
Edited by M. Houck
85 **Advanced textiles for wound care**
Edited by S. Rajendran
86 **Fatigue failure of textile fibres**
Edited by M. Miraftab
87 **Advances in carpet technology**
Edited by K. Goswami
88 **Handbook of textile fibre structure Volume 1 and Volume 2**
Edited by S. J. Eichhorn, J. W. S. Hearle, M. Jaffe and T. Kikutani
89 **Advances in knitting technology**
Edited by K-F. Au
90 **Smart textile coatings and laminates**
Edited by W. C. Smith
91 **Handbook of tensile properties of textile and technical fibres**
Edited by A. R. Bunsell

92 **Interior textiles: Design and developments**
Edited by T. Rowe
93 **Textiles for cold weather apparel**
Edited by J. T. Williams
94 **Modelling and predicting textile behaviour**
Edited by X. Chen
95 **Textiles, polymers and composites for buildings**
Edited by G. Pohl
96 **Engineering apparel fabrics and garments**
J. Fan and L. Hunter
97 **Surface modification of textiles**
Edited by Q. Wei
98 **Sustainable textiles**
Edited by R. S. Blackburn
99 **Advances in yarn spinning technology**
Edited by C. A. Lawrence
100 **Handbook of medical textiles**
Edited by V. T. Bartels
101 **Technical textile yarns**
Edited by R. Alagirusamy and A. Das
102 **Applications of nonwovens in technical textiles**
Edited by R. A. Chapman
103 **Colour measurement: Principles, advances and industrial applications**
Edited by M. L. Gulrajani
104 **Fibrous and composite materials for civil engineering applications**
Edited by R. Fangueiro
105 **New product development in textiles: Innovation and production**
Edited by L. Horne
106 **Improving comfort in clothing**
Edited by G. Song
107 **Advances in textile biotechnology**
Edited by V. A. Nierstrasz and A. Cavaco-Paulo
108 **Textiles for hygiene and infection control**
Edited by B. McCarthy
109 **Nanofunctional textiles**
Edited by Y. Li
110 **Joining textiles: Principles and applications**
Edited by I. Jones and G. Stylios
111 **Soft computing in textile engineering**
Edited by A. Majumdar
112 **Textile design**
Edited by A. Briggs-Goode and K. Townsend
113 **Biotextiles as medical implants**
Edited by M. W. King, B. S. Gupta and R. Guidoin
114 **Textile thermal bioengineering**
Edited by Y. Li
115 **Woven textile structure**
B. K. Behera and P. K. Hari

116 **Handbook of textile and industrial dyeing. Volume 1: Principles, processes and types of dyes**
Edited by M. Clark
117 **Handbook of textile and industrial dyeing. Volume 2: Applications of dyes**
Edited by M. Clark
118 **Handbook of natural fibres. Volume 1: Types, properties and factors affecting breeding and cultivation**
Edited by R. Kozłowski
119 **Handbook of natural fibres. Volume 2: Processing and applications**
Edited by R. Kozłowski
120 **Functional textiles for improved performance, protection and health**
Edited by N. Pan and G. Sun
121 **Computer technology for textiles and apparel**
Edited by J. Hu
122 **Advances in military textiles and personal equipment**
Edited by E. Sparks
123 **Specialist yarn and fabric structures**
Edited by R. H. Gong
124 **Handbook of sustainable textile production**
M. I. Tobler-Rohr
125 **Woven textiles: Principles, developments and applications**
Edited by K. Gandhi
126 **Textiles and fashion: Materials design and technology**
Edited by R. Sinclair
127 **Industrial cutting of textile materials**
I. Viļumsone-Nemes
128 **Colour design: Theories and applications**
Edited by J. Best
129 **False twist textured yarns**
C. Atkinson
130 **Modelling, simulation and control of the dyeing process**
R. Shamey and X. Zhao
131 **Process control in textile manufacturing**
Edited by A. Majumdar, A. Das, R. Alagirusamy and V. K. Kothari
132 **Understanding and improving the durability of textiles**
Edited by P. A. Annis
133 **Smart textiles for protection**
Edited by R. A. Chapman
134 **Functional nanofibers and applications**
Edited by Q. Wei
135 **The global textile and clothing industry: Technological advances and future challenges**
Edited by R. Shishoo
136 **Simulation in textile technology: Theory and applications**
Edited by D. Veit
137 **Pattern cutting for clothing using CAD: How to use Lectra Modaris pattern cutting software**
M. Stott

138 **Advances in the dyeing and finishing of technical textiles**
M. L. Gulrajani
139 **Multidisciplinary know-how for smart textiles developers**
Edited by T. Kirstein
140 **Handbook of fire resistant textiles**
Edited by F. Selcen Kilinc
141 **Handbook of footwear design and manufacture**
Edited by A. Luximon
142 **Textile-led design for the active ageing population**
Edited by J. McCann and D. Bryson
143 **Optimizing decision making in the apparel supply chain using artificial intelligence (AI): From production to retail**
Edited by W. K. Wong, Z. X. Guo and S. Y. S. Leung
144 **Mechanisms of flat weaving technology**
V. V. Choogin, P. Bandara and E. V. Chepelyuk
145 **Innovative jacquard textile design using digital technologies**
F. Ng and J. Zhou
146 **Advances in shape memory polymers**
J. Hu
147 **Design of clothing manufacturing processes: A systematic approach to planning, scheduling and control**
J. Gersak
148 **Anthropometry, apparel sizing and design**
D. Gupta and N. Zakaria
149 **Silk: Processing, properties and applications**
Edited by K. Murugesh Babu
150 **Advances in filament yarn spinning of textiles and polymers**
Edited by D. Zhang
151 **Designing apparel for consumers: The impact of body shape and size**
Edited by M.-E. Faust and S. Carrier
152 **Fashion supply chain management using radio frequency identification (RFID) technologies**
Edited by W. K. Wong and Z. X. Guo
153 **High performance textiles and their applications**
Edited by C. A. Lawrence
154 **Protective clothing: Managing thermal stress**
Edited by F. Wang and C. Gao
155 **Composite nonwoven materials**
Edited by D. Das and B. Pourdeyhimi
156 **Functional finishes for textiles: Improving comfort, performance and protection**
Edited by R. Paul
157 **Assessing the environmental impact of textiles and the clothing supply chain**
S. S. Muthu
158 **Braiding technology for textiles**
Y. Kyosev
159 **Principles of colour appearance and measurement**
Volume 1: Object appearance, colour perception and instrumental measurement
A. K. R. Choudhury

160 **Principles of colour appearance and measurement**
Volume 2: Visual measurement of colour, colour comparison and management
A. K. R. Choudhury

161 **Ink jet textile Printing**
C. Cie

162 **Textiles for Sportswear**
Edited by R. Shishoo

163 **Advances in Silk Science and Technology**
Edited by A. Basu

164 **Denim: Manufacture, Finishing and Applications**
Edited by R. Paul

165 **Fabric Structures in Architecture**
Edited by J. Ignasi de Llorens

166 **Electronic Textiles: Smart Fabrics and Wearable Technology**
Edited by T. Dias

167 **Advances in 3D Textiles**
Edited by X. Chen

168 **Garment Manufacturing Technology**
Edited by R. Nayak and R. Padhye

169 **Handbook of technical textiles Second edition Volume 1**
Edited by A. R. Horrocks and S. C. Anand

170 **Handbook of technical textiles Second edition Volume 2**
Edited by A. R. Horrocks and S. C. Anand

171 **Sustainable Apparel**
Edited by R. S. Blackburn

172 **Handbook of Life Cycle Assessment (LCA) of Textiles and Clothing**
Edited by S. S. Muthu

Introduction

Life cycle assessment (LCA) is one of the most commonly used tools to analyze the environmental impacts of a product from its cradle-to-grave stages. Variants of these stages include cradle to gate and cradle to cradle. LCA is also used to analyze the environmental impacts of processes and services. The International Organization for Standardization (ISO) has a dedicated series of standards for LCA. LCA is a four-step process, namely goal and scope definition, life cycle inventory, life cycle impact assessment and life cycle interpretation. LCA is gaining momentum these days and its applicability keeps increasing in many industrial sectors. With LCA, one can measure any environmental impact indicator such as carbon footprint, ecological footprint, water footprint, eutrophication, acidification and human toxicity. In addition, as a methodological point of view LCA is constantly growing and to date there have been numerous important developments. LCA has crossed many milestones and now is a well-defined tool.

While I was writing my previous book for the Woodhead Publishing Series in Textiles entitled *Assessing the Environmental Impacts of Textiles and the Clothing Supply Chain*, I had this idea of bringing out a dedicated handbook for LCA in textiles. Though the previous book dealt with the environmental impacts of textiles keeping LCA as a crux, still there is a dire need to have a dedicated handbook on LCA of textiles and clothing and hence the reason for this book. Textiles and clothing products having a lengthy supply chain demand LCA to assess the environmental impacts made by them in different phases of their life cycle. However, due to the complexity of a lengthy supply chain and accounting for all the elements pertaining to life cycle inventory, data collection, modelling aspects, etc., it is a bit challenging to conduct an LCA study for a clothing product taking into account the entire supply chain and again the entire life cycle. There are many other difficulties added to the above such as the availability of data and databases dedicated to textiles, know-how on LCA models to apply in textiles, and so on, which were all discussed in detail in my earlier book. Notwithstanding all these, LCA practitioners still make earnest efforts to conduct LCA studies in textiles.

There are a lot of product sectors under the umbrella of textiles and even with each product sector, there are umpteen numbers of developments happening day by day. There are numerous process innovations and product innovations constantly evolving in the field of textiles. One can see an ample amount of new fibres, new materials and innovative processes to attain sustainable development in the field of textiles. Considering the volume of all these innovations day by day and counting even the existing (traditional) materials, processes and product sectors, the current number of LCA studies in the world of textiles and clothing is minimal and there is a long way to go. In my view, every innovation should have an LCA element to appreciate the differences in terms of environmental impacts from the conventional ones.

As I have been constantly saying in all my publications for the past 7 years, there is a dearth of LCA studies considering the entire spectrum of the textile world and I am sure I need not obviate the need to have an LCA study for clothing products here. Some of the challenges pertaining to data availability and so on will be addressed if there are a plenty of studies on the shelf on LCA of textiles. This book in my list of books is also an attempt towards it. When I framed the contents, I framed with still more chapters than this book presently has. However, due to the difficulty in locating the experts to write chapters, many chapters were dropped. This again indicates the scarcity of experts/researchers working in this important area. Albeit this difficulty, this book is sprouting out with very important elements of LCA in textiles with detailed chapters split into two parts namely Key indicators and processes in LCA and Applications of LCA in the textile industry. Each chapter is detailed enough and covering all the aspects pertaining to LCA of textiles and clothing.

The first part of the book presents key indicators and processes in LCA, consisting of methodological, modelling aspects of LCA in textiles. The main areas covered include carbon footprints in the textile industry, energy footprints in the textile industry, ecological footprints in the textile industry, measuring the reusability of textile products, environmental impacts of the use phase of the clothing life cycle, open- and closed-loop recycling of textile products, life cycle assessment and certification of textiles and clothing and environmental impact assessment methods for textiles and clothing.

The second part of the book presents applications of LCA in the textile industry, consisting of case studies of LCA in many textile products/textile sectors. Chapters deal with the applications, methodological issues, challenges and future outlook of LCA in different sectors of textiles and clothing. There is also discussion of LCA of cotton textiles and clothing, LCA of woolen textiles and clothing, LCA of silk textiles and clothing, comparative LCA of natural and synthetic textiles, LCA of cotton shopping bags, LCA of fibre-reinforced composites, life cycle assessment and the environmental and social labels in textile and clothing industry, and clothing disposal phase – environmental benefits of reuse and recycling.

I wish to take this opportunity to thank all the contributors who have made this handbook an important reference in this field. I hope that this book will be of great benefit to all the students, researchers, industrialists and others who work in LCA/ sustainability of the textiles and clothing sector.

S.S. Muthu
Global Sustainability Services, SGS Hong Kong Limited,
Hong Kong

Part One

Key indicators and processes in LCA

Carbon footprints in the textile industry

1

G. Peters[1], M. Svanström[1], S. Roos[1,2], G. Sandin[1,3], B. Zamani[1]
[1]Chalmers University of Technology, Gothenburg, Sweden; [2]Swerea IVF, Mölndal, Sweden; [3]SP Technical Research Institute of Sweden, Borås, Sweden

1.1 Introduction

Anthropogenic climate change has damaged overall agricultural productivity, has pushed species out of their normal habitats, and is expected to accelerate species extinctions on our planet (IPCC, 2014). If political activity is an indication, climate change is considered to be one of the most important environmental problems of the last 100 years. Few issues associated with the chemistry of our environment have motivated the same level of international engagement, with some exceptions like the efforts in relation to the protection of the stratospheric ozone layer (resulting in the Montreal Protocol) and the elimination of some persistent chemical compounds (resulting in the Stockholm Convention).

It is no wonder then that assessing textiles and other products with respect to their contributions to climate change is increasingly popular (Shen and Patel, 2008; Peters et al., 2014). Carbon footprinting is the central method for doing this, and it takes into account the relative importance of different greenhouse gases (discussed further below). Another key feature of carbon footprinting is that it typically attempts to take a perspective of emissions that encompasses the entire life cycle of a product or service. So in the case of a T-shirt, this means considering the greenhouse gas emissions all the way from the cultivation of cotton through to the ultimate disposal of the T-shirt (see Muthu, 2014, for more detail on the elements of the life cycle).

Carbon footprinting is a simplified form of product environmental footprint (PEF) calculation, and both tools are ultimately based on ISO14040, the global standard for life cycle assessment (LCA). Carbon footprinting is simplified in the sense that only one impact category (climate change) is considered, while a PEF or LCA will typically consider other resource, environmental and human health categories, like energy consumption, impacts on habitat and the emission of carcinogens. International standards are discussed further in Section 1.3 of this chapter. Baumann and Tillman (2004) produced an easily read, classic guide to LCA, which is a good place to learn the basics of carbon footprinting. What follows in the remainder of the introduction is a very brief description of the method.

ISO14040 describes LCA as a four-step process as shown in Figure 1.1. Goal and scope definition is a process for making clear the intent, audience and

Handbook of Life Cycle Assessment (LCA) of Textiles and Clothing. http://dx.doi.org/10.1016/B978-0-08-100169-1.00001-0

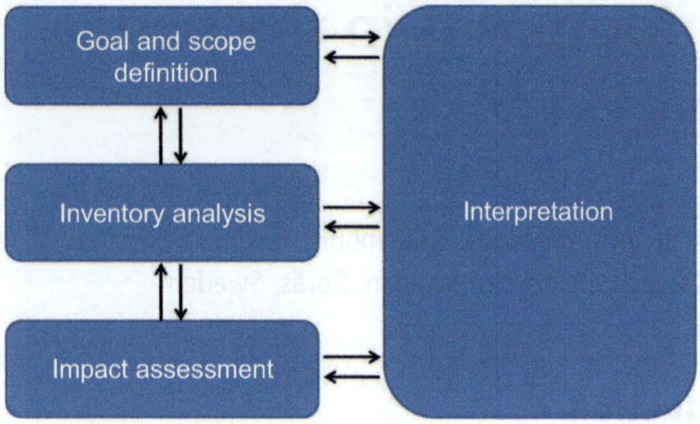

Figure 1.1 Overview of carbon footprint process.

boundaries of the study. A key element of this first step in carbon footprinting is definition of the 'functional unit' – a quantitative description of the benefit a product is expected to provide – which enables two products to be compared on a fair basis. For example, given that some T-shirts in the market are very poorly made and only last a few washes, a fair basis for comparing the carbon footprint of two T-shirts might be a functional unit of 'wearing a T-shirt daily for two years'. In this case, a heavier, more durable fabric might cost the environment more per shirt than the production of a cheaper shirt, but if the more durable T-shirt lasts three times longer than the cheaper one, the carbon footprint results will favour the more durable shirt. Section 1.2.2.1 of this chapter describes some of the challenges of goal and scope definition.

Inventory analysis is the most time-consuming step in carbon footprinting. Typically, this involves the foreground product system defined in the previous step (large dark box in Figure 1.2), carefully examining each life cycle element (labelled boxes) and identifying what greenhouse gases (black arrows) may be expected to pass out of the product system.

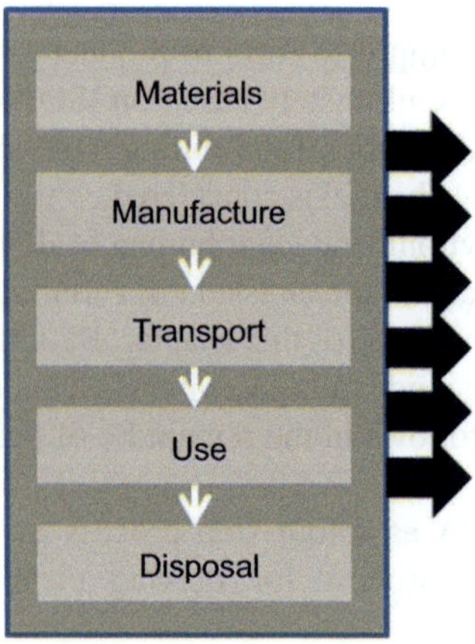

Figure 1.2 The idea of inventory analysis.

Section 1.2.2.2 describes some of the challenges of doing inventory analysis, while Sections 1.4 and 1.5 discuss some practical examples.

The third step of carbon footprinting, impact assessment, is when the differences between the greenhouse gases are considered. Carbon footprinting typically considers all the greenhouse gases identified by the Intergovernmental Panel on Climate Change: carbon dioxide, methane, nitrous oxide, hydrofluorocarbons, perfluorocarbons and sulphur hexafluoride. In the atmosphere, a kilogram of each of these gases causes a different degree of insulation, and each has a different residence time. Carbon footprinting typically uses 100 years as a standard time horizon for analysis – this is discussed further in Section 1.2.2.3.

The interpretation phase of a carbon footprint study is a time to formally reflect on the meaning of outcomes of the preceding three steps, and probably to do parts of them again. This iterativeness is the symbolic intent of the double arrows in Figure 1.1 – in fact at any step in the overall process the analyst will frequently learn about the effect of decisions made in the other steps, which may point out inconsistencies that need to be removed or opportunities for product improvement that deserve consideration in additional scenarios.

1.2 Key issues in calculating carbon footprints

In the third step of a carbon footprint calculation, global warming potentials (GWPs) are used to calculate an aggregated indicator in units of 'kilograms of carbon dioxide equivalent'. These units differentiate the footprint concept in this context from other uses, as 'footprint' otherwise commonly refers to a surface area. A GWP is based on the potential for a gas to cause radiative forcing (to change the balance of energy entering and leaving the atmosphere) in comparison to the potential for carbon dioxide (CO_2) to do this, over a set period of time. A carbon footprint can therefore be called a 'midpoint indicator' in LCA. This means a numerical value along the cause–effect chain somewhere between the mass of an emission and an estimated impact on human health, ecological quality or resources, rather than an 'end-point indicator' that describes impacts on one of these latter three protection objects in more meaningful detail. So carbon footprint is a proxy indicator for the impact of stressors that affect the climate, and it is meaningful in comparisons of the impact on climate of alternative products and processes.

Sometimes, only carbon-containing greenhouse gases (CO_2 and methane), or sometimes even only CO_2, have been included when the carbon footprint is calculated. Other variations of scope have included whether biogenic CO_2 emissions and/or emissions related to land transformation are considered. Even other climate impacts, such as a changed albedo of the earth's surface, may be translated into CO_2 equivalents and included. It is important for readers to understand how the carbon footprint was calculated when interpreting numbers found in the literature, and consequently, it is very important to be explicit about how it was calculated when publishing carbon footprint results. The most common life cycle impact assessment (LCIA) methods, CML and ReCiPe, rely, as suggested by the ILCD handbook (JRC, 2010), mainly on greenhouse gases and GWP data listed in the Fourth Assessment Report of the IPCC from 2007 (100-year time horizon; Forster et al., 2007).

1.2.1 Methodological issues

1.2.1.1 The goal and scope – which system to assess

A key challenge in estimating the carbon footprint (as in LCA in general) is to determine the scope of the system to be assessed, that is what activities should be included in the assessment. Depending on the character of the study and the question(s) that it is intended to answer, many different choices are available, some more relevant than others for a given situation.

One such important consideration is whether **an attributional or a consequential approach** should be applied. While an attributional approach is more straightforward and primarily answers questions like 'What is the carbon footprint of this product manufactured at this site?', a consequential approach takes into account changes in the surrounding system that may be caused, for example by initiated or changed production and may answer questions like 'How does the carbon footprint change if this new product is introduced to the market on this scale?' The choice between an attributional or a consequential approach may determine both the scope of the system to be studied and what type of data are to be used in the inventory of included activities. For the textile industry, this may, for example relate to how the energy supplies and land use activities should be addressed. In an attributional approach, greenhouse gas emissions from an energy mix relevant for the geographical scope can be used, but in a consequential study, the peak load energy source is often a more relevant choice; for electricity, this is often coal. For land use activities, greenhouse gas emissions from, for example, cultivation of cotton, could be included in an attributional study. On the other hand, in a consequential study, emissions from indirect land use change (ILUC) are relevant, for example if increased cultivation of cotton pushes other agricultural activities into native forests.

Another generic challenge is how to deal with **allocation of environmental impacts when multifunctional processes are involved**. When several coproducts (or services) are generated in a single process included in the system, or when an input to a single process in the system comes from a process (or service) that generated several coproducts, the analyst must decide how much greenhouse gas should be allocated to each flow. If the single process cannot be divided into subsystems, the allocation can be handled either by partitioning or by substitution. Partitioning means that emissions are allocated based on, for example mass, volume or market price. Substitution means that if emissions are avoided elsewhere in the global economy on account of the coproducts, the avoided emissions can be subtracted from the multifunction process to leave only the amount associated with the main product. The choice of allocation method may have a very large impact on the results (see, e.g. Sandin et al., 2015) and therefore needs to be done in a relevant way. Sandin et al. (2015) provide some advice on how the decision at hand should influence this choice.

1.2.1.2 Inventory analysis – emission data

A major challenge in any LCA is to find relevant **inventory data**. Carbon footprinting can be applied at many scales, from individual products (e.g. a T-shirt) to corporations (e.g. Nike) and national industries (e.g., the United States' textile industry). The methodologies and standards relevant to a study depend upon the scale of the study and its purpose. There are two fundamental methodologies in use for obtaining inventory data for carbon footprinting: process analysis and input–output analysis. Process analysis is an approach made popular in the field of LCA, and it is the basis of the majority of publications in the *International Journal of Life Cycle Assessment*. It typically involves quantifying physical inputs and outputs, and any energy flows, at the level of a product, process or business. Input–output analysis has historically been used at the regional or national level. For carbon footprinting, it typically requires the performance of matrix algebra on national economic statistics and data on the greenhouse gas emissions of each industrial segment of the national economy. Process analysis has the advantage of specificity, but the downside that it is relatively time consuming. Input–output analysis has the advantage that it requires little effort for the analyst to include data for small flows, and the risk of truncation error (i.e. lost system components) is reduced. It has the disadvantage that by using average emissions for each segment of the economy, the environmental performance of the worst polluters in a segment is underestimated and environmental leaders are punished with overestimates. The different strengths and weaknesses of the methods have resulted in a number of attempts to hybridise them (e.g. Alvarez-Gaitan et al., 2013), using process analysis for the most important elements of the system under study, and input–output analysis for the services that support those systems. In this way, both approaches can be used to analyse textile products, manufacturing plants and companies.

It is important to obtain high-quality data for the foreground system (those environmental stressors that may be affected by the decision at hand) and for parts of the life cycle that make up important contributions to the total impact (which can often be identified during the work because of the iterative approach in LCA). Greenhouse gas emission databases are available for most common industrial processes (e.g. in the Ecoinvent database; Ecoinvent Centre, 2010) and emissions can also often be fairly easily measured or estimated when the emission is from a point source and a result of a technical process. However, for some human activities, such as different land use activities (e.g. forestry and agriculture), relevant greenhouse gas emission inventories or emission factors may be hard to find. Due to work spurred on by the need to account for national greenhouse gas emissions within the Kyoto Protocol and its continuation, this field is advancing rapidly (see e.g. searchable database for IPCC emission factors on http://www.ipcc-nggip.iges.or.jp/EFDB/main.php). For textile fibres and materials, a particular challenge arises in the fact that value chains may be long and complex. Therefore, relevant inventory data may be hard to find.

Another key challenge is in making sure that **all relevant greenhouse gas emissions are included**, which may be an active choice but most often is merely a result of which data were included in inventory data sets that were employed in the study (and sometimes also a result of which gases are listed in the LCIA method applied). Data sets should be selected based on their relevance and evaluated for their completeness, possibly in an iterative process where the most highly impacting activities are first selected for a quality check and then improved if needed.

A special case is **whether or not the net carbon balance is considered** in determining which CO_2 emissions should be included in the inventory. It has been common to include only CO_2 from nonbiogenic sources (thus excluding emissions from biogenic sources) with the motivation that the long-term carbon balance will not be affected by these as the emissions can be assumed to soon be involved in new biomass generation, but this approach is often criticised today (Gunn et al., 2012; Johnson, 2009). The dynamics of the carbon balance are affected by the soil carbon content, land management practices, crop selection, harvesting and regrowth rates and many other parameters. Furthermore, it has been suggested (and applied in a few cases) that when carbon from biogenic sources becomes tied up in long-lived products or in soil (sequestered), this may contribute to a net decrease in atmospheric greenhouse gases and should therefore be included as a negative emission (Sandin et al., 2014). If the carbon tied up in a long-lived product is from a nonbiogenic source, it could be treated as a delayed emission. It is thus a challenge to know how to handle even a simple and well-known gas as CO_2.

1.2.1.3 Impact assessment – choice of method

Carbon footprint studies typically assume a 100-year **time perspective**, abbreviated 'GWP100'. This is often an arbitrary choice (Reap et al., 2008) and introduces the value-based standpoint that what happens 100 years from now is not the responsibility of the decision-maker in the study. It relies on the argument that 100 years from now, we should have solved this problem and that greenhouse gas emissions will be under control. With a 100-year perspective, long-lived greenhouse gases (having a lifetime over 100 years) get a lower value than they would with a longer time perspective. Other fairly common alternatives are a 20-year and a 500-year time perspective. A 20-year perspective would perhaps better address the urgency of current impacts on climate, whereas the 500-year perspective also considers what happens in the more distant future.

The timing of emissions should be seen in the light of this. It has been suggested that the carbon footprint should take into account **when emissions take place** (Sandin et al., 2014), both in terms of the carbon balance (as discussed earlier) and in terms of when emissions will actually happen, for example in end-of-life activities. If a greenhouse gas emission will not take place until 50 years from now, it may seem unreasonable that all of this impact should be included if a 100-year perspective is applied. This is thus a matter of whether the temporal dynamics of emissions and impact should be allowed to influence the results or not.

1.2.1.4 *Holistic assessments may require complementary indicators*

Finally, the carbon footprint of a product or a process can only rarely reflect all the important environmental aspects of a product or process choice. Therefore, other environmental impact categories or even technical, social and economic assessments are necessary complements. The carbon footprint has often been the only environmental indicator assessed, due to lack of data or well-developed impact assessment methodology for other impacts, or sometimes because of ignorance of the potential risks constituted by the other environmental aspects. The choice of making a carbon footprint assessment or not and how to complement it with other evaluations must be guided by the decision at hand.

1.3 Carbon footprint standards

The original standardisation process for carbon footprinting was the development of the ISO1404x series of standards. ISO14040 (Environmental management – Life cycle assessment – Principles and framework) was first published in 1997. ISO14040 is broader than carbon footprinting, providing a basis for assessing products against other LCIA indicators like those for contributions to ozone layer depletion and pollution of oceans, but the assessment of greenhouse gas emissions has been central to LCA from the beginning. ISO/TS14067 (ISO, 2013) has emerged as an important technical standard for carbon footprinting after the original proponents paused in their drive to turn it into an ISO standard, though it is possible it may be converted into a full ISO standard after it is reviewed in 2016.[1]

Two other important standardisation processes have developed in parallel to the ISO/TS14067 process. One is the Greenhouse Gas Protocol (GHG Protocol) development by the World Resources Institute and the World Business Council for Sustainable Development (WRI/WBCSD, 2004). The other is the Publically Available Standard 2050 (PAS2050 see BSI, 2011) of the British Standards Institute. These and other technical guidance documents of greatest relevance to the carbon footprinting of textile products in the English-speaking world are shown in Table 1.1. PAS 2395:2014 sets out supplementary requirements that tailor PAS2050 to the assessment of greenhouse gas emissions from the life cycle of textile products. There are some common features across all of the standards produced by these three institutional groupings and some differences. As is the case with the ISO1404x standards, ISO/TS14067 is more general than the GHG Protocol or PAS2050. There are also many other ISO standards and other initiatives of relevance to greenhouse gas accounting, for example the French BPX30-323 ‘General principles for an environmental communication on mass market products’. At the more detailed level, the International Environmental Product Declaration System contains more detailed Product Category Rules (PCRs) that

[1] http://www.pef-world-forum.org/initiatives/international-standards/iso-14067/.

Table 1.1 Key standard and technical guides in international use for carbon footprinting

Source	Label	Name	Latest version	Scale
International Standardization Organization	ISO14040	Environmental management – Life cycle assessment – Principles and framework	2006	Generic
	ISO14044	Environmental management – Life cycle assessment – Requirements and guidelines	2006	Generic
	ISO14025	Environmental labels and declarations – Type III environmental declarations – Principles and procedures	2006	Generic
	ISO14064	Greenhouse gases – Part 1: Specification with guidance at the organization level for quantification and reporting of greenhouse gas emissions and removals	2009	Organisation
	ISO/TS 14067	Carbon footprint of products – Requirements and guidelines for quantification and communication	2013	Product
World Resources Institute/World Business Council for Sustainable Development	GHG Protocol	Greenhouse gas protocol: A corporate reporting and accounting standard	2004	Organisation
		Greenhouse gas protocol: The GHG protocol for project accounting	2005	Project
		Greenhouse gas protocol: Product life cycle accounting and reporting standard	2011	Product

Table 1.1 Continued

Source	Label	Name	Latest version	Scale
Global Reporting Initiative	GRI	Sustainability reporting guidelines – Reporting principles, standard disclosures	2014	Organisation
British Standards Institute	PAS2050	Specification for the assessment of the life cycle greenhouse gas emissions of goods and services	2011	Product
	PAS 2395	Specification for the assessment of greenhouse gas (GHG) emissions from the whole life cycle of textile products	2014	Product
European Commission	PEF	Product environmental footprint guide	2013	Product
	OEF	Organisation environmental footprint guide	2013	Organisation

provide guidance on the assessment of climate and other impacts, specific to a particular textile product (e.g. nonwoven cleaning cloths).[2]

1.3.1 Standards for product assessment

On many issues, the principal standards are in agreement. For example, for allocation of multifunction processes, the standards are all in agreement that avoiding allocation by substitution is preferable and that allocation based on economic values is the last resort. The standards are united in disallowing the inclusion of emission offsets in the calculations. Another issue is whether to include emissions from land use change. ISO14044 does not discuss this, but PEF, ISO/TS 14067, PAS 2050 and the GHG Protocol (WRI/WBCSD, 2011a) all instruct analysts to include emissions associated with direct land use change. PAS2050 excludes ILUC and the others defer to future revisions of the standards for agreement regarding this topic.

One of the key methodological issues product carbon footprint standards have to deal with is whether capital goods should be included in the calculations. In some

[2] http://environdec.com/en/What-is-an-EPD/#.VC6Zak0cRlY.

industries that involve transport of heavy materials, for example the water industry, capital goods are often unimportant elements of the total carbon footprint – analytical time is better spent on improving estimates of operational energy consumption than estimates of how much concrete was used in a wastewater treatment facility. On the other hand, if the analytical task is to compare options including one with a large amount of distributed infrastructure, capital goods may be significant. Capital goods are specifically excluded in PAS2050 but they are considered optional in ISO14040.

A related and more general difference concerns the treatment of cut-offs: when is it acceptable to exclude small components of the total carbon footprint? Cut-offs are ruled out in PEF (EU, 2013) but acceptable in ISO14044 depending on their total mass or environmental significance. PAS2050 sets a quantitative threshold: 95% of the total 'anticipated' emissions must be included along with 'any one source' that contributes at least 1% of the anticipated total.

1.3.2 Standards for organisational assessment

Perhaps the most important issue confronting organisational carbon footprints is the question of whether to include 'scope 3', which refers to indirect emissions from a company's supply chain, whereas scope 1 includes all direct on-site emissions and scope 2 includes emissions from energy purchases. There is an element of pragmatism here because, thanks to normal financial accounting systems, energy purchases are usually the easiest causes of greenhouse gas emissions for analysts to identify. Consequently, all the organisational carbon footprint standards require the reporting of scope 1 and 2 emissions. Where they diverge is the treatment of scope 3. The ISO standards consider scope 3 optional. Other standards like the OEF and GRI (GRI, 2014) focus on the question of control: does the organisation under study have financial or operational control of the scope 3 emission? The GRI defers to the GHG Protocol regarding scope 3 but is broader in its coverage. For example, GRI allows that if stakeholders (such as customers, suppliers, investors or civil society) consider a particular emission to be material to the organisation, then it should be considered in the carbon footprint. The World Resources Institute and the World Business Council for Sustainable Development have published the most extensive guidance on scope 3 in the form of a dedicated 'Corporate Value Chain (Scope 3) Accounting and Reporting Standard' to supplement the GHG Protocol (WRI/WBCSD, 2011b).

1.4 Calculating carbon footprints of textile manufacturing processes

Textile manufacturing plants select processes for transforming raw materials to products depending on what is appropriate for their products' end uses. The three main end uses are clothing, home furnishing and industrial use (European Commission, 2003).

The manufacturing processes can include fibre production, yarn production, fabric production (nonwoven, knitting, weaving), wet treatment, finishing and sewing. A textile manufacturing plant that performs all the processes from yarn production to product is commonly termed a 'vertical plant'. The materials include well-known ones like cotton, polyester, viscose, lyocell, wool and silk. Calculations of the carbon footprint of textile manufacturing plants are generally made within the context of an EMS, an organisational carbon footprint or for the purpose of a product LCA. If it is the first or second of these, the functional unit can be simply defined as one year of operation of the textile plant, regardless of which processes are present. If the goal is a product LCA, allocation procedures are required, and the extent to which different processes are applied to the specific product under investigation needs to be considered. Throughout the production processes for textiles, material is wasted, which means that the carbon footprint of the product has to be adjusted to consider the material losses from previous steps (Collins and Aumônier, 2002). This means that carbon footprint data published about textile processes in the context of a product carbon footprint are not readily converted into carbon footprints of textile manufacturing processes or textile manufacturing plants unless background data are provided detailing the losses at different process steps. Applications and examples on how carbon footprints of textile manufacturing plants are calculated are described below. More information about differences in production technology with major implications on the carbon footprint is found in Section 1.5, 'Further information and advice'.

'Cradle-to-gate' is a common expression in LCA indicating that raw materials are included but end use and disposal are excluded. This is similar to 'cradle-to-plate' used in food LCA and is in contrast to 'cradle-to-grave' indicating an all-inclusive system analysis. Figure 1.3 shows the cradle-to-gate carbon footprints from each of the textile manufacturing processes for a mixed material garment (a jacket) as calculated in the Mistra Future Fashion programme (Roos et al., 2015). In the figure, the contribution

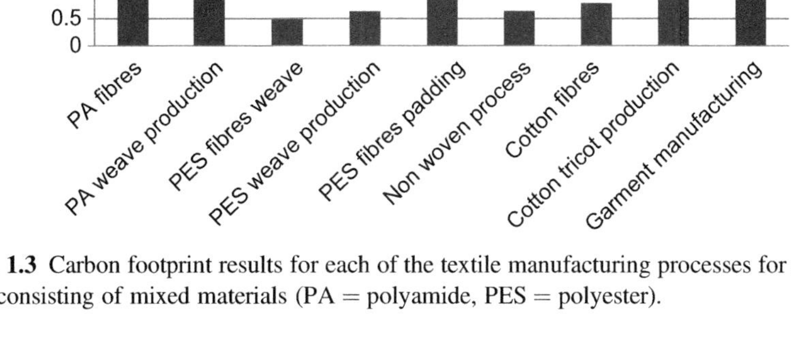

Figure 1.3 Carbon footprint results for each of the textile manufacturing processes for a single jacket consisting of mixed materials (PA = polyamide, PES = polyester).

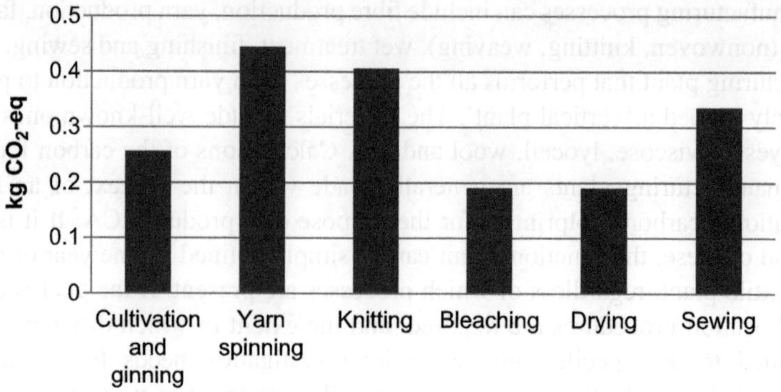

Figure 1.4 Carbon footprint results for each of the textile manufacturing processes for a T-shirt consisting of pure cotton.

of each manufacturing process is related to the fraction of the jacket that each material represents.

In contrast to Figure 1.3, in Figure 1.4 a monomaterial garment is examined (a cotton T-shirt). This makes the relative scale of the carbon footprint of each manufacturing step easier to grasp.

As Figure 1.4 indicates, yarn and fabric production, wet treatment, and sewing processes at manufacturing plants can make up a considerable proportion of the cradle-to-gate carbon footprint of a garment, compared to the cultivation step. Table 1.2 shows a set of published carbon footprint studies of textile manufacturing plants, with a description of the method used for carbon footprinting, the studied reference flow (most often a product) and the calculated carbon footprint. In some studies, sensitivity or scenario analysis provides a range of results – in such cases, results shown here are for the baseline scenario.

Other studies from which guidance to carbon footprints of textile processes can be retrieved include the latest IMPRO report (Beton et al., 2014), the publication on spin dyeing of modal fibres by Terinte et al. (2014), the study on recycling gains by Muthu et al. (2012), the study on pad dyeing by Yuan et al. (2012), the overview provided by Business for Social Responsibility (2009) and the French jeans study by Ademe (Labouze et al., 2006). These reports are not listed in the table above because greenhouse gas emissions of processes are only reported in relation to the final product, and adjustments in relation to process losses need to be made if data is to be used for the manufacturing process or plant carbon footprint. The specifications on the equipment used, quality of product outcome and energy sources are also lacking in general, which makes comparisons of different figures difficult. Publications of other product-related carbon footprint calculations include (Bhurtun et al., 2006; Grace, 2009; Nagel, 2010; Roos et al., 2015; Smith and Barker, 1995; Steinberger et al., 2009). Further, several studies that report energy consumption of textile processes are available (Barber and Pellow, 2006; Beton et al., 2014; Bevilacqua et al., 2011; Bidoki and Wittlinger, 2006; Browne et al., 2005; Collins and Aumônier, 2002; European Commission,

Table 1.2 Overview of recent studies of the carbon footprint of textile manufacturing plants

Publication	Method	Studied item (in kg) (*origin*)	Results (kg CO_2 eq.)
Peer-reviewed scientific papers			
Van der Velden et al. (2014) (uses data from ecoinvent (Swiss Centre for Life Cycle Inventories (Ecoinvent Centre, 2010)) and IDEMAT (Vogtländer, 2012))	GWP100 (IPCC, 2007; 100 years)	Cotton fibres, cultivation and ginning[a] (*Ecoinvent*)	3.47
		Polyethylene terephthalate, granulate, amorphous (*Ecoinvent*)	2.70
		Ring[a] spinning cotton yarn 45 dtex[b] (*IDEMAT*)	11.32
		Ring[a] spinning cotton yarn 300 dtex (*IDEMAT*)	1.70
		Texturing polymer fibres (*IDEMAT*)	0.51
		Circular[a] knitting 83 dtex (*IDEMAT*)	0.26
		Circular[a] knitting 300 dtex (*IDEMAT*)	0.07
		Weaving 45 dtex (*IDEMAT*)	16.60
		Weaving 45 dtex (*IDEMAT*)	2.49
		Airjet[a] dyeing of polyester, excluding chemicals (*IDEMAT*)	2.25
		Heat setting and washing synthetic fabrics (*IDEMAT*)	0.91
Astudillo et al. (2014)	IPCC (2013) GWP 100a v.1.02	Silk production, farm practices	80.90
		Silk production, recommended practices	52.50
		Nylon 66 polymer production – (Swiss Centre for Life Cycle Inventories (Ecoinvent Centre, 2010)	8.00

Continued

Table 1.2 Continued

Publication	Method	Studied item (in kg) (*origin*)	Results (kg CO_2 eq.)
Other publications			
Cherrett et al. (2005)[c]	Wackernagel et al. (2004)	Spun fibres of organic cotton, US	0.0024
		Spun fibres of conventional cotton, US	0.0059
		Spun fibres of hemp	0.0045
		Spun fibres of polyester, Europe	0.0072
		Spun fibres of polyester, US	0.0095
Murugesh et al. (2013)	Not reported	Exhaust dyeing of knitted cotton using softflow dyeing machine, colour 1	12.43
		Exhaust dyeing of knitted cotton using softflow dyeing machine, colour 2	16.88
		Cold pad batch dyeing of knitted cotton, colour 3	6.59
		Squeezer and softener pad	0.076
		Relax drying	0.072
		Stenter drying	0.113
		Compacting	0.135

[a]The marked text is an assumption regarding technology when data are lacking in the source document.
[b]dtex or decitex is the mass in grams per 10,000 m.
[c]The results shown here are based on the values per tonne of fibres as stated in that report, but it seems likely that the original report data is actually per kg of fibre.

2003; Kalliala and Nousiainen, 1999; Koç and Çinçik, 2010; Laursen et al., 2007). In fact, primary energy consumption was found to be the most commonly reported impact for LCAs of textiles in a recent overview from Oakdene Hollins (Chapman, 2010).

1.5 Calculating carbon footprints of textile products

In addition to the textile manufacturing processes described in the previous subsection, the full life cycle of textile products encompasses processes related to raw material extraction (e.g. cultivation, forestry, extraction of oil), distribution, retail, use (e.g. consumer transportation to and from the store, laundry) and end-of-life handling. Typically, manufacturing and use (consumer transportation as well as laundry) are, because of high energy intensity, the most important contributors to the carbon footprint (Roos et al., 2015). Depending on methodology, raw material extraction may also be an important contributor.

There are few published carbon footprinting studies of textile products compared to product categories such as foodstuffs or fuels (Peters et al., 2014). Table 1.3 lists some key studies, academic as well as other publications. The list includes studies of products that are available to consumers (e.g. garments and fabrics) but excludes studies of textile fibres. Also, the list excludes studies limited to a part of the manufacturing process (e.g. finishing). The table suggests that GWP100 is the dominant metric for carbon footprinting but that further methodological details (e.g. concerning the handling of biogenic CO_2 emissions and soil carbon changes, or the use of a carbon footprinting standard) are seldom clearly stated in published studies. This most likely means that existing standards are often not considered and that less traditional methodological aspects are seldom accounted for (which means, among other things, that biogenic CO_2 emissions most often are assumed to be climate neutral).

In terms of results, the table suggests that the cradle-to-gate carbon footprints of fabrics range from about 0.7 to 36 kg CO_2 equivalents per kg and that the cradle-to-grave carbon footprints of garments range from about 1 to 18 kg CO_2 equivalents per garment (as a comparison, the carbon footprint of 1 kg beef is about 14–32 kg CO_2 equivalents (de Vries and de Boer, 2010), and the carbon footprint of 100 km of car transport is about 10 kg CO_2 equivalents). Note that there are large differences between seemingly similar garments: a wool garment is among the studied items with both lowest and highest carbon footprint (about 8 kg vs 46 kg CO_2 equivalents per kg of garment in different studies). Differences between the calculated carbon footprints are most often not because of differences in the applied carbon footprinting methodology but rather because of other assumptions, for example regarding user behaviour (washing temperatures, number of washes before end of life, etc.) and technology (manufacturing practices, electricity mixes, etc.). One exception is the study on black modal knitted fabrics, which show considerably lower carbon footprints per kg of fabric compared to other studies – this is most likely because biogenic carbon stored in the fibres is considered a carbon sink, which results in a negative GWP score. This shows that the choice of carbon footprint methodology is, although seldom emphasised in carbon footprinting studies of textile products, potentially very important for the calculated carbon footprint because of the presence of biobased materials (Levasseur et al., 2010; Brandão et al., 2013; Helin et al., 2013; Røyne et al., in press). Thus, LCA practitioners and decision-makers must be very careful in extracting carbon footprint results from studies and using them out of context. To interpret results, the specific methodology and assumptions of each study must be considered.

Table 1.3 **Recently published carbon footprint studies of textile products**

Publication	Method	Studied item and scope	Results[a] (kg CO_2-e)
Peer-reviewed scientific papers			
Terinte et al. (2014)	GWP100. CML 2001 LCIA method. Biogenic carbon stored in fibres considered negative emissions	Black modal knitted fabric, 1 kg (conventionally dyed), cradle-to-gate	5.6
		Black modal knitted fabric, 1 kg (spun-dyed), cradle-to-gate	2.07
		Black modal knitted fabric, 1 kg (undyed), cradle-to-gate	2.05
Van der Velden et al. (2014)	GWP100	Cotton fabric, 1 kg, cradle-to-gate (the study also covers other life cycle phases and fabrics, for which numeric results are not shown in the report)	~9–22.5
De Saxce et al. (2012)	GWP100. CML 2001 LCIA method	One year use of cotton bed sheet, 240 × 300 cm, cradle-to-use	~3–5
		One year use of cotton/polyester bed sheet, 240 × 300 cm, cradle-to-use	~4
Muthu et al. (2012)	GWP100	Cotton fabric, 1 kg, cradle-to-gate	0.66
Walser et al. (2011)	GWP100	Polyester T-shirt, 130 g, cradle-to-gate	2.55
		Polyester T-shirt, 130 g, cradle-to-grave	~15
Bevilacqua M et al. (2011)	GWP100. PAS 2050 was taken into account	Wool sweater, 265 g, cradle-to-grave	1.95
Other publications			
Roos et al. (2015)	GWP100. Biogenic CO_2 emissions considered climate neutral. ReCiPe characterisation framework	Cotton T-shirt, 110 g, cradle-to-grave	3
		Cotton/elastane jeans, 477 g, cradle-to-grave	14

Table 1.3 **Continued**

Publication	Method	Studied item and scope	Results[a] (kg CO_2-e)
		Polyester dress, 478 g, cradle-to-grave	10
		Polyamide/polyester jacket, 444 g, cradle-to-grave	18
		Cotton/polyester hospital uniform, 340 g, cradle-to-grave	11
Beton et al. (2014)	GWP100. Biogenic CO_2 emissions considered climate neutral. ReCiPe LCIA method	Nine different fabrics (viscose, flax, silk, wool, cotton, polyester, PA6, acrylic, polypropylene), 1 kg, cradle-to-gate	14.9–35.7
Thomas et al. (2012)	GWP100. Carbon change due to land use excluded	Cotton garment (average), 1 kg, cradle-to-gate	27.7
		Wool garment (average), 1 kg, cradle-to-gate	46.3
		Silk garment (average), 1 kg, cradle-to-gate	25.4
		Flax/linen garment (average), 1 kg, cradle-to-gate	15.0
		Viscose garment (average), 1 kg, cradle-to-gate	30.1
		Polyester garment (average), 1 kg, cradle-to-gate	21.3
		Acrylic garment (average), 1 kg, cradle-to-gate	38.4
		Polyamide garment (average), 1 kg, cradle-to-gate	24.3
		Polyurethane/polypropylene garment (avg), 1 kg, cradle-to-gate	19.8
		Tops (average), cradle-to-grave	6.5

Continued

Table 1.3 Continued

Publication	Method	Studied item and scope	Results[a] (kg CO_2-e)
		Underwear, nightwear and hosiery (average), cradle-to-grave	1.9
		Bottoms (average), cradle-to-grave	8.7
		Jackets (average), cradle-to-grave	13.3
		Dresses (average), cradle-to-grave	14.8
		Suits and ensembles (average), cradle-to-grave	16.8
		Gloves (average), cradle-to-grave	1.1
		Sportswear (average), cradle-to-grave	6.1
		Swimwear	1.8
		Scarves, shawls, ties etc.	1.8
Jungmichel (2010)	GWP100	Cotton white long shirt, 222 g, cradle-to-grave	10.8
		Sweat jacket	13.4
		Acrylic children jacket	13.7
Laursen et al. (2007)	GWP100. EDIP LCIA method	Cotton T-shirt, nylon/cotton jogging suit, polyester/cotton work jacket, viscose/nylon/elastane blouse, cotton table cloth, nylon/polypropylene floor, all cradle-to-grave (numeric results are not shown in the report)	–
Allwood et al. (2006)	GWP100. EDIP LCIA method	Cotton T-shirt, 250 g, cradle-to-grave	4 kg CO_2 eq.

[a]In some studies, sensitivity or scenario analysis provides a range of results – in such cases, the result shown here is on the baseline scenario.

1.6 Carbon footprint labelling of textile products

Product carbon footprint labelling has been initiated in countries like the United Kingdom, the Netherlands and Japan in order to educate consumers in making informed purchasing decisions (Tan et al., 2014). From a corporate perspective, carbon footprint labelling offers the potential to gain competitive advantage via green marketing that taps into increasing global concerns about carbon emissions (Vanclay et al., 2010). From a non-governmental organisation perspective, the information provided by carbon footprint labels can assist customers thinking about which product to buy (Upham et al., 2011), and endorsing companies with carbon footprint labels can encourage them to publicly commit to reducing their carbon emissions (Boardman, 2008).

Consumers' awareness about environmental impacts of products' life cycles is being increased by communication of quantitative values on carbon footprint labels, backed by strategies and methods developed for quantifying greenhouse gas emissions of products and services (Tan et al., 2014).

There are a number of labels for textiles and other products that include a carbon footprint component. One of the first was the Carbon Reduction Label, which was first introduced in the UK in 2006 by the Carbon Trust (see www.carbontrust.com). The original labels licensed by the Trust for use on products stated the amount of CO_2 and other greenhouse gases emitted per product or per use of the product. Today, two different types of labels are offered: a 'Reducing CO_2' label and a 'CO_2 Measured' label. The Reducing CO_2 label is a communication tool to show the company's commitment for carbon reduction and can also appear with a statement about the size of the carbon footprint of the labelled product or service. Certified applicants are committed to reduce their measured carbon footprint over two years or lose the label – results will be assessed by a third party after 2 years to verify the correct use of the label (Carbon Trust, 2015). The CO_2 Measured label indicates that a product's carbon footprint has been calculated but makes no commitment in relation to reducing that footprint. The carbon footprint assessment is based on PAS2050 or the GHG Protocol. This means that the whole life cycle of the product including production, use and disposal is included in measuring the carbon footprint.

Another carbon footprint label is Carbon Care® Asia, which was developed by a Hong Kong–based company. The Carbon Care® labelling system evaluates companies' and organisations' efforts in reducing carbon footprint in three steps: measuring, reducing and offsetting carbon footprints. An applicant who wishes to use the label needs to provide a verified audit report of its emission inventory according to the Carbon Care® label protocol. Six different labelling options are allowed based on the level of emission reduction of the applicant compared with its emission baseline. The awarded labels show the level of achievement in carbon reduction as 5%, 20%, 40%, 60% or 80% compared with the measured emission baselines. A golden label is awarded to applicants that can achieve carbon neutrality (CarbonCareAsia, 2014).

Carbonfund.org is an American nonprofit organisation that developed the Carbon-Free Product Certification to make consumers aware of the companies that are compensating for their greenhouse gas emissions including CH_4, N_2O, SF_6, HFCs, PFCs and biomass CO_2 emissions (Carbonfund, 2012). To acquire the CarbonFree certification, an LCA needs to be performed to measure the greenhouse gas emissions throughout the product's life cycle. Components of the carbon footprint of the product's life cycle that cannot be reduced are compensated with third-party validated renewable energy, energy efficiency and forestry carbon offset projects. The measurement protocols are based on ISO 14044, PAS 2050 and the GHG Protocol (Carbonfund, 2012).

1.7 Future trends

The textile industry has been relatively late in engaging with carbon footprinting, but its engagement with this aspect of life cycle thinking, measured in terms of literature output, is now growing exponentially (Peters et al., 2014). In this context, it is to be expected that more textile products will be labelled with some kind of carbon-related claim that has to be backed up with carbon footprint calculations. If carbon footprint calculations are not the basis of such claims, the claims risk being labelled greenwash. On account of this risk, there is also an increased emphasis on transparency in presenting data, which is leading textile companies to invest more in understanding their upstream supply chains. In the past, it was not necessary for companies to know exactly who their subcontractors' subcontractors were, but initiatives like the Sustainable Apparel Coalition (SAC) emphasise the need for this. The SAC has prepared spreadsheet templates for communication with subcontractors to facilitate the dialogue necessary to allow transparent reporting of greenhouse gas emissions and other sustainability issues (www.apparelcoalition.org). Initiatives such as this will make it easier to do the carbon footprint calculations necessary for the implementation of carbon labelling schemes.

Looking at the progress of standardisation indicated in Table 1.1, it is apparent from the dates of the various standards that there is a trend toward increasing specificity in the standardisation process. This trend is also apparent in the timing of the development of PCRs for various products including textile products. At the time of writing, there were 12 PCRs on the Environmental Product Declaration EPD website under the category 'textile and leather products'[3], all of them published since 2010. It is hoped that in this context of increasing standardisation that the potential for greater unity will arise regarding the appropriate degree to which scope 3 emissions are included in carbon footprints. It should also be hoped that efforts such as those of the SAC will put more information in the hands of clothing designers, to give them the information tools necessary to create sustainable fashion. Nevertheless, given the rapid rate of growth in the number of different eco-labelling schemes, it may be some time before there is a universal, global carbon label.

[3] http://environdec.com/en/EPD-Search/.

This is not just a consequence of institutional diversity but also the methodological diversity underlying the calculation of carbon footprints. As described in this chapter, in any carbon footprint there are important questions to be answered around attribution and consequence, allocation, the inclusion of biogenic carbon and the selection of temporal aspects. Beginners attempting their first carbon footprint are prone to underestimate the importance of goal and scope definition in the process, but it has been identified in the standards as a key element of the analysis, because of the powerful role it plays in determining the results and giving them meaning.

1.8 Conclusions

Carbon footprint analysis of textiles and clothing suggests that many possible improvements to the product life cycle can be achieved because of large differences between possible alternatives. Selecting low-emission fibres is a worthwhile intervention, as suggested by the data in Table 1.2, but these data, along with Figure 1.3 and Figure 1.4, also indicate that the spinning and weaving processes are important variables in the overall carbon footprint of textile products. A meta-analysis of various carbon footprint studies (Muthu, 2014) has suggested that important gains can be made in the user phase of the textile life cycle, in no small part due to the use of energy to heat laundry water and dry clothing. We should also bear in mind that recent laboratory work has shown how laundering damages fibres at the molecular level, reducing their life length and their potential for reuse or mechanical recycling (Palme et al., 2014). So improving the quality of the products made by the textile industry has an important role to play in reducing the impacts of life cycle phases before and after the use phase (Zamani et al., 2014). Commercial interests also play a role in creating consumers' fashion expectations, with the consequence that most textile products are not used for their whole technical life length. Assumptions about the use phase must be made carefully in carbon footprinting in this context (one may easily overestimate the number of washing and drying cycles per new garment). It can be argued that the current trend towards 'fast fashion', in which both product quality and fashion expectations cause shorter product lives, must be reversed if the textile and clothing industry is to significantly improve environmental outcomes (Peters et al., 2014).

The diversity of methodological questions that need to be answered when calculating a carbon footprint mean that consumers of this information must be on their guard. Without any intent to mislead, it is feasible for different analysts to provide aggregated carbon footprint results that differ considerably on account of the complexity of such calculations. Practical decisions are necessary concerning the choice of functional unit, system boundaries, land use change, biogenic carbon and allocation. We have seen in Section 1.3 of this chapter the importance of many parts of the textile supply chain in terms of their contributions to the total carbon footprint of textile products, but we should be careful when extrapolating from such results. Results that are perfectly correct in their original context may be inappropriate when borrowed by writers outside the field of LCA, for example to try and illustrate some related question of consumer textile policy. Standardisation has gathered pace over the last 20 years, and the key standards bear many signs of common ancestry and purpose. Increasingly, we will see comparable results, but the residual diversity among the standards means that readers of carbon footprint information must read carefully.

Far from decrying diversity, we should bear in mind that a diversity of standards and approaches is necessary given the many different purposes to which carbon footprinting can be put – from product labelling to policy analysis and process design. The context of each carbon footprint calculation will have to influence the necessary and appropriate inclusions and choices. Finally, let us bear in mind that despite hundreds of years of global trade, and the increasing intensity of globalisation in the last century, there are still different shoe size scales in different countries. Therefore, it is to be expected that a diversity of carbon footprint standards will persist for years to come.

1.9 Further information and advice

In addition to the references provided in Tables 1.1–1.3, the text below provides a brief description of each production stage and key publications that will assist the reader interested in calculating carbon footprints of textile products and processes.

1.9.1 Fibre production

For synthetic and regenerated fibres, the fibre production (primary spinning) is mainly made by melt spinning (polyester, nylon etc.), dry spinning (elastane etc.) or wet spinning (viscose, lyocell etc.). Lenzing AG has produced carbon footprints of viscose and lyocell fibres and compared these with other fibres (Shen and Patel, 2008; Shen et al., 2010; Terinte et al., 2014; Van der Velden et al., 2014). The production of natural fibres via agriculture or forestry has been studied elsewhere (e.g. Sandin et al., 2013).

1.9.2 Yarn production

Textile yarn spinning is in general an energy intensive process, and energy consumption depends on the yarn size and the spinning technique. Thinner yarns require higher energy consumption (Van der Velden et al., 2014). Ring spinning gives a smooth yarn with good pilling resistance and high strength but entails the highest energy consumption, while open-end (rotor) spinning gives a rougher yarn with less strength but at lower energy consumption. Air-jet spinning is a growing technology with both good performance and low energy consumption. The waste in the yarn spinning process depends on material and fibre quality and can vary between 0.5% and 22% (Roos, 2012).

1.9.3 Fabric production

Textile fabric is constructed either as nonwoven, tricot or weave. Weaving is the most energy intensive process using both electricity and compressed air, and energy consumption depends on the type of equipment (Koç and Çinçik, 2010). Knitting is described in the EDIPTEX report (Laursen et al., 2007). Nonwoven fabrics have relatively low energy consumption. For plants with low energy consumption of the actual technical processes, the supplementary processes such as ventilation, air conditioning and personnel facilities are usually the main contributors to the carbon footprint (Elle, 2010). This is problematic when it comes to explaining the differences between the values in the Ecoinvent database (Ecoinvent Centre, 2010), which has a consistent structure of including also infrastructure such as machinery and buildings versus the IDEMAT database (Vogtländer, 2012) where documentation is lacking regarding the inclusion of supplementary processes and infrastructure.

1.9.4 Wet treatment

Textile wet treatment technology is very diverse, but the Best Available Techniques Reference Document (BREF) for the Textile Industry (European Commission, 2003) provides a useful overview. Energy consumption is in general linked to the water ratio (kg water needed per kg textile) and the number of processes performed. However, continuous processing, with lower water ratios compared to batch processing, has in general a higher chemical consumption, and a broader perspective than only carbon footprint impact should be applied to take environmental impacts of chemicals into account. The carbon footprint of wet treatment plants is to a large extent dependent on the percentage of output that is dyed 'right first time'. Also, the waste water treatment is crucial for the environmental performance although this also contributes to the carbon footprint (Roos et al., 2015).

1.9.5 Finishing and sewing

Finishing can include dry coating, wet coating, printing, tearing and spraying to give the finished product the desired visual and technical properties. This is seldom described in the literature. Cutting, ironing and sewing are described in several publications (Laursen et al., 2007). For plants with low energy consumption of the actual technical processes, the supplementary processes such as ventilation, air conditioning and personnel facilities are usually the main contributors to the carbon footprint (Elle, 2010).

Acknowledgement

This work was financed by MISTRA – the Swedish Foundation for Strategic Environmental Research.

References

Allwood, J.M., Laursen, S.E., Malvido de Rodríguez, C., Bocken, N.M.P., 2006. Well Dressed? The Present and Future Sustainability of Clothing and Textiles in the United Kingdom. University of Cambridge, Institute for Manufacturing, Cambridge, UK.

Alvarez-Gaitan, J.P., Peters, G.M., Rowley, H.V., Moore, S., Short, M.D., 2013. A hybrid life cycle assessment of water treatment chemicals: an Australian experience. International Journal of Life Cycle Assessment 18, 1291–1301.

Astudillo, M.F., Thalwitz, G., Vollrath, F., 2014. Life cycle assessment of Indian silk. Journal of Cleaner Production 81, 158–167.

Barber, A., Pellow, G., 2006. Merino Wool – Life Cycle Assessment: New Zealand Merino Industry. Total Energy Use and Carbon Dioxide Emissions. The AgriBusiness Group, Pukeeohe, New Zealand.

Baumann, H., Tillman, A.-M., 2004. The Hitch Hiker's Guide to LCA. Studentlitteratur, Malmö, Sweden.

Beton, A., Dias, D., Farrant, L., Gibon, T., Le Guern, Y., Desaxce, M., Perwueltz, A., Boufateh, I., Wolf, O., Kougoulis, J., Cordella, M., Dodd, N., 2014. Environmental Improvement Potential of Textiles (IMPRO Textiles). Publications Office of the European Union, European Union, Luxembourg. http://dx.doi.org/10.2791/52624.

Bevilacqua, M., Ciarapica, F.E., Giacchetta, G., Marchetti, B., 2011. A carbon footprint analysis in the textile supply chain. International Journal of Sustainable Engineering 4 (1), 24–36.

Bhurtun, C., Kistamah, N., Chummun, J., 2006. Energy saving strategies in textile industry: the case of Mauritius. In: Proceedings of Industrial and Commercial Use of Energy Conference 2006, pp. 53–57. Cape Town, South Africa.

Bidoki, S., Wittlinger, R., 2006. Eco-efficiency analysis of textile coating materials. Journal of the Iranian Chemical Society 3 (4), 351–359.

Boardman, B., 2008. Carbon labelling: too complex or will it transform our buying? Significance 5 (4), 168–171.

Brandão, M., Levasseur, A., Kirschbaum, M., Weidema, B., Cowie, A., Jørgensen, S., Hauschild, M., Pennington, D., Chomkhamsri, K., 2013. Key issues and options in accounting for carbon sequestration and temporary storage in life cycle assessment and carbon footprinting. International Journal of Life Cycle Assessment 18, 230–240.

Browne, M., Rizet, C., Anderson, S., Allen, J., Keïta, B., 2005. Life cycle assessment in the supply chain: a review and case study. Transport Reviews 25, 761–782.

BSI, 2011. PAS 2050:2011 Specification for the Assessment of the Life Cycle Greenhouse Gas Emissions of Goods and Services. British Standards Institute. http://shop.bsigroup.com/en/forms/PASs/PAS-2050/ (accessed October 2014).

CarbonCareAsia, 2014. CarbonCare® Label. http://www.carboncareasia.com/eng/CCL/CCL.php (accessed May 2015).

Carbonfund, 2012. CarbonFree® Product Certification Carbon Footprint Protocol. Version 4.0. http://www.carbonfund.org/site/2012-03-15%20-%20Product%20Certification%20Protocol%20-%20Final.pdf (accessed May 2015).

Carbon Trust, 2015. Carbon footprint labels from the Carbon Trust. http://www.carbontrust.com/client-services/footprinting/footprint-certification/carbon-footprint-label (accessed May 2015).

Chapman, A., 2010. Mistra Future Fashion – Review of Life Cycle Assessments of Clothing. Mistra – The Foundation for Strategic Environmental Research, Stockholm, Sweden.

Cherrett, N., Barrett, J., Clemett, A., Chadwick, M., Chadwick, M.J., 2005. Ecological Footprint and Water Analysis of Cotton, Hemp and Polyester. Stockholm Environment Institute, Sweden. ISBN 91 975238 2 8.

Collins, M., Aumônier, S., 2002. Streamlined Life Cycle Assessment of Two Marks & Spencer plc Apparel Products. Environmental Resources Management Ltd, Oxford, UK.

De Saxce, M., Pesnel, S., Perwuelz, A., 2012. LCA of bed sheets – some relevant parameters for lifetime assessment. Journal of Cleaner Production 37, 221–228.

De Vries, M., de Boer, I.J.M., 2010. Comparing environmental impacts for livestock products: a review of life cycle assessments. Livestock Science 128 (1–3), 1–11.

EC, 2013. 2013/179/EU: Commission Recommendation of 9 April 2013 on the Use of Common Methods to Measure and Communicate the Life Cycle Environmental Performance of Products and Organisations. European Commission. http://ec.europa.eu/environment/eussd/smgp/dev_pef.htm (accessed October 2014).

Ecoinvent Centre, 2010. Ecoinvent Data v2.2. Final reports ecoinvent 2010. Swiss Centre for Life Cycle Inventories, Dübendorf.

Key Results. In: Elle, M. (Ed.), 2010. ENGINE, Energy Efficiency in SME. ENGINE Consortium, Target GmbH, Hannover, Germany.

European Commission, 2003. Integrated Pollution Prevention and Control (IPPC) Reference Document on Best Available Techniques for the Textiles Industry. European IPPC Bureau, Seville, Spain.

Forster, P., Ramaswamy, V., Artaxo, P., Berntsen, T., Betts, R., Fahey, D.W., Haywood, J., Lean, J., Lowe, D.C., Myhre, G., Nganga, J., Prinn, R., Raga, G., Schulz, M., Van Dorland, R., 2007. Changes in atmospheric constituents and in radiative forcing. In: Climate Change 2007: The Physical Science Basis IPCC 2007. Contribution of Working Group I to the Fourth Assessment Report of the Intergovernmental Panel on Climate Change.

Grace, P., Gane, M., Garcia, F., 2009. Life Cycle Assessment of a 100% Australian Cotton T-shirt. Queensland University of Technology, Brisbane, Australia.

GRI, 2014. Sustainability Reporting Guidelines – Implementation Manual. Global Reporting Initiative. https://www.globalreporting.org/resourcelibrary/GRIG4-Part2-Implementation-Manual.pdf (accessed October 2014).

Gunn, J.S., Ganz, D., Keeton, W., 2012. Biogenic vs geologic carbon emissions and forest biomass energy production. GCB Bioenergy 4, 239–242.

Helin, T., Sokka, L., Soimakallio, S., Pingoud, K., Pajula, T., 2013. Approaches for inclusion of forest carbon cycle in life cycle assessment – a review. GCB Bioenergy 5 (5), 475–486.

IPCC, 2007. Climate Change 2007: The Physical Science Basis. In: IPCC Fourth Assessment Report: Climate Change 2007. Intergovernmental panel on climate change (IPCC). Geneva, Switzerland. Retrieved from: https://wwwipcc.ch/publications_and_data/ar4/wg1/en/ch2s2-10-2.html.

IPCC, 2013. In: Stocker, T.F., Qin, D., Plattner, G.-K., Tignor, M., Allen, S.K., Boschung, J., et al. (Eds.), Climate Change 2013: The Physical Science Basis. Working Group I Contribution to the Fifth Assessment Report of the Intergovernmental Panel of Climate Change. Cambridge University Press, Cambridge, United Kingdom and New York, NY, USA. Available at: http://www.ipcc.ch/report/ar5/wg1/ (accessed October 2014).

IPCC, 2014. Climate Change 2014 Synthesis Report – Summary for Policymakers. Intergovernmental Panel on Climate Change. http://www.ipcc.ch/pdf/assessment-report/ar5/syr/SYR_AR5_SPMcorr1.pdf.

ISO, 2013. ISO/TS 14067:2013 Greenhouse Gases – Carbon Footprint of Products – Requirements and Guidelines for Quantification and Communication. International Standardization Organization. http://www.iso.org/iso/home/store/catalogue_tc/catalogue_detail.htm?csnumber=59521 (accessed October 2014).

Johnson, E., 2009. Goodbye to carbon neutral: getting biomass footprints right. Environmental Impact Assessment Review 29, 165–168.

JRC, March 2010. Joint Research Centre – Institute for Environment and Sustainability: International Reference Life Cycle Data System (ILCD) Handbook – General Guide for Life Cycle Assessment – Detailed Guidance, first ed. Publications Office of the European Union, Luxembourg. EUR 24708 EN.

Jungmichel, N., 2010. The Carbon Footprint of Textiles. Systain Consulting. http://www.ci-romero.de/fileadmin/media/informieren-themen/gruene_mode/Jungmichel._Systain.pdf (accessed September 2014).

Kalliala, E., Nousiainen, P., 1999. Environmental profile of cotton and polyester-cotton fabrics. Autex Research Journal 1, 8–20.

Koç, E., Çinçik, E., 2010. Analysis of energy consumption in woven fabric production. Fibres and Textiles in Eastern Europe 79 (2), 14–20.

Labouze, E., Le Guern, Y., des Abbayes, C., 2006. Analyse de Cycle de Vie d' un Pantalon en Jean Sommaire. Ivry-sur-Seine, France. Retrieved from: http://www.biois.com/wp/wp-content/uploads/2011/02/BIO_Intelligence_Service_ACV_jean_ADEME.pdf.

Laursen, S.E., Hansen, J., Knudsen, H.H., Wenzel, H., Larsen, H.F., Kristensen, F.M., 2007. EDIPTEX – Environmental Assessment of Textiles. Danish Environmental Protection Agency. Working Report No. 24. Retrieved from: http://www2.mst.dk/udgiv/publications/2007/978-87-7052-515-2/pdf/978-87-7052-516-9.pdf.

Levasseur, A., Lesage, P., Margni, M., Deschenes, L., Samson, B., 2010. Considering time in LCA: dynamic LCA and its application to global warming impact assessments. Environmental Science and Technology 44, 3169–3174.

Murugesh, K., Selvadass, M., 2013. Life cycle assessment for the dyeing and finishing process of organic cotton knitted fabrics. Journal of Textile and Apparel, Technology and Management 8 (2), 7.

Muthu, M., 2014. Assessing the Environmental Impact of Textiles and the Clothing Supply Chain. Woodhead Publishing. ISBN 9781782421047.

Muthu, S.S., Li, Y., Hu, J.Y., Ze, L., 2012. Carbon footprint reduction in the textile process chain: recycling of textile materials. Fibers and Polymers 13 (8), 1065–1070.

Nagel, S., 2010. Exploring ways to reduce greenhouse gas emissions in the textile supply chain. In: Bachelors of Industrial Engineering Thesis, Faculty of Engineering, Built Environment and Information Technology. University of Pretoria.

Palme, A., Idström, A., Nordstierna, L., Brelid, H., 2014. Chemical and ultrastructural changes in cotton cellulose induced by laundering and textile use. Cellulose 2 (6), 4681–4691.

Peters, G., Granberg, H., Sweet, S., 2014. The role of science and technology in sustainable fashion. Chapter 18. In: Fletcher, K., Tham, M. (Eds.), The Handbook of Sustainable Fashion. Routledge, UK. ISBN 978-0-415-82859-8.

Reap, J., Roman, F., Duncan, S., Bras, B., 2008. A survey of unresolved problems in life cycle assessment, Part 2: impact assessment and interpretation. International Journal of Life Cycle Assessment 13, 374–388.

Roos, S., Posner, S., Jönsson, C., Peters, G.M., 2015. Is unbleached cotton better than bleached? Exploring the limits of life cycle assessment in the textile sector. Clothing and Textiles Research Journal. http://dx.doi.org/10.1177/0887302X15576404.

Roos, S., 2012. Livscykelanalys av Tencelfiber. Swerea IVF Report 23497, Swerea IVF AB, Mölndal, Sweden.

Roos, S., Sandin, G., Zamani, B., Peters, G.M., 2015. Environmental assessment of Swedish fashion consumption. Five garments - sustainable futures. A Mistra Future Fashion Project 2 Report. Chalmers University of Technology, Gothenburg, Sweden.

Røyne, F., Peñaloza, D., Sandin, G., Svanström, M., Berlin, J. Climate impact assessment in life cycle assessments of forest products: Implications of method choice for results and decision-making. Journal of Cleaner Production, in press.

Sandin, G., Peters, G.M., Svanström, M., 2013. Moving down the cause-effect chain of water and land use impacts: an LCA case study of textile fibres. Resources, Conservation and Recycling 73, 104–113.

Sandin, G., Peters, G.M., Svanström, M., 2014. Life cycle assessment of construction materials: the influence of assumptions in end-of-life modeling. International Journal of Life Cycle Assessment 19, 723–731.

Sandin, G., Røyne, F., Berlin, J., Peters, G.M., Svanström, G.M., 2015. Allocation in LCAs of biorefinery products: implications for results and decision making. Journal of Cleaner Production 93, 213–221.

Shen, L., Patel, M.K., 2008. Life cycle assessment of polysaccharide materials: a review. Journal of Polymers and the Environment 16 (2), 154–167.

Shen, L., Worrell, E., Patel, M.K., 2010. Environmental impact assessment of man-made cellulose fibres. Resources, Conservation and Recycling 55, 260–274.

Smith, G.G., Barker, R.H., 1995. Life cycle analysis of a polyester garment. Resources, Conservation and Recycling 14, 233–249.

Steinberger, J.K., Friot, D., Jolliet, O., Erkman, S., 2009. A spatially explicit life cycle inventory of the global textile chain. International Journal of Life Cycle Assessment 14, 443–455.

Tan, M.Q.B., Tan, R.B.H., Khoo, H.H., 2014. Prospects of carbon labelling – a life cycle point of view. Journal of Cleaner Production 72, 76–88.

Terinte, N., Manda, B.M.K., Taylor, J., Schuster, K.C., Patel, M.K., 2014. Environmental assessment of coloured fabrics and opportunities for value creation: spin-dyeing versus conventional dyeing of modal fabrics. Journal of Cleaner Production 72, 127–138.

Thomas, B., Fishwick, M., Joyce, J., van Santen, A., 2012. A Carbon Footprint for UK Clothing and Opportunities for Savings. Final Report. WARP. http://www.wrap.org.uk/sites/files/wrap/Appendix%20IV%20-%20Carbon%20footprint%20report.pdf (accessed September 2014).

Upham, P., Dendler, L., Bleda, M., 2011. Carbon labelling of grocery products: public perceptions and potential emissions reductions. Journal of Cleaner Production 19 (4), 348–355.

Van der Velden, N.M., Patel, M.T., Vogtländer, J.G., 2014. LCA benchmarking study on textiles made of cotton, polyester, nylon, acryl, or elastane. International Journal of Life Cycle Assessment 19, 331–356.

Vanclay, J.K., Shortiss, J., Aulsebrook, S., Gillespie, A.M., Howell, B.C., Johanni, R., Maher, M.J., Stewart, M.D., Yates, J., 2010. Customer Response to Carbon Labelling of Groceries. Journal of Consumer Policy 34 (1), 153–160.

Vogtländer, J.G., 2012. IDEMAT 2012 Database. Retrieved at: www.ecocostsvalue.com.

Wackernagel, M., Monfreda, C., Moran, D., Goldfinger, S., Deumling, D., Murray, M., 2004. National Footprint and Biocapacity Accounts 2004: The Underlying Calculation Method. Global Footprint Network, Oakland, CA, USA.

Walser, T., Demou, E., Lang, D.J., Hellweg, S., 2011. Prospective environmental life cycle assessment of nanosilver T-shirts. Environmental Science and Technology 45 (10), 4570–4578.

WRI/WBCSD, 2004. The Greenhouse Gas Protocol: A Corporate Reporting and Accounting Standard, Revised edition. World Resources Institute/World Business Council for Sustainable Development. http://www.ghgprotocol.org/files/ghgp/public/ghg-protocol-revised.pdf. (accessed October 2014).

WRI/WBCSD, 2011a. Greenhouse Gas Protocol Product Life Cycle Accounting and Reporting Standard. World Resources Institute/World Business Council for Sustainable Development. http://www.ghgprotocol.org/files/ghgp/public/Product-Life-Cycle-Accounting-Reporting-Standard_041613.pdf (accessed October 2014).

WRI/WBCSD, 2011b. Corporate Value Chain (Scope 3) Accounting and Reporting Standard – Supplement to the GHG Protocol Corporate Accounting and Reporting Standard. World Resources Institute/World Business Council for Sustainable Development. http://www.ghgprotocol.org/files/ghgp/public/Corporate-Value-Chain-Accounting-Reporing-Standard_041613.pdf (accessed October 2014).

Yuan, Z.-W., Zhu, Y.-N., Shi, J.-K., Liu, X., Huang, L., 2012. Life-cycle assessment of continuous pad-dyeing technology for cotton fabrics. International Journal of Life Cycle Assessment 18 (3), 659–672.

Zamani, B., Svanström, M., Peters, G., Rydberg, T., 2014. A carbon footprint of textile recycling: case study – Sweden. Journal of Industrial Ecology. http://dx.doi.org/10.1111/jiec.12208.

Energy footprints in the textile industry

2

S. Palamutcu
Pamukkale University, Denizli, Turkey

2.1 Introduction: key issues in calculating energy footprints

2.1.1 Energy and energy footprint

Energy is defined as the ability or the capacity to do work, and it can be converted from one form to another form. The root of the word *energy* is in the late Latin word *energia* that means "activity and operation." A broader meaning of the word is "power" in the English language as it is in most other languages (Etymonline). There are many forms of energy, including mechanical, solar, thermal, chemical, electrical, electromagnetic, and nuclear. Conversions of different forms of energy from one to the other are possible based on the thermodynamic laws of "energy conservation". Conversion of energy most of the time requires multiple transitions, for example, for nuclear power generation the split of nuclear fuels into their atoms is required to be able to release nuclear energy and thermal energy correspondingly. Subsequently generated heat energy may later be captured in steam form and utilized to drive any turbine generator or any other system or machine requiring kinetic energy. Kinetic energy may be used to spin a magnetic field around a conductor, creating electrical energy. Successively created energy types of heat or electric may commonly be utilized in most of the manufacturing industries, including the textile industry.

Measurement of energy is the most important issue from the viewpoints of efficiency and cost. To measure the energy, "heating value," of a fuel, a certain amount of "consumed fuel" and certain amount of "converted heat" data are utilized. Following are units and conversion factors that are commonly used (Tables 2.1 and 2.2).

The textile industry has become one of the industries concerning about sustainable initiatives in the world. Global textile markets have lately been involved with environmental approaches that take into account consumption of water, energy, and other natural sources during production and utilization stages of consumers. The goal is that textile production stages, consumer utilization stages, and recycling stages should cause minimal negative impact to the environment. The relevant terminology of energy efficiency, carbon emissions, water, chemical, and energy footprint terms have become very familiar to both the textile manufacturing sector and consumers.

Energy footprint is the term used to explain the costs of climate change due to carbon dioxide (CO_2) releases of energy production and consumption of any human

Handbook of Life Cycle Assessment (LCA) of Textiles and Clothing. http://dx.doi.org/10.1016/B978-0-08-100169-1.00002-2

Table 2.1 Commonly used energy units (World Energy Outlook, 2013)

Mtce:	Million tons of coal equivalent (equals 0.7 Mtoe)
Boe:	Barrels of oil equivalent
Toe:	Ton of oil equivalent[a]
ktoe:	Kilotons of oil equivalent
Mtoe:	Million tons of oil equivalent
MBtu:	Million British thermal units
kcal:	Kilocalorie (1 calorie*10^3)
Gcal:	Gigacalorie (1 calorie*10^9)
MJ:	Megajoule (1 joule*10^6)
GJ:	Gigajoule (1 joule*10^9)
TJ:	Terajoule (1 joule*10^{12})
PJ:	Petajoule (1 joule*10^{15})
EJ:	Exajoule (1 joule*10^{18})
kWh:	Kilowatt-hour
MWh:	Megawatt-hour
GWh:	Gigawatt-hour
TWh:	Terawatt-hour

[a]Term of ton equivalent oil "toe" is equal to the amount of energy that results from burning one ton of oil.

Table 2.2 General conversion factors for energy (World Energy Outlook, 2013)

Convert to	TJ	Gcal	Mtoe	mBtu	GWh
From:	Multiply by:				
TJ	1	238.8	2.388*10^{-5}	947.8	0.2778
Gcal	4.1868*10^{-3}	1	10^{-7}	3.968	1.163*10^{-3}
Mtoe	4.1868*10^4	10^7	1	3.968*10^7	11 630
MBtu	1.0551*10^{-3}	0.252	2.52*10^{-8}	1	2.931*10^{-4}
Gwh	3.6	860	8.6*10^{-5}	3 412	1

activities. *Unit of energy footprint* is defined as the measure of land required to absorb the CO_2 emissions (Gdrc). This approach focuses on the outcome of energy use, which is CO_2 emissions, to understand the environmental cost of the problem. The environmental cost of CO_2 emissions keeps increasing as a result of rising urbanization rates, intensive industrial activity, climate change, and transitions in land use and agricultural practices.

The earliest approaches about ecological footprint are defined in the study of Wackernagel and Rees (1996), where land need for fossil energy can be calculated in three ways that all lead to similar results (Stoglehner, 2003).

In the first method, the footprint of fossil fuels is expressed by the land need of an energy carrier produced by agriculture or forestry, that is, the area that is necessary for the production of the same amount of energy as generated by fossil fuels.

The second method that is commonly used for ecological footprint calculation suggests calculating the land need for fossil energy consumption by assessing the forest area that is necessary to absorb the carbon dioxide emissions generated by burning fossil fuels.

In the third method, the land need for fossil energy is estimated by the area needed to compensate the amount of energy burned from fossil fuels with renewable resources at the same rate at which the fossil energy is depleted. For instance, forests with the same energy content as the burned fossil fuels have to be planted. According to this philosophy, the energy contained in the earth should not decrease.

2.1.2 Energy consumption of world textile industry

The textile and clothing industry, which will be addressed here as the *textile industry*, is one of the most globalized, labor intensive and consumer-oriented industries in the world. The textile sector involves the manufacturing of natural and man-made fibers and fiber-based textile products. Production and consumption of fiber-based products have been increasing as a result of population growth of emerging markets, rising life standards, age structure in the developing countries, and increases of textile-based product types and amounts of their consumption.

Per capita fiber consumption has been increased by nearly 39%, from 8.5 kg in 2000 to 12.0 kg, roughly 82 million tons in total for the year 2011 (Icac; Saskflax; Textileworld). Annual fiber market growth has been varying between 1.5% and 3.0%, correlating with growth in population and increase in real per capita incomes. Changing economic conditions in various regions of the world affect the fiber and textile market growth, however, on average, 120,000–240,000 tons of additional textile fibers in quantity terms are consumed in the world every year (Saskflax). Increasing trends in the fiber demand are shown in Figure 2.1.

According to the records of the World Trade Organization (WTO), annual revenue of the textile industry associated with clothing is about USD 710 billion representing about 1.9% of world trade volume for the year 2012 (WTO, 2013). In 2015, the global market for textiles is forecast to have a value of USD 1.6 trillion, with an increase of 32.5% since 2010 (Lenzing). It can therefore be emphasized that the textile industry is one inevitable component of the global manufacturing sector

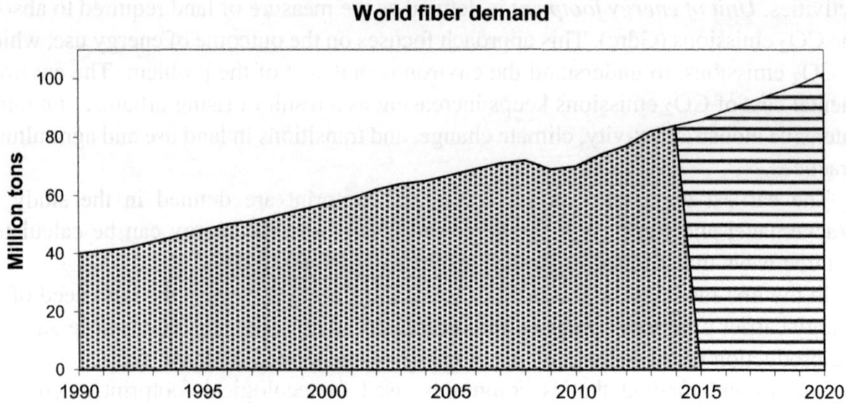

Figure 2.1 World fiber demand, million tons.

and is a definite contributor to the global warming problem with its manufacturing processes, distribution and transportation of products to stores and customers, and recycling stages.

Energy consumption volume in the textile industry is increasing along with the rising demand of textile-based product types. Between the years 1971 and 2004, final energy use in the world textile sector doubled from 4.7 ExaJoule (EJ)/year to 9.0 EJ/year (International Energy Agency, 2008), which is about 2% of total world energy use, where total fiber consumption was about 60 million tons/year. Energy requirements of the textile industry with utilization of over 80 million tons fiber in 2012 are estimated about 1.9% (Energy [R]Evolution, 2012) of total world energy consumption.

Total world energy consumption estimations vary according to different sources: Ma et al. (2012) report total world energy consumption at 475 EJ with 2005 data; according to the report of Energy [R]Evolution (Energy [R]Evolution, 2012), consumption was 304 EJ in 2009 and expected to grow to 350 EJ in 2015; according to the report of BP (BP, 2014), consumption was 550 EJ at the primary energy level in 2010, and the current energy demand of the world textile industry is raised to the 13,000 Mtoe, which is equivalent to 544 EJ (Theoildrum) (Figure 2.2); and 14,920 Mteo, which is equivalent to 625 EJ, of estimated primary energy demand for 2035 (International Energy Agency, 2013).

Regarding gathered data about world annual energy consumption, and the amount of total fiber demand, a unit energy requirement of 1 kg textile product can be calculated using the simple approach of division. When the world annual energy consumption of the textile industry is taken at 10.33 EJ (1.9% of world industrial energy consumption of 544 EJ/year) and produced textile products are 82 million tons as rough numbers, unit energy consumption of 1 kg of textile on average can be calculated as 126.05 MJoule (MJ) as a rough estimation.

Related to energy utilization, CO_2 emissions occur as an undesired by-product of energy consumption. CO_2 is naturally present in the atmosphere as part of the earth's carbon cycle (the natural circulation of carbon among the atmosphere, oceans, soil,

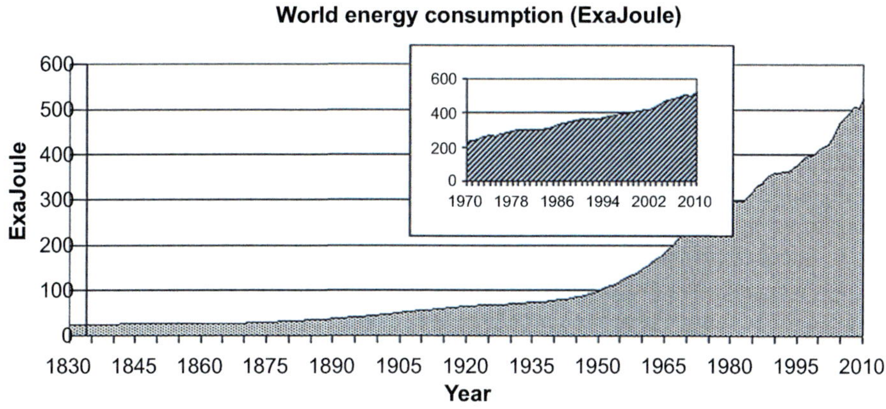

Figure 2.2 World energy consumption (ExaJoule), 1830–2010 (Source: Theoildrum).

plants, and animals), and it is the primary greenhouse gas emitted through human activities. The increasing volume of human activities, starting with the industrial revolution, has dramatically added CO_2 to the atmosphere.

Change in the total world CO_2 emissions in million tonness is shown in Figure 2.3, for the years 1965–2013.

Sources of human activity that result in CO_2 emissions are electricity, transportation, industry, residential–commercial, and nonfossil fuel combustion. The share of industrial purposed energy demand, including textile industry, is about one-third of total global energy demand that is consumed largely by chemicals-petrochemicals, iron-steel, cement, paper, and aluminum industries. CO_2 emission of the textile industry has been reported as about 2% of total industrial CO_2 emission of the world, estimated to be about 0.18 gigatons/year (Ngai et al., 2012).

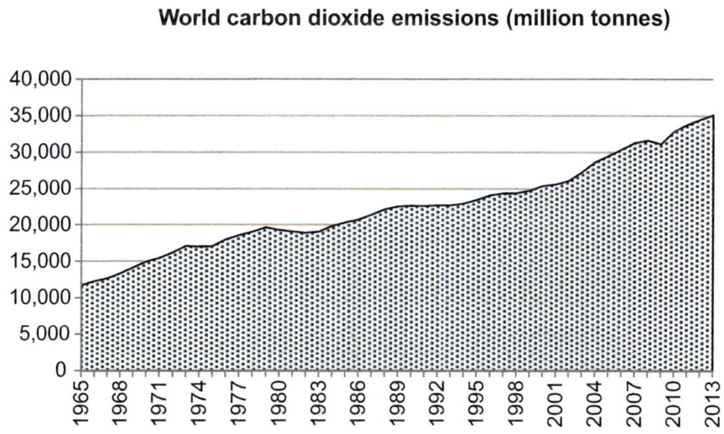

Figure 2.3 World CO_2 emissions (BP, 2014).

2.1.3 Textile production stages

The textile manufacturing process is divided into five main substages of fiber processing, spinning, surface structuring of weaving, knitting, and nonwoven processing, wet processing, clothing or any other consumer-oriented product-associated textile manufacturing stages. General production path line of a woven textile product is given in Figure 2.4. The dotted boxes 1 and 2 represent primary and secondary energy consuming applications in an ordinary textile production plant. The boxes labeled A, B, and C represent the primary energy consuming production lines of spinning, sizing, weaving; wet processing; and clothing manufacturing. Heating, cooling, sanitation, transportation, exterior, and other energy-consuming applications are classified as secondary energy-consuming applications (EMS-Energy).

Textile fiber types are mainly classified into two categories of natural and man-made fibers. Natural fiber, mainly cotton, is an agricultural product, and during its cultivation large amounts of water, insecticides, pesticides, and artificial fertilizers are consumed besides a relatively low amount of energy. Man-made fiber processing requires higher amounts of energy compared to natural fiber. Adding to the high energy consumption, certain manufacturing stages of man-made fibers release toxic by-products.

Spinning processes are mostly involved with fiber separation, ventilation, and guidance that require mechanical and air movements. Production purposed, air conditioning and ventilation purposed machinery and systems of spinning lines are reported as electric energy-intensive processes.

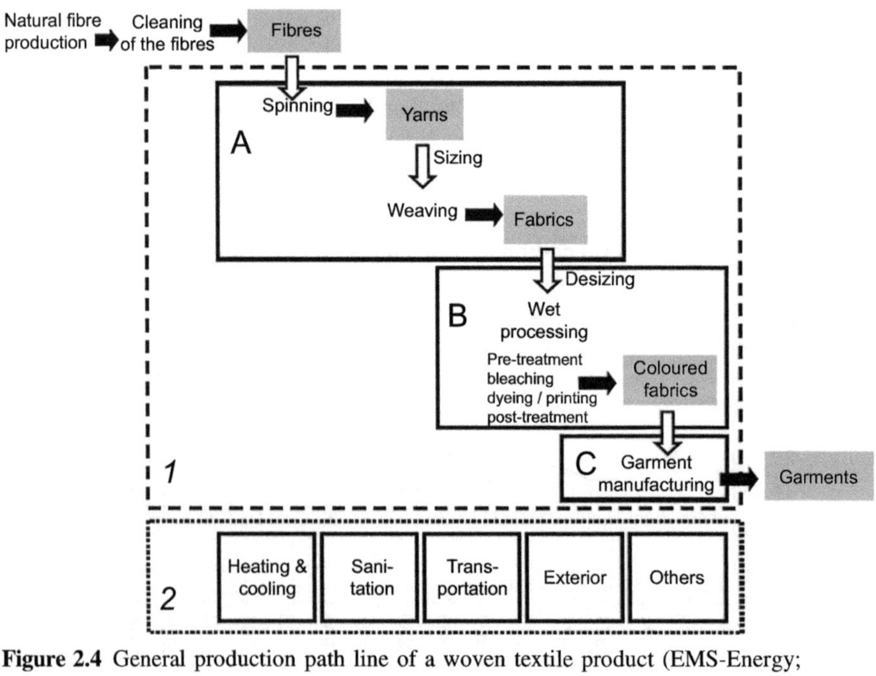

Figure 2.4 General production path line of a woven textile product (EMS-Energy; Palamutcu, 2010).

Textile surface structuring processes, which are basically classified as weaving, knitting, and nonwoven, vary based on their processing steps and machinery types. They are mostly dry operations and mostly consume electric energy for machine driving and air conditioning purposes.

Wet processing steps can be roughly classified as pretreatment, bleaching, washing, dyeing/printing, and posttreatments. The sizing process of warp yarns before weaving processes can also be classified as wet processing. All of the wet processing steps require hot water, steam, and/or hot dry air, which are mainly warmed up using fossil fuels or electric energy.

Clothing production stages are labor-intensive processes. Cutting, sewing, and ironing purposed machineries are utilized and electric energy is consumed.

Consumed energy type of traditional cotton textile production line is given in Table 2.3. As a whole, processing steps of fiber opening, yarn processing, textile surface generation, any kind of wet processing, and apparel manufacturing operations; energy costs of electricity and heat constitute about 8–10% of total production cost (Palamutcu, 2010).

2.1.4 Energy types used in textile production stages

Textile and clothing manufacturing processes have their own energy consumption characteristics. Consumed energy types and energy intensity of each process vary; some processes require more electricity while others requires more heat energy. The share of electricity and fuels within the total final energy use of any one country's textile sector depends on the structure of the textile industry in that country (Hasanbeigi, 2010).

Electric energy is the dominant energy type comparing the fuel-based heat energy; the highest electric energy consumption in the textile production stages belongs to the spinning together with twisting and texturing processes (Figure 2.5). Weaving, knitting, and clothing processes are also electric energy-intensive production processes, while man-made fiber production and dyeing (wet processing) stages highly consume fuel-based heat energy.

It is estimated from the UNIDO report (UNIDO, 1992) that electricity consumption rates in the total consumed energy for individual textile production stages are 93% for spinning, 85% for weaving, 43% for wet processing, and 65% for clothing manufacturing. The rest of the energy that is used in the textile processing plants is supplied by other energy sources of fuel, natural gas, and coal.

In a study by Ongüt, the share range of energy costs in the total production cost of Turkish textile production stages are given as 5–10% for yarn production, 5–8% for weaving processes, and 5–8% for clothing processes (Ongüt, 2007).

Cost of energy consumption shares of a man-made fiber-based textile production stages are shown for the Indian textile industry in Figure 2.6.

The dyeing process is reported as consuming the highest share of energy (25.6%). Man-made fiber production is in second place (21.0%) concerning the energy consumption share in total energy consumption of the textile sector in India. Following the energy consumption share of man-made fiber production, spinning processes spend

Table 2.3 **Energy map of a cotton fabric manufacturing machinery line in general**

Name of production process	Machine name	Consumed energy type
Cotton spinning	Opening-blending line	Electricity
	Carding machine	
	Drawframes	
	Roving machine	
	Ring spinning machine	
	Winding machine	
Warping and sizing	Serial warping machine	Electricity
	Conical warping machine	Electricity
	Sizing machine	Electricity + steam
Weaving or knitting	Weaving machines	Electricity
	Knitting machines	
	Quality control machinery	
Wet processing	Prefinishing line	Electricity + steam
	Washing line	Electricity + steam
	Drying line	Electricity + hot oil
	Dyeing machine	Electricity
	Finishing line	Electricity
	Volume machine	Electricity + hot oil
	Opening and rewinding machines	Electricity
	Quality control machinery	Electricity
Cutting-sewing and ironing	Long side cutting machine	Electricity
	Long side hemming machine	
	Sewing machine	
	Overlock machine	
	Ironing	

a high share of energy (19.7%). The energy consumption share of weaving, clothing, and knitting processes in the Indian textile industry is 11.4%, 9.2%, and 6.1%, respectively (Figure 2.6).

Total energy consumption of man-made fiber textile products is generally accepted as higher than those textile items that are produced using natural fibers.

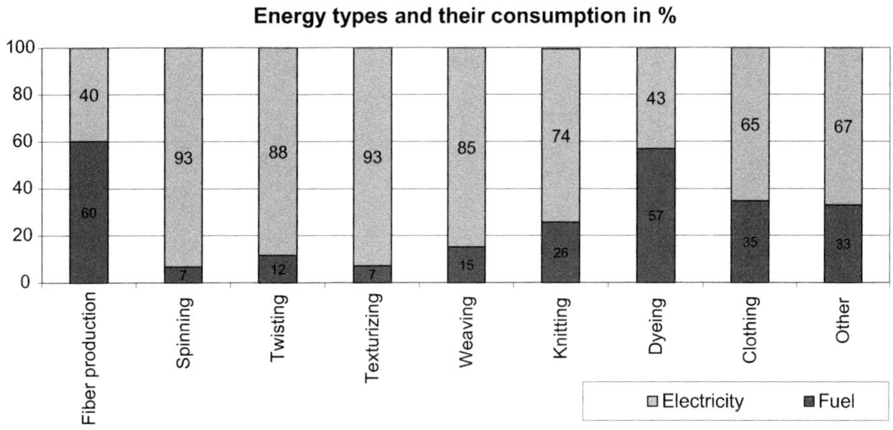

Figure 2.5 Energy types and their share in textile processing stages (Source: S. Sharma).

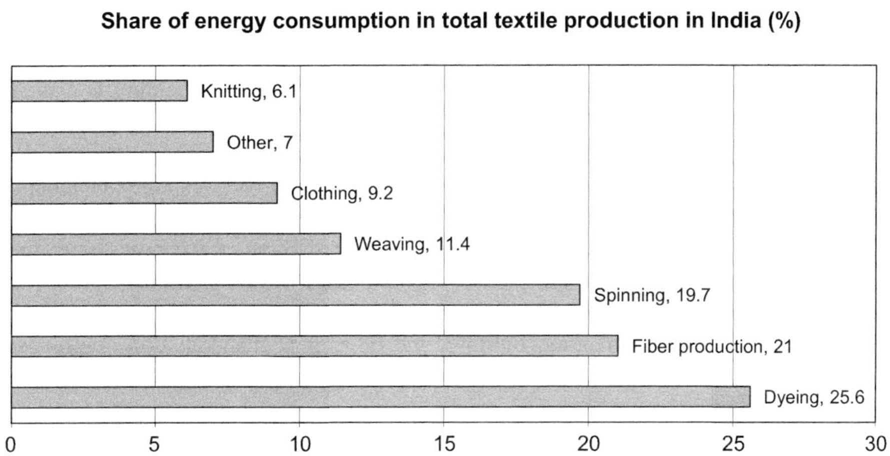

Figure 2.6 Share of energy consumption in total energy consumption of textile industry in India (Data from S. Sharma).

Such an amount of consumption is not encountered in the cost of natural fiber-based textile products. In the case of cotton fiber, energy consumption share is comparatively low and belongs to the agricultural activities, transportation, and ginning process.

Breakdown of final energy use by end use in the US textile industry is given in Figure 2.7, where steam production, motor driving purposed energy consumption has the highest share of 28%. Industrial process heating purposed energy consumption's share and facility-based consumption's share are 20% and 18%, respectively. The lowest level of 4% and 2% energy consumptions, respectively, belong to the process cooling and other utilization purposed consumptions.

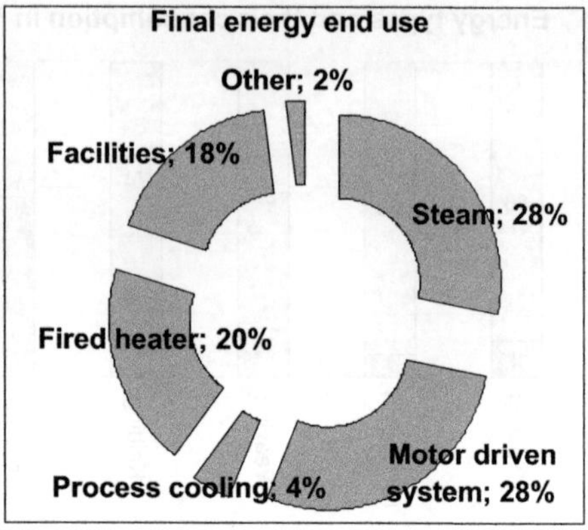

Figure 2.7 Final energy end use in the US textile industry (U.S. DOE, 2004; Sathaye et al., 2005).

Electricity and thermal energy consumption rates of different production departments in an integrated Indian textile plant that has spinning, weaving/knitting, and wet processing (preparation, dyeing/printing, finishing) all on the same site is given in Figure 2.8. According to the study of Sathaye et al. (2005), spinning consumes the greatest share of electricity (41%), followed by weaving (weaving preparation and weaving; 18%). Wet-processing preparation (desizing, bleaching, etc.) and finishing together consume the greatest share of thermal energy (35%). A significant amount of thermal energy is also lost during steam generation and distribution (35%).

Electric energy is intensively used for machine driving, air conditioning, and ventilation in the man-made fiber processing, spinning, weaving-knitting-nonwoven, and clothing production stages. Besides machine driving purposed consumption of electric energy, cooling, heating, lighting, and other miscellaneous consumption of electricity in the textile plants are encountered.

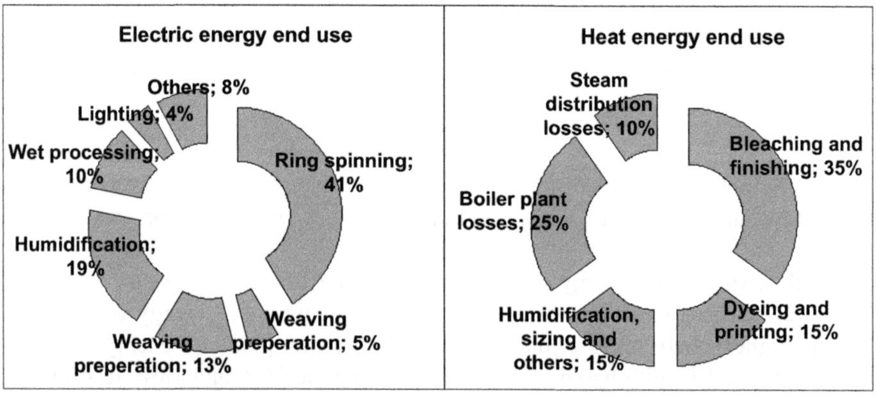

Figure 2.8 Breakdown of electric and heat use in an integrated textile production plant (Sathaye et al., 2005; Energymeasuretosave).

Heat energy is mainly acquired by burning of any fuel material and is consumed in two basic operations in an ordinary textile plant: heating water and removing soaked water from the textile structure. Besides dyeing and other wet-processing utilization, water and oil heating, steam production, and ironing purposed consumption of heat energy are encountered in an ordinary textile production plant.

2.2 Methodologies and standards

Textile production, utilizations, and recycling steps are complicated operations with variable machinery parks, different energy types, and varying energy intensities. Methodologies and standards used to define and characterize the energy consumption and energy footprint of textile products also vary, which threatens the consistency of a considerable amount of data about the textile sector. Regulations, reports, scientific works, and studies about energy consumption measurements, environmental affects evaluation, and life cycle assessment (LCA) works for textile products and production methods have recently increased as result of wild energy cost policies and raising environmental awareness worldwide. Methods, models, approaches, and standards that are used to designate energy footprints in the textile industry are summarized below.

2.2.1 *Energy balance approach*

Energy balance approach together with material balance approach is used in the examination of the various stages of any process, over the whole process, and even extending over the total production system from the raw material to the finished product in textile production to maximize product yields and minimize costs. The energy given into a unit operation can be balanced with the energy going out and the energy stored.

$$\text{Energy}_{\text{in}} = \text{Energy}_{\text{out}} + \text{Energy}_{\text{stored}}$$

Energy balances can be calculated, using SI system with the only unit of joule, on the basis of external energy used per kilogram of product or raw material processed. The energy consumed in textile production includes *direct energy*, which is fuel and electricity used in the production plant, and in transport and in factories, and in storage, selling, etc.; and *indirect energy*, which is used to actually build the machines, to make the packaging, and to produce the electricity and the oil (Bureau of Energy Efficiency).

Energy balancing works are unique operations for each textile processing plant and each different product. Process parameters of a traditional textile production may become complicated considering differences in fiber properties, yarn properties, fabric properties, wet-processing plans, and clothing-processing steps. Adding to the differences in product-processing parameters, energy conversion laps may vary among electricity, fuel, steam, thermal, or mechanical energy types, where their quantities must be balanced at the end.

2.2.2 Sankey diagrams

Sankey diagrams are a visual tool to represent the entire inlet and outlet energy flow in energy-consuming equipment, system, or process, in which the width of the arrows is shown proportionally to the flow quantity. They are typically used to visualize energy or material or cost outputs and losses between processes.

Energy flow diagram for traditional knitting plant as given by Ashok (Energymeasuretosave) as an example of Sankey diagram for textile processing. Supplied amount of energy is consumed by light and fan installation, air conditioning system, computers, and water pumping system, and resulting energy is consumed by motor of production machine.

2.2.3 Exergy analysis

Exergy is formally defined as the maximum amount of useful work that can be obtained as the system is brought to equilibrium with the environment. Exergy analysis is an efficient method to analyze the performance of thermal systems by using the second law of thermodynamics as well as the mass and energy balances. Exergy analysis is used to detect the inefficiencies of a system, and the outputs of the analysis can be suitable for the improvement of systems in terms of efficient energy use. Wet processing groups in textile production stages are suitable systems that can be analyzed via exergy analysis method.

Applications of exergy analysis in the textile sector were studied by Cay et al. (2012, 2009, 2007, 2010), where thermal energy consumption of printing, drying, and stenting machinery and processes are analyzed and the exergy destructions in each of the components of the overall systems are determined for measured average values of processing parameters.

2.2.4 Specific energy consumption

Specific energy consumption (SEC) calculation method is one of the basic approaches to calculate unit energy consumption of a product. In the calculation of SEC the number of energy consuming unit (J) that are associated with energy consumption, the average amount of energy (E_t) that is used during period (t), and the quantity of production (P_t) that is produced during period (t) are involved.

$$SEC = J * E_t / P_t$$

Close tracking of the SEC value for production processes is substantial to follow the energy use per unit product in severe circumstances of global market competition and increasing environmental compulsions over the manufacturing industry. The textile industry as part of the global competitive manufacturing sector is one suitable ground for SEC calculation. Estimated SEC data of a textile product can easily be transformed to the energy footprint of a unit (kg, meter, or piece) textile product. SEC data can also be utilized both for operating profit margins per product and also for indication of machine safety.

2.2.5 Carbon emission and carbon emissions signature

Carbon emissions (CEs) signature is specific to an electrical energy grid and can be used by manufacturers or anyone using energy that has been generated by fossil fuels. A CE label is proposed as a way of showing how much carbon has been emitted during the manufacturing period of a product (Jeswiet and Kara, 2008). Jeswiet developed a method that connects the electrical energy used in manufacturing directly to the CE.

2.2.6 ISO 14031 environmental standard

ISO 14031 is one of the earlier environmental standards; it was introduced in 1998 and revised in 2013 as ISO 14031:2013. Standard explains the external and internal environmental performance evaluation (EPE) method (ISO-14031, 1998) and defines guiding information. The task of measuring and evaluating environmental performance of technology and/or activities is accomplished by a set of environmental performance indicators (EPIs) and environmental condition indicators (ECIs). ECIs are basically the outcomes of normalization and aggregation of EPIs. EPIs and ECIs can be used not only for EPE but also help a company in selecting environmentally friendly techniques, benchmarking, environmental reporting, and establishing an environmental management system (EMS) (Ren, 2000). EMS has found wide usage in the textile and fashion industry as a result of rising concerns over product sustainability. Some worldwide-known textile and apparel brands have even developed their own standards for environmental compliance and conducted their own audits to determine the level of compliance of their suppliers (Hamner, 2006). Rising environmental concerns of conscious consumers and stakeholder groups push the textile and apparel manufacturers to settle environmental management in their production chains. High amounts of energy and other natural source consumption of textile manufacturing results in high emission of pollutants to the environment. Adoption of EMSs to the textile production chain has significant impact on operational performance of the production facility and helps to reduce their energy footprint.

2.2.7 The Carbon Trust Standard

The Carbon Trust Standard is the world's leading independent certification of an organization's impact with respect to the three primary constituents of environmental sustainability:

- Energy usage and greenhouse gas (CO_2e) emissions
- Water usage, management and effluent
- Waste management and disposal

Awarded to recognize best practices and real achievements in reduction, the Carbon Trust Standard helps organizations to measure, manage, and reduce their environmental impact, whilst improving their resource management and operational sustainability (Carbontrust).

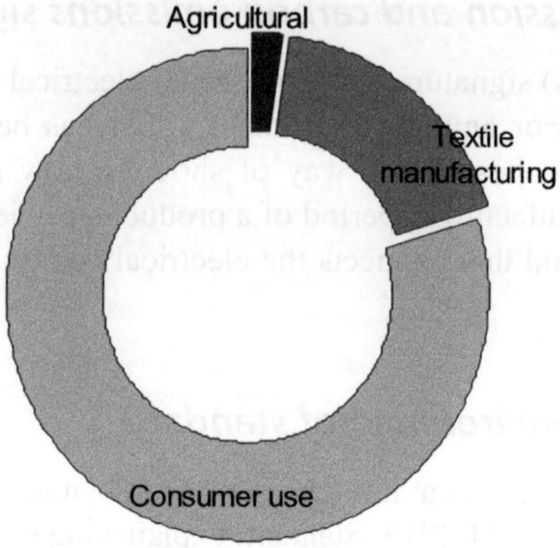

Figure 2.9 Energy-based environmental impacts of cotton textiles.
Derived from data of Anonymous (2011).

According to the report of International Carbon Flows Clothing (Anonymous, 2011), annual global clothing consumption results in about 330 Mt CO_2 emissions; added to this is 530 Mt CO_2 emissions of clothing use, which are washing, drying, ironing, and dry cleaning in orderly manner. A simple T-shirt has about 15 kg CO_2 emissions over its lifetime, and more than half of the CO_2 emissions are during its use phase.

Energy demand on the cotton-based textile industry is divided into three categories: agricultural impact, textile manufacturing impact, and consumer using impact. The rate of each energy-consuming category is shown in Figure 2.9; the highest energy consumption rate belongs to the consumer use phase.

2.2.8 The Carbon Reduction Label

The Carbon Reduction Label (Environmentalleader) for textiles was launched by UK-based "Continental Clothing" company as participant of an action plan conducted by the Department for Environment, Food and Rural Affairs (Defra). Cooperation of BSI British Standards, the Carbon Trust and Defra has shaped the carbon reduction labeling standard to help businesses assess the carbon footprint of textile goods and services. The standard measures the green house gas (GHG) emissions in goods and services throughout the entire life cycle of textile products and will provide the carbon footprint data of the entire lifecycle of the clothes, from raw materials and manufacturing through to consumer use and disposal.

2.3 Calculating energy footprints of textile products (applications and examples)

The term *energy footprint* has been used for about 20 years in the global textile and clothing industry worldwide. Before then, energy costs, efficient production, environmental issues, and green production terms were not main issues for the sector.

Table 2.4 Specific energy consumption (Report of Bureau of Energy Efficiency)

	SEC
Average specific electricity consumption	1.2 kWh/kg (best observed value – 0.95 kWh/kg)
Average specific thermal energy consumption	15000 kCal/kg (best observed value – 10932 kCal/kg)

Research about energy footprint analysis in the textile and clothing industry started long after the major energy crisis in the world in the 1970s.

Tarakcioglu published one of the earliest studies in Turkey about energy consumption in textile processes, that electrical energy consumption of 1 kg of yarn changes between 2.7 kWh/kg and 4 kWh/kg. Additionally, it was pointed out that thermal energy alternating between 1.1 MJ/kg and 4.7 MJ/kg is necessary for processes such as fixation besides electrical energy needs.

Koc and Kaplan (2007) estimated the SEC of electricity for 20 tex combed ring yarn as 3.32 kWh/kg and 20 tex open-end rotor yarn as 2.95 kWh/kg (Kaplan and Koç, 2010), which are relevant to the available literature.

Koç and Çinçik (2010) studied the specific energy consumption for woven fabric, where SEC for the chosen fabric has been estimated as 5.06 kWh/kg for electrical energy and 9.85 kJ/kg for thermal energy. The values calculated have been found to be compatible with the available literature.

Report of Bureau of Energy Efficiency mentions the average specific electric energy consumption and average specific thermal energy consumption for the polyester wet-processing stages in the area of Pali Textile cluster in India (Table 2.4). Average SEC for electrical energy and thermal energy is given with the best observed values.

Palamutcu (2010) has studied average SEC values for electricity in spinning, weaving, wet-processing, and clothing stages in a cotton textile production plant in Denizli, Turkey. Average SEC values are found as 3.24–3.47 kWh/kg for yarn spinning plant (compatible with the current literature); 0.0073 kWh/kg for warping-sizing plant; 1.58–2.24 kWh/kg for the weaving plant (lower than the current literature); 0.79–1.05 kWh/kg for the wet processing (compatible with the current literature); and 0.065–0.195 kWh/kg for the clothing production plant.

Hasanbeigi et al. (2011) studied the energy intensity of textile plants in five major subsectors, that is, spinning, weaving, wet processing, worsted fabric manufacturing, and carpet manufacturing in Iran. Results of the study showed that spinning plant electricity intensity varies between 3.6 MWh/ton yarn and 6.6 MWh/ton yarn, while fuel intensity ranges between 6.7 MBtu/ton yarn and 11.7 MBtu/ton yarn. In weaving plants, electricity intensity ranges from 1.2 MWh/ton fabric to 2.2 MWh/ton fabric, while fuel intensity was 10.1 MBtu/ton fabric and 16.4 MBtu/ton fabric for the two plants studied. In three wet-processing plants, the electricity intensity was found to be between 1.5 MWh/ton finished fabric and 2.5 MWh/tons finished fabric, while the fuel intensity was between 38.2 MBtu/tons finished fabric and 106.3 MBtu/tons finished fabric.

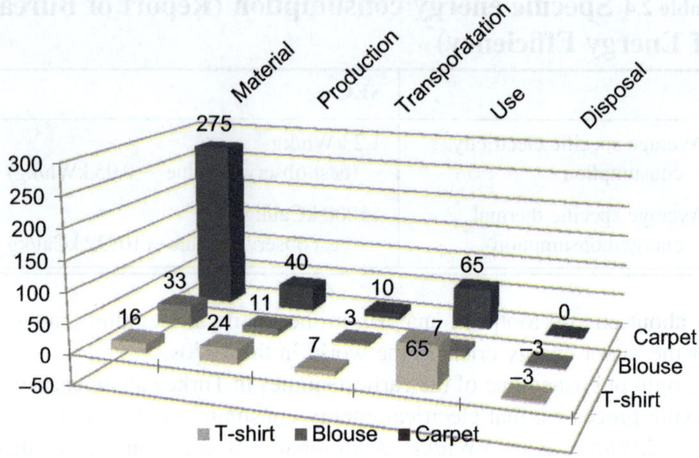

Figure 2.10 Primary energy profile for piece of T-shirt and blouse and square meter of carpet, MJ. Derived from data of CB2 1RX.

Energy consumption per unit kg of cotton textile is given at about 180 MJ/kg of cotton textile in total, at 15 MJ for the farming and ginning processes, and 165 MJ for textile processing (Anonymous, 2011).

Reports about the UK textile industry (CB2 1RX) energy profile for the T-shirt, blouse, and carpet are given in Figure 2.11 from the view of primary energy breakdown in the four major phases of transportation, material, production, disposal, and use. In Figure 2.10, estimated total energy consumption is approximately 109 MJ per T-shirt, 51 MJ per blouse, and 390 MJ per square meter of carpet.

The breakdown of energy for piece of cotton T-shirt shows that transportation, material, production, and disposal phases are dwarfed by the use phase. The use phase includes 25 washes at 60 °C, followed by tumble drying and ironing, that consume 60% of the total energy. The disposal phase includes incineration in which heat is generated and used so the net energy consumption is negative in this phase.

The breakdown of energy for piece of viscose blouse is a strikingly different shape compared to that of the T-shirt. The energy consumption of the material phase dominates at about 65% of the total energy. Approximately 50% of the energy consumed in the production of viscose is from the renewable origin ("from the basic feed-stock wood") and the output from the viscose factory is filament yarn resulting in a shortage of energy consumption in traditional short-staple spinning processes. The production phase energy requirement is 11 MJ, which is less than that of T-shirt. The use phase of the blouse is much less energy intensive, consuming only 7 MJ (14% of the total) compared to 65 MJ for the T-shirt. This is primarily a result of the much less extensive maintenance, such as low temperature washing, 40 °C, and without tumble drying or ironing.

The energy profile of the carpet is given per square meter of carpet. Energy breakdown of carpet shows similar proportions to that of the blouse. The carpet weighs approximately 10 times more than either the T-shirt or the blouse. The material production phase represents approximately 71% of the total energy. This is partly the result of the relatively large energy consumption in the production of the synthetic fiber polyamide—approximately 160 MJ per kg—compared to about 50 MJ per kg for cotton. The use phase includes vacuum cleaning only. Total energy consumption for the carpet amounts to approximately 390 MJ per square meter or 150 MJ per kg.

2.4 How to reduce energy footprint in textile and clothing products

Energy consumption level and size of sector's energy footprint have been assessed in many different academic works and sectorial reports. It is now known that the textile and clothing industry is an energy-intensive industry with many appropriate saving possibilities and opportunities during their manufacturing phases and consumer use phase. The energy-intensity ratio of any textile product varies depending on the fiber type, manufacturing steps, and consumer behaviors during the use phase of the product. It is clear that many clothing products have a higher energy intensity at the stage of consumer use than in the manufacturing stages.

Having a high level of energy intensity and large size of footprint, energy saving, and efficiency in the textile sector has become a main concern in the sector. In the extensive work that is conducted in UK and reported by WRAP (Final Report, 2012; Dynamics) (Material Change for a Better Environment Report) the carbon emissions of textile products through the product life cycle period is assessed annually in a country based on size. Textile product is assessed following the processing route of fiber, yarn, fabric processing, clothing manufacturing, retailing, consumer use time, and recycling stages. Besides carbon emissions analysis and carbon footprint of the textile materials and clothing products, the report also presents detailed information about the reduction possibilities of carbon dioxide emission and energy-saving possibilities during manufacturing, consumer utilization period, and recycling stages of textile and clothing products.

According to the results of the study, the total annual carbon footprint of all garments, both new and existing, in use in the UK in 2009 was approximately 38 million ton of CO_2 ($\sim$0.6 tons per person per year), representing about 2% of the UK's total direct carbon footprint. The carbon footprint of each individual garment in use in the UK in 2009 ranges between 1 and 17 kg CO_2 and the per person carbon footprint of all personal clothing use, both new and existing, is estimated around 0.6 ton of CO_2.

In this study, LCA methodology of a full cradle-to-grave or business-to-consumer carbon footprint approach is used to draw the general map of energy consumption, and later any possible savings opportunities are analyzed and appropriate action plans are structured. In Table 2.5 life cycle stages that are included and also excluded in the conducted work are listed.

Unit electric energy consumption (kWh/kg), unit waste creation (kg/kg), transport distance of, sea, air, and road (km), unit cost of washing detergent, electric energy consumption of washing machines for different energy rates—at different temperatures per load of washing (kW/h per load), electric energy consumption of drying

Table 2.5 Life cycle stages in the footprint assessment of a textile product in the UK

Included life cycle stages	Excluded life cycle stages
• Extraction of raw materials required for the production of fibers; • Processing of materials (e.g., production of synthetic polymer resin); • Production of fibers (either at farm or factory); • Production of yarn; • Production of fabric; • Treatment of fabric (e.g., bleaching, dyeing etc.); • Production of garments; • Packaging of garments; • Transportation of materials and goods to and from production locations; • Waste at all stages of production; • Transportation of garments to the UK; • Storage at regional distribution center (RDC) in the UK; • Transportation from RDC to retail outlets; • Storage at retail outlets in the UK; • Use of clothing (e.g., washing [energy, water, and detergent use], tumble drying, ironing); • End of life of clothing (e.g., reuse, recycling, landfill and incineration)	• Transportation of consumers to and from the point of retail purchase; • Packaging of packaging used at all life cycle stages; • Fabric softeners, color catches, etc. or other material inputs used during washing; • Water use for ironing; • Preparation for reuse burdens; • Stain removers used during the use phase. • Capital goods (e.g., the manufacture of weaving looms, washing machines, irons etc.); • Human energy inputs to processing; and • Animals providing transport services

The table is arranged using the data from report of WRAP (Final Report, 2012).

(kWh/kg), and ironing hours per kg of garment are taken into consideration to assesses the footprint of textile product.

Estimated footprints are then reassessed to find any possible "hotspots" in the life cycle, where the highest component of footprint comes. Defined hotspot areas are accepted highest possible saving achievement areas. Determined saving areas are then inspected to narrow the potential reduction or saving measures. The reduction measures are then identified by consultation of WRAP and selected stakeholders of the Sustainable Clothing Action Plan members; and the 10 reduction options considered for analysis are infiltrated. A list of reduction measures, phase of textile life cycle stage, name of responsible group for reduction, expected output message of measures, baseline saving rates, and best possible level of saving of each measure are given in Table 2.6. Cumulative savings of all measures are estimated as a 20.9% reduction rate for baseline saving scenario and 70.5% reduction rate for best level of possible saving scenario of total current CO_2 emissions.

Table 2.6 Reduction measures in the life cycle of textiles in UK, in 2009

Responsible group	Phase of textile	Expected output	Reduction measure	Achieved saving rate of CO_2e, %	
				Base	Best
Design and production	Manufacturer/ retailer	Lean production	**1-Eco efficiency across supply chain (production, distribution and retail)**	−4,1	−24,5
	Manufacturer/ retailer/consumer	Longer product lifetime	**2-Design for durability (and product lifetime optimization)**	−7,7	−27,3
		Buy differently	**3-Shift in market to higher proportion of synthetic fibers**	−0,4	−1,7
In use	Consumer/retailer	Reduce consumer footprint through behavioral change	**4-Clean clothing less**	−2,6	−3,9
			5-Wash at lower temperature	−1,4	−2,7
			6-Increase size of washing and drying loads	−1,4	−2,7
			7-Use the tumble dryer less	−1,1	−2,8
Reuse and recycling	Consumer/retailer	Reuse more	**8-Dispose less and reuse more**	−0,7	−2,1
		Recycle more	**9-Dispose less and recycle more (open loop)**	−0,5	−1,0
Reuse and recycling/Design and production	Manufacturer/ retailer/consumer	Recycle more	**10-Start closed loop recycling of synthetic fibers**	−0,9	−1,8
			TOTAL REDUCTION	**−20,9**	**−70,5**

The table is part of the work that was published in report of WRAP (Final Report, 2012).

Besides the above-defined LCA approach used, country-based sectorial analysis, energy footprint analysis studies have also been practically conducted in individual textile manufacturing plants. One recent successful example was presented by "Sustainable Apparel to Help a Planet in Crisis" of a Swedish-based "earth positive" initiation named "Textile Supply Chain 10 Steps to Green". According to the study, energy, water, and carbon footprints of a cotton apparel product are defined as subdivisions of a greening approach over the production steps of farming, ginning, spinning, knitting, dyeing, finishing, cutting, sewing, packaging, and shipping. Advised action steps are organic production, reduction of cotton-processing waste, utilization of green energy, ethical trade practices, reduce–reuse–recycle, decreased transportation costs, water care, lowered water footprint, reduced carbon footprint, and informed customer.

As a market phase example of reduced carbon footprint textile product, Rapanui (Rapanuiclothing) trademark has a group of T-shirt collections that are promoted as low carbon dioxide clothing. In the energy analysis of T-shirts, it has been reported that CO_2 emissions per product have been reduced by 80% compared to any other average T-shirt. As shown in Figure 2.11, energy consumption of a T-shirt during its life time is the sum of material-based consumption, production-phase consumption, transportation consumption, use-customer utilization-phase consumption, and disposal consumption.

Energy consumption amount on material, manufacturing, and transportation phases of a T-shirt can be decreased by:

- use of organic fabrics that consume less carbon-intensive nitrogen fertilizer and utilize less machinery
- utilization of renewable energy sources during manufacturing processes
- use of a new print technology that is reducing waste massively
- concretely shortened transportation routes.

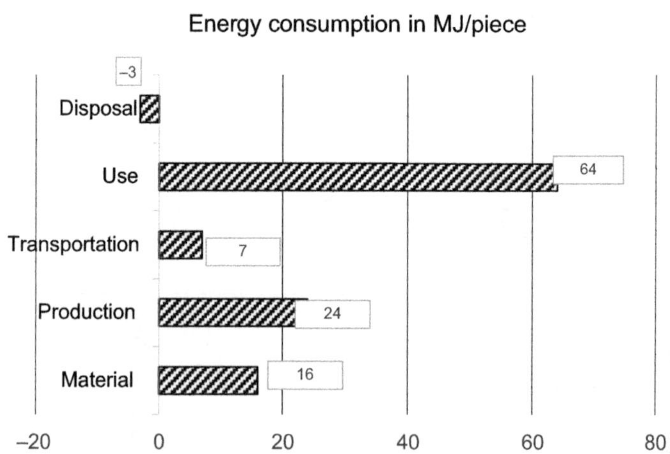

Figure 2.11 Primary energy profile of a T-shirt during its lifetime.
Graphic has been arranged using data given in Rapanuiclothing.

As it is seen from Figure 2.11, the highest consumption amount belongs to the consuming phase where washing, drying, and ironing processes are practically repeated several times. In order to reduce the energy consumption at this phase of the T-shirt, the following listed actions can be utilized (Rapanuiclothing):

- appropriately labeled product, informing costumers about postpurchase care
- informing customers about the need to wash cool and hang dry
- informing consumers about right stuff to wash with and together
- informing consumers about material recovery program.

WRAP (Final Report, 2012; Dynamics) research group has also focused on the consumer intervention phase of textile products where the highest environmental influences are added to the energy footprint and carbon footprint chains. In the study, a series of consumer interventions were tested for their influence on carbon footprint results. Reducing measures to change consumer behavior during the use phase of clothing are listed as follows:

- Increased size of washing and drying loads
- Decreased number of cloth cleaning
- Decreased number of tumble dryer use
- Washing at lower temperature, 30°
- Washing at lower temperature, 30°, and periodic biofouling "service" wash 90°
- Washing at lower temperature, 30°, periodic biofouling "service' wash 90°, and descaling/detergent removal
- Enhanced energy efficiency of washing—improved appliance rating of washing machines or achieved by more use of existing economy wash setting for washing machine
- Increased spin drying cycle speed of washing machine
- Reduced detergent use
- Increased lifetime of clothing with better washing and drying behavior.

Company-based carbon footprint analysis has been done by the Otto Group that is partner of a venture project in Germany. In the project (Carbon Footprint Study, 2009), 100% cotton, 220 g/piece, size 40–42, white color, long shirt has been selected for the analysis.

Analysis route of the long shirt: started from the cotton field in the US to the production units in Bangladesh to shipment of the product to Germany and finally to the customer, who ordered the article. This distance amounts to more than 35,000 km—almost one trip around the world, and constitutes the transportation-based CO_2 emissions of 3% in the annual total CO_2 emissions of shirt (Figure 2.12).

The highest CO_2 emission is linked to the consumer use phase of the shirt (31%), where 55 repeated washing processes are counted. The carbon footprint for the use phase of the shirt is found to depending significantly on the concrete washing temperature, the energy efficiency of washing machine and dryer, the load degree when washing and drying, and the frequency of automatic drying.

Another product-based footprinting analysis has been conducted by Continental Clothing (Environmentalleader), where it has been revealed that as much as half of the total footprint of clothing can come from consumers' home laundry activities.

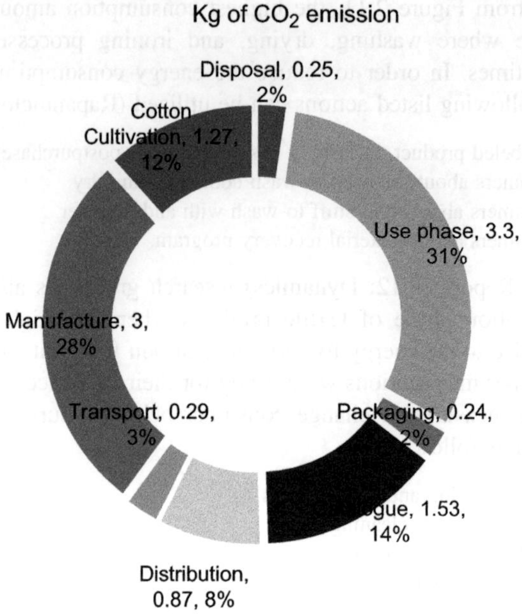

Figure 2.12 CO_2 emissions of 100% cotton, white color, long shirt, size 40–42, 220 g. Graph is arranged using the data given in Carbon Footprint Study (2009).

2.5 How to reduce energy footprint in textile and clothing supply chain

The supply chain is one fact of global textile and clothing market. Management of the supply chain from the view of environment and reduction of carbon dioxide release are major subjects in the textile and clothing industry. Definition of supply chain is stated as "all the activities involved in delivering a product from raw material through to the customer including sourcing raw materials and parts, manufacturing and assembly, warehousing and inventory tracking, order entry and order management, distribution across all channels, delivery to the customer and the information systems necessary to monitor all of these activities" in the study of Jimmy and Postle (Lam and Postle, 2006).

The textile and clothing supply chain mainly involves energy consumption phases of international delivery that require fossil fuel consumption; manufacturing and assembly that require processing and machine driving purposed heat or electric energies; office work that requires electric energy; and customer distribution that requires fossil fuel consumption. Reduction of energy footprint possibilities can only become possible in case of focused evaluations, analysis, restructuring, and continued tracking of every energy-consuming phase.

Main applicable tools and methods on the method of energy footprint reduction of clothing supply chain can be listed as follows:

- increased energy efficiency in the machinery use

- efficient processing technology utilization on the manufacturing stages
- efficient auxiliary utilization on the manufacturing stages
- route planning for decreased transportation distances
- selection of less energy-intensive transportation types among, air, marine, truck, or railway transportation
- utilization of renewable energy consumption.

In Muthu's recent book by Woodhead publications (Muthu, 2014), he comprehensively reviews the methods used to calculate environmental impact of textile and clothing supply chain, including product carbon footprints, ecological footprints, and life cycle assessment. The book presents information about the textile supply chain, its environmental impact, and an overview of the methods (greenhouse gas emissions, the water and energy footprints of the industry, and depletion of resources, as well as the use of LCA) used to measure the overall environmental impact of the textile industry.

In the work of WRAP (Final Report, 2012; Dynamics), eco-efficiency across the supply chain of textile and clothing product consumption in the UK over one year is analyzed, consisting only of the production, distribution, and retailing phases. It is seen that in the baseline scenario a 4.1% reduction on current carbon dioxide emission is possible; and if the highest possible reduction scenario occurs, the rate of reduction is estimated as 24.5%. Detailed information about the estimation of reduction in sector-based supply chain carbon dioxide emission in the UK can be found in the report.

Continental clothing (Environmentalleader) company has also worked on the impact of low carbon economy throughout the entire supply chain of textile products in the Carbon Trust's pilot program. As a result of the conducted low carbon economy program, the clothing company has managed to reduce its CO_2 emissions in its manufacturing operations in total by 90% through the use of a low footprint production approach and use of renewable energy.

2.6 Energy analysis in the textile industry

Academic studies that involve energy efficiency in the textile industry are not at a desired level compared to the volume of the sector in the world. Most of the published studies focus on the energy savings, measurements, comparison of current methods and technologies, potential benefits and energy savings in plant size and national size.

Tang and Mohanty (1996) have presented an efficient method of electrical and thermal energy cogeneration in a plant with a heat recovery system. Palanichamy et al. (2001) have published results of an in-plant study that substantial energy savings become readily achievable by taking action about problems found during an energy audit in production-related and production-supporting facilities. Ozturk (2004) has presented energy conservation measures for implementing efficient energy using rules, identifying losses, and reducing such losses in textile factories. Palanichamy and Sundar (2005) have also presented results of adopting conservation measures in the textile industry, showing the reduction in energy and the estimated annual reduction of greenhouse gases. Kocabas et al. (2009) have presented soft application of best-available technology measures in a textile wet-processing plant to manage reduced water and energy consumption. Palamutcu (2010) investigated electrical energy consumption of cotton textile processing steps of yarn processing, weaving, wet processing, and apparel manufacturing, and has stated that actual electricity consumption per unit textile is higher than the estimated amount of electricity. Ngai et al. (2012) have presented a soft systems methodology (SSM) to identify management support system opportunities for managing energy and utility usage, and related to the SSM approach they introduce system utilization of energy and utility management support system in textile manufacturing processes.

Country-based sector analysis concerning energy consumption and efficiency is published in several studies, for the Indian textile industry (Palanichamy and Sundar Babu, 2005; Sivaramakrishnan et al., 2009), for the Taiwanese textile industry (Hong et al., 2010), for the German and Columbian textile industry (Martínez, 2010), for the Malaysian industrial sectors (Masjuki et al., 2009), for the Iranian textile industry (Hasanbeigi et al., 2012), for the Thailand textile industry (Tang and Mohanty, 1996), and for the Turkish textile industry (Pulat et al., 2009; Palamutcu, 2010; Rapanuiclothing).

Adding to the mentioned scientific papers, new reports, guiding documents, and program improvement initiations are being studied and published all around the world.

2.7 Energy analysis from the view of environmental approaches

Rising environmental awareness worldwide has some positive influences on textile production and consumer behaviors. Energy savings, reduced energy consumption, energy watch programs, and renewable energy use in textile industry are carried out under the sustainable production programs. There have been active industry-driven programs and academic research about energy in a sustainable textile production frame. Reuse of energy, reduction of energy, and renewable energy-based papers in the textile industry have been carried out by Kiran-Ciliz (2003), Dayem and Mohamad (2001), Niinimäki and Hassi (2011), Zabaniotou and Andreou (2010), Muneer et al. (2006), and Pulat et al. (2009).

Manufacturing, logistics, and transportation-purposed energy consumption of textile products in global markets are some of the main concerns of sustainable production in the textile industry in countries. Many of the textile manufacturing companies have not yet managed to start implementation of even cost-effective applications in their production sites because of limited information about energy efficiency in the textile industry. The main reason for such low interest concerning the energy efficiency issues and less participation in such programs in textile plants can be explained with the category of the textile companies themselves, which mostly are small- and medium-sized enterprises (SMEs) with local and limited resources, and comparatively low amounts of energy consumption compared to the high-energy-consuming sectors of chemicals, metal processing, cement, pulp, and paper industries.

2.8 Future trends

2.8.1 Energy-based trends

Energy has continued to be one of the major concerns of the world. Management of current fuel-based energy sources, energy saving approaches and technologies, and new and renewable energy sources are strengthening their positions in textile industry.

Reduce, reuse, recycle, up-cycle, and efficiency terms are already placed into the textile and energy literature, textile production and applications stages, and textile-using lifetime by consumers and even recycling stages.

Feasible global scenarios of energy reports define new ground in the energy debate that world's energy supply must be provided by renewable and sustainable sources by mid-century (Panda, 2014). As part of the global energy market and production chain, energy-consuming expectations in the global textile industry, energy supply and energy footprint values are expected to change parallel to the world's energy debates. The dominant form of energy in the 2050s is expected to be electricity, whose easy transportation and relatively easy applications are possible with only the need for improved, intelligent energy grids of efficient electricity transportation.

Current state of the art in textile production technologies, consumer-based energy consumption contributions, and recycling approaches can fully transform the electricity-consuming processes by the year 2050.

Production of electric energy that is planned to be consumed in the textile sector can also be mostly supplied using renewable energy sources of sun or wind. There are already many academic studies and practical application of renewable energy use in textile industries and they are expected to increase after establishment of appropriate infrastructure for energy supply. Current high costs of renewable energy systems will become cheaper compared to the continued fossil fuel system. Adding to the reduction in energy cost, lessening impacts of climate change gained via renewable energy transformation will even double the cost reduction in total. As an optimistic approach, energy consumption trends in the textile sector are part of the manageable efforts of energy-transforming programs for the benefit of all.

2.8.2 Textile-based trends

Textile production is increasing and it is estimated that world fiber demand will sustain its rising trend for the next several decades. Reasons for this trend include the increasing world population, increasing market of technical textile products, and rising personal consumption of fast fashion trends.

Consumer awareness about environmental influences of textile products is also increasing in both the developed and developing countries in the world. Pressure from consumers and legislation is likely to drive increasing demands for environmentally sensitive production technologies, product types focusing on the use of chemicals, energy consumption, and also extend to include reuse of materials and substitution of alternative materials. Improvement in the environmental performance of the textile sector will continue to be material specific and depend on the energy profile of the material. For conventional cotton products, the requirement for energy is driven by number of laundry and consumed energy volume. Energy requirements for cotton garments are dominated by washing, drying, and ironing. Novel treatments may provide resistance to odors, reduce the total number of washes, or allow faster drying with less ironing.

Recycling, reusing, up-cycling, and reducing approaches in used textile products will continue to increase to manage a positive impact on the energy footprint of a single textile product. Success of the sustainability in the textile sector will continue depending on the consumer response and choices.

Future trends of consumer behaviors may include buying second-hand textiles where possible; buy fewer and more durable textile products; for new product shopping select the ones made with least energy and also toxic chemicals; lease clothing, where possible; wash textiles less often at lower temperatures with appropriate detergents; dry in open air; avoid ironing where possible; and repair the available textile products. Possible consumer opposition to the above-mentioned precautions can be changed with education about a sustainable textile approach.

Technological improvements, government policies, consumer behaviors, and sensitive approaches of the fashion world would help to decrease environmental influences and the energy footprint in textile sector in the near future.

Renewable energy types of solar, wind, and geothermal energies are starting to be used in textile and clothing industry as in the other energy-intensive manufacturing industries. The simple and the most traditional method in the frame of renewable energy utilization in textile supply chain is drying processes, where air drying can easily be replaced over tumble drying. Wind energy, solar energy, or geothermal energy-based production of heat and electric energy utilization in any manufacturing phase of a textile product will help to reduce the energy footprint of textile supply chain.

2.9 Conclusions

Textile products have an energy footprint like many other industrial products. All manufacturing operations, retail operations, total energy consumption of textile items utilization, and recycling phases are indispensable energy consumption areas. There is no possibility of zero impact or entirely erasing the energy footprint left behind from a textile product. Improved tools that have already started to be used in several companies and brands are energy-concerned management of supply chains, renewable energy utilizations, implication of efficiency in manufacturing phases, increased consumer awareness on textile consumption phase, and improved recycling–reusing–reducing approaches in society. Current technological developments offer several opportunities to take action and to build technical solutions and models for reduction of energy footprint in textile industry.

Local, national, and international projects and campaigns about energy management, energy saving and sustainable production in textile processing steps are continuously increasing all over the developed and developing countries in the world.

The textile and clothing sector has complicated issues from the viewpoint of raw material variations, processing steps, machinery parks, production components, and consumed energy types and amounts. Fragmented and nonhomogeneous structure of the sector is dominated by SMEs where energy management systems are either not well managed yet or even not present. Energy, as one of the main cost components of a textile product, should be one main concern both from the view of manufacturing efficiency and environmental approaches.

2.10 Additional readings about energy in the textile industry

- Carbon Trust's report (Carbontrust, 1997)
- The Canadian Industry Program for Energy Conservation (CIPEC) (NRCAN)
- United Nations Industrial Development Organization (UNIDO)'s report (UNIDO, 1992)
- Environmental Protection Agency (EPA) of United States (EPA)
- Report of European Commission by European Intelligent Energy according to the EIE/04/113/S07.38648 Grand Agreement program (Energy Management Review Report)
- Japan Energy Conservation Center report of energy saving programs in textile industry (ECCJ, 2007)
- Report of Centre for the Analysis and Dissemination of Demonstrated Energy Technologies (CADDET) in Netherland textile industry (Caffal, 1995)
- Research Report of China's Textile Industry by China Research and Intelligence (CRI) (CRI, 2009)
- European Commission, Reference document on best available techniques for the textile industry (European Commission, 2003)
- Report of the Natural Resources Defense Council (NRDC) for textile mills to save money and reduce pollution (Greer et al., 2010)
- Report of EU COST Action 628:life cycle assessment (LCA) of textile products, eco-efficiency and definition of best available technology (BAT) of textile processing (Nieminen et al., 2007)

References

Anonymous, Longevity: Tackling the Carbon Crisis, August–September 2011. Ecotextilenews, p. 66–68.

BP Statistical Review of World Energy, 2014 (access date 15.09.14.). http://www.bp.com/en/global/corporate/about-bp/energy-economics/statistical-review-of-world-energy.html.

Bureau of Energy Efficiency. http://www.em-ea.org/Guide%20Books/Book-1/1.4%20MATERIAL%20%20AND%20ENERGY%20BALANCE.pdf.

Caffal, C., 1995. Energy Management in Industry. Centre for the Analysis and Dissemination of Demonstrated Energy Technologies (CADDET), Sittard, the Netherlands.

Carbon Footprint Study, 2009. Final Summary, Systain Consulting GmbH.

Carbon Trust Cutting Your Energy Costs–a Guide for the Textile Dyeing and Finishing Industry, 1997. http://www.carbontrust.co.uk/Publications/pages/publicationdetail.aspx?id=GPG168.

Cay, A., Tarakcioglu, I., Hepbasli, A., 2007. Exergetic performance assessment of a stenter system in a textile finishing mill. International Journal of Energy Research 31 (13), 1251–1265.

Cay, A., Tarakcioglu, I., Hepbasli, A., 2009. A study on the exergetic analysis of continuous textile dryers. International Journal of Exergy 6 (3), 422–439.

Cay, A., Tarakcioglu, I., Hepbasli, A., 2010. Exergetic analysis of textile convective drying with stenters by subsystem models: Part 2—parametric study on exergy analysis. Drying Technology: An International Journal 28 (12), 1368–1376.

Çay, A., Özgüney, A.T., Yavaş, A., 2012. Application of exergy analysis to textile printing process. Fibres & Textiles in Eastern Europe 6A (95), 37–42.

Well Dressed? the Present and Future Sustainability of Clothing and Textiles in the United Kingdom, University of Cambridge Institute for Manufacturing Mill Lane, Cambridge CB2 1RX, UK, ISBN:1-902546-52-0.

China Research and Intelligence (CRI), 2009. Research Report of China's Textile Industry. Available at: http://www.researchandmarkets.com/reports/1053813/research report of chinas textile industry.pdf.

Dayem, A.M., Mohamad, M.A., 2001. Potential of solar energy utilization in the textile industry a case study. Renewable Energy 23, 68–694.

Energy Conservation Center, Japan (ECCJ), 2007. Overview of Energy Saving Technologies in Textile Industry. http://www.aseanenergy.org/download/projects/promeec/2007-2008/industry/eccj/ECCJ SW03%20Overview%20of%20energy%20saving%20technology TH.pdf.

ISO-14031 Environmental Management-Environmental Performance Evaluation Guidelines, 1998, 2013.

Energy Management Review Report, EMS-textile Project: Promotion of Energy Management Practices in the Textile Industries of Greece, Portugal, Spain and Bulgaria.

EMS-Energy Management Standard in Textile Project –EC European Intelligent Energy Report, Section 4 – Energy Consumption.

European Commission, 2003. Reference Document on Best Available Techniques for the Textiles Industry. http://eippcb.jrc.ec.europa.eu/reference/brefdownload/download TXT.cfm.

Final Report a Carbon Footprint for UK Clothing and Opportunities for Savings, July 2012. http://www.wrap.org.uk/sites/files/wrap%20/Appendix%20IV%20-%20Carbon%20footprint%20.

Greer, L., Egan Keane, S., Lin, Z., 2010. NRDC's Ten Best Practices for Textile Mills to Save Money and Reduce Pollution. http://www.nrdc.org/international/%20cleanbydesign/files/rsifullguide.pdf.

Hamner, B., 2006. Effects of green purchasing strategies on supplier behaviour. In: Sarkis, J. (Ed.), Greening the supply chain. Springer-Verlag London Limited, Germany.

Hasanbeigi, A., Hasanabadi, A., Abdorrazaghi, M., 2011. Energy Efficiency Technologies and Comparing the Energy Intensity in the Textile Industry. ACEEE Summer Study on Energy Efficiency in Industry, Vol. No.- (2011).

Hasanbeigi, A., Hasanabadi, A., Abdorrazaghi, M., 2012. Comparison analysis of energy intensity for five major sub-sectors of the textile Industry in Iran. Journal of Cleaner Production 23, 186–194.

Hasanbeigi, A., September 2010. Energy-efficiency Improvement Opportunities for the Textile Industry. LBNL-3970E.

Hong, G.B., Su, T.L., Lee, J.D., Hsu, T.C., Chen, H.W., November 2010. Energy conservation potential in Taiwanese textile industry. Energy Policy 38 (11), 7048–7053.

International Energy Agency, 2013. World Energy Outlook.

International Energy Agency, 2008. World Energy Outlook. OECD/IEA, Paris, ISBN 978-92-64-04560-6.

Jeswiet, J., Kara, S., 2008. Carbon emissions and CESTM in manufacturing. CIRP Annals - Manufacturing Technology 57, 17–20.

Kaplan, E., Koç, E., 2010. Investigation of energy consumption in yarn production with special reference to open-end rotor spinning. Fibres & Textiles in Eastern Europe 18 (2), 7–13.

Kiran-Ciliz, N., 2003. Reduction in resource consumption by process modifications in cotton wet processes. Journal of Cleaner Production 11, 481–486.

Koç, E., Çinçik, E., 2010. Analysis of energy consumption in woven fabric production. Fibres & Textiles in Eastern Europe 18 (2(79)), 14–20.

Koc, E., Kaplan, E., October/December 2007. An investigation on energy consumption in yarn production with special reference to ring spinning. Fibres & Textiles in Eastern Europe 15 (4(63)).

Kocabas, A.M., Yukseler, H., Dilek, F.B., Yetis, U., 2009. Adoption of European Union's IPPC directive to a textile mill: analysis of water and energy consumption. Journal of Environmental Management 91, 102–113.

Lam, Jimmy K.C., Postle, R., 2006. Textile and apparel supply chain management in Hong Kong. International Journal of Clothing Science and Technology 18 (4), 265–277.

Ma, L., Allwood, J.A., Cullen, J.M., Li, Z., 2012. The use of energy in China: tracing the flow of energy from primary source to demand drivers. Energy 40, 174–188.

Martínez, C.I.P., 2010. Energy use and energy efficiency development in the German and Colombian textile industries. Energy for Sustainable Development 14 (2), 94–103.

Masjuki, H.H., Mekhilef, S., Ping, H.W., Jamaluddi, M.F., 2009. End-use energy analysis in the Malaysian industrial sector. Energy 34, 153–158.

Muneer, T., Maubleu, S., Asif, M., 2006. Prospects of solar water heating for textile industry in Pakistan. Renewable and Sustainable Energy Reviews 10, 1–23.

Muthu, S.S., 2014. Assessing the Environmental Impact of Textiles and the Clothing Supply Chain. Woodhead Publishing.

Ngai, E.W.T., To, C.K.M., Ching, V.S.M., Chan, L.K., Lee, M.C.M., Choi, Y.S., Chai, P.Y.F., 2012. Development of the conceptual model of energy and utility management in textile processing: a soft systems approach. International Journal of Production Economics 135, 607–617.

Nieminen, E., et al., 2007. EU COST Action 628: life cycle assessment (LCA) of textile products, eco-efficiency and definition of best available technology (BAT) of textile processing. Journal of Cleaner Production 15, 1259–1270.

Niinimäki, K., Hassi, L., 2011. Emerging design strategies in sustainable production and consumption of textiles and clothing. Journal of Cleaner Production 13, 1876–1883.

Ongüt, C.E., 2007. Turkısh Textile and Clothing Industry's Adaptation to the Changing Competitive Conditions Report of State Planning Organization (in Turkish).

Ozturk, H.K., 2004. Energy use and cost in textile industry: a case study for Turkey. Energy 33, 2424–2446.

Palamutcu, S., 2010. Electric energy consumption in the cotton textile processing stages. Energy 35, 2945–2952.

Palanichamy, C., Sundar Babu, N., 2005. Second stage energy conservation experience with a textile industry. Energy Policy 33, 603–609.

Palanichamy, C., Sundar, N., 2005. Second stage energy conservation experience with a textile industry. Energy 33, 603–609.

Palanichamy, C., Nadarajan, C., Naveen, P., Sundar, N., 2001. Budgetconstrained energy conservation—an experience with a textile industry. IEEE Transactions on Energy Conversion 16 (4), 340–345.

Pulat, E., Etemoglu, A.B., Can, M., 2009. Waste-heat recovery potential in Turkish textile industry: case study for city of Bursa. Renewable and Sustainable Energy Reviews 13, 663–672.

Report of Bureau of Energy Efficiency, Detailed Project Report on Installation of System for Control of Excess Air In Boiler Textile SME Cluster, Pali, Rajasthan (India), Detail Project Report No.: PAL/TXT/EAC/05, (access date 20.08.14.).

Ren, X., 2000. Development of environmental performance indicators for textile process and product. Journal of Cleaner Production 8, 473–481.

Sathaye, J., Price, L., de la Rue du Can, S., Fridley, D., 2005. Assessment of Energy Use and Energy Savings Potential in Selected Industrial Sectors in India. Report No. 57293, Lawrence Berkeley National Laboratory, Berkeley.

Sharma, S., Energy management in textile industry. International Journal of Power System Operation and Energy Management ISSN (PRINT) 2(1,2) 2231–4407. http://interscience.in/IJPSOEM_Vol2Iss1-2/45-49.pdf.

Sivaramakrishnan, A., Muthuvelan, M., Ilango, G., Alagarsamy, M., 2009. Energy Saving Potential in Spinning, Weaving, Knitting, Processing, and Garmenting. Available at: http://www.emtindia.net/Presentations2009/3L 2009Aug8 Textile/06-SITRA.pdf.

Stoglehner, G., 2003. Ecological footprint—a tool for assessing sustainable energy supplies. Journal of Cleaner Production 11, 267–277.

Tang, O., Mohanty, B., 1996. Industrial energy efficiency improvement through cogeneration: a case study of the textile industry in Thailand. Energy 21 (12), 1169–1178.

United States Department of Energy (U.S. DOE), 2004. Energy Use, Loss and Opportunities Analysis: U.S. Manufacturing & Mining. Available at: https://www.eecbg.energy.gov/industry/intensiveprocesses/pdfs/energy_use_loss_opportunities_analysis.pdf.

United Nations Industrial Development Organization (UNIDO), 1992. Textile Industry. Output of a Seminar on Energy Conservation in Textile Industry. www.unido.org/fileadmin/import/userfiles/puffk/textile.pdf.

World Energy Outlook, 2013. Annexes.

Wackernagel, M., Rees, W.E., 1996. Our Ecological Footprint: Reducing Human Impact on the Earth. New Society Publishers, Gabriola Island, BC.

Energy [R]Evolution a Sustainable World Energy Outlook, report fourth ed., 2012 world energy scenario.

World Trade Organization International Trade Statistics, 2013. www.wto.org/statistics.

Zabaniotou, A., Andreou, K., 2010. Development of alternative energy sources for GHG emissions reduction in the textile industry by energy recovery from cotton ginning waste. Journal of Cleaner Production 18, 784–790.

http://www.etymonline.com/index.php?term=energy.

http://www.gdrc.org/uem/footprints/energy-footprint.html (access date 22.09.14.).

https://www.icac.org/cotton_info/publications/statistics/world-apparel-survey/FAO-ICAC-Survey-2013-Update-and-2011-Text.pdf.

http://www.saskflax.com/worldfibermarket_overview.html.

http://www.textileworld.com/Articles/2012/April/Textile_News/The_Fiber_Year_To_Release_2012_Textiles_And_Nonwovens_Survey.

http://www.lenzing.com/fileadmin/template/pdf/konzern/prasentationen/CMD_Management_Board_Presentation.pdf.

http://www.theoildrum.com/node/8936 (access date 12.08.14.).

http://www.energymeasuretosave.com/yahoo_site_admin/assets/docs/%20TEXTILE_KNITTING_ENERGY_%20CONSERVATION_2183906.345121918.pdf.

http://www.carbontrust.com/.../footprint-certification.

http://www.environmentalleader.com/2009/03/27/uk-launches-first-carbon-footprint-label-for-retail-clothing/#ixzz3OtfsZLXk.

http://www.dynamics.se/pdf/Textile%20Supply%20Chain%2010%20steps%20to%20go%20Green.pdf

https://rapanuiclothing.com/carbon-in-clothing/ (access date 19.01.15.).

http://wwf.panda.org/what_we_do/footprint/climate_carbon_energy/energy_solutions22/renewable_energy/sustainable_energy_report/ (access date 10.10.14.).

http://www.nrcan.gc.ca/sites/www.nrcan.gc.ca/files/oee/pdf/industrial/technical-info/benchmarking/ctwp/BenchmarkingCanadianTextilesWetProcessing-March2007.pdf (access date 10.10.14.).

http://www.epa.gov/region5/waste/solidwaste/p2pages/pdfs/tb-energy.pdf (access date 10.10.14.).

Ecological footprints in the textile industry

3

E. Roca[1], D.M. Herva[2]

[1]Sustainable Processes and Products Engineering Group (PRODES), School of Engineering, University of Santiago de Compostela, Santiago de Compostela, Spain; [2]Torres Quevedo Program at Cetus Group, Cerceda, A Coruña, Spain

3.1 Introduction

The ecological footprint (EF) indicator was mainly founded on the carrying capacity concept, which refers to the number of individuals who can be supported in a given area within natural resource limits and without degrading the natural, social, cultural, and economic environment for present and future generations (Kratena, 2008). Originally, the EF was advocated to assess the level of sustainability of the urban development, lifestyles, or regions. A more appropriate definition for the corporate level is that the EF determines the space required to support an activity by means of the area needed to provide the resources consumed and to absorb the wastes generated (Wackernagel and Rees, 1996; Monfreda et al., 2004; Kitzes et al., 2007; Venetoulis and Talberth, 2008). Thus, the most appealing characteristic of the EF is its integrative nature. Expressing all environmental aspects in a single score facilitates the understanding and communication of results (Ferguson, 1999).

In contrast, the term biocapacity refers to the capacity of ecosystems to produce useful biological materials and to absorb the waste generated by humans using current management schemes and extraction technologies (Kitzes et al., 2007). A comparison between the EF and the biocapacity reveals whether existing natural capital is sufficient to support consumption and production patterns (Monfreda et al., 2004). The ecological deficit occurs when the EF exceeds the available biocapacity.

There are some limitations that were acknowledged for this methodology, even though active development on EF methodology poses to continuous new proposals to overcome core critiques (Venetoulis and Talberth, 2008; Kitzes et al., 2009; Herva et al., 2010, 2012b). Van den Bergh and Verbruggen (1999) were pioneers in publishing an evaluation of the EF, but many of the ideas they exposed were further debated by different authors during the last decade (Fiala, 2008; Kitzes et al., 2009; Wiedman and Barret, 2010). The main critiques refer to the fact that there is no distinction between sustainable and unsustainable use of land (intensive production increases waste, land depletion, and soil degradation), the measure refers to virtual land area and therefore it cannot be compared to real biocapacity, it reflects neither relative scarcity changes over time nor variation over space, the quality and quantity of renewable resource use is missed, the evaluation of energy use is focused on emissions and

Handbook of Life Cycle Assessment (LCA) of Textiles and Clothing. http://dx.doi.org/10.1016/B978-0-08-100169-1.00003-4

not in the effects of the consumption of resources, and the aggregation using physical weight is considered unfair (it refers to world average technology).

On the contrary, the most appealing characteristic of the EF is its integrative nature. Expressing all environmental aspects in a single score facilitates the understanding and communication of results (Ferguson, 1999). Further, the comparison with the available biocapacity is quite straightforward; hence, the EF is often regarded as an indicator of sustainability since it states limits for the consumption of resources (materials and energy). The physical areas considered in EF accounting are weighted according to their relative productivity to obtain a final figure expressed in global hectares. Aggregation is conducted using weighting coefficients based on the relative productivity of the different area types (Kitzes and Wackernagel, 2009) rather than on the relative importance derived from the subjective opinion of experts or decision makers.

The EF has been used for different products and processes in environmental appraisals since its development (Kratena, 2008; Mamouni Limnios et al., 2009), like in aquaculture processes (Kautsky et al., 1997; Muir, 2005), seafood production (Folke et al., 1997), a water supplier company (Lenzen et al., 2003), mobile phones (Frey et al., 2006), biofuels (Stoeglehner and Narosdoslawsky, 2009), wine production (Niccolucci et al., 2008), construction of residential buildings (Solís-Guzmán et al., 2013), particleboard production (Saravia-Cortez et al., 2013), and textile and footwear industry (Herva et al., 2008, 2011, 2012a). Also, the European Union has shown particular interest in evaluating the EF capability to measure sustainable use of resources (ECOTEC, 2001; EUROSTAT, 2006; Best et al., 2008).

In spite of the increasing interest shown by the stakeholders due to the proper characteristics of the EF, currently the application of EF to evaluate environmental behavior of processes and products is still quite scarce, especially those applications to textile industries. The textile and clothing supply chain demands a lot of resources and material and chemical requirements, and each stage of the life cycle chain for textile products contributes to the EF (from fiber production to waste disposal) (Muthu, 2014). The application of EF to textiles showed that the type of material used is a key issue in the sustainability of this activity. A number of works considering different types of fibers (conventional and organic cotton, wool, hemp, synthetic stitch, and polyester among others) were done (Cherrett et al., 2005; Herva et al., 2008, 2012a; Muthu, 2014). In summary, it can be stated that these studies pointed out that conventional cotton is clearly the material with a highest EF value due to water and chemical consumption and agriculture land needs; wool also presents an important EF due to pastureland needs. In contrast, hemp poses a low EF value together with organic cotton, while synthetic fibers are in an intermediate position in terms of sustainability and resource depletion.

In this chapter, the EF methodology is described considering its suitability to appraise products and processes. The methodology is applied to two different case studies of textile products. Firstly, a case study of the EF calculation of a textile manufacturing plant is shown. The example selected with this aim is a dressmaking factory process. In this plant, cotton and synthetic jackets are produced. This example is considered to visualize the influence of resources, energy, wastes, and emissions flows in the EF obtained. After that, the EF will be applied to evaluate four different types of children's footwear. This product is consists of a more complex type of elements than other textile products, which will help to better show the interest in the application of this indicator.

3.2 Methodologies and standards

The EF methodology determines the space required to support an activity considering the following major land use types (Table 3.1): cropland (crops for food, animal feed, fiber, oil, etc.); grazing land (raising animals for meat, hides, wool, milk, etc.); fishing grounds (harvesting fish and other marine products); forest area (harvesting timber products and fuelwood); and built-up land (infrastructure for housing, transportation, industrial production, etc.). Additionally, there is a component accounting for the carbon land, that is, the area required to absorb carbon dioxide emissions derived from energy consumption. Many different ecosystem types have the capacity for long-term storage of CO_2, such as cropland or grassland. However, since most terrestrial carbon uptake in the biosphere occurs in forests, and to avoid overestimations, carbon uptake land is assumed to be forestland. For this reason, forest for timber and fuelwood is currently not separated from forest for carbon uptake (Ewing et al., 2010). There is also 12% that must be reserved for biodiversity, as stated in the Brundtland Report (WCED, 1987).

Two EF approaches can be distinguished: compound and component methods (Chambers et al., 2000; Monfreda et al., 2004). While the compound method is based on national statistics of input–output flows (production, import, export), the component method is based on life cycle data for each individual component involved in calculations. In the case of application of EF to textile products, as in general for any type

Table 3.1 The different categories of land area used in ecological footprint calculations

Land types (different EF components)	Definition
Bioproductive sea and land (Forest, pasture, agriculture)	Biologically productive land and water are categorized into five types of surface area: cropland, grazing land, forest and fishing ground.
Built-up land	Land area dedicated to buildings and constructions.
Carbon land	Land area required to sequester CO_2 emissions from fossil fuel combustion. Typically that of unharvested forests needed to absorb the fraction of fossil CO_2 that is not absorbed by the ocean.
Biodiversity	Amount of biocapacity set aside to maintain representative ecosystem types and viable populations of species. How much needs to be set aside depends on biodiversity management practices and the desired outcome.

of product, the application of the component method is preferred for the EF assessment of textile products, since it is based on the real-life cycle data of the manufacture process. Furthermore, two perspectives, namely additive or mutually exclusive use of land, can be adopted. The former means that the same area can be used for different purposes at a time (e.g., the forest used to harvest timber can be the same as the one used for carbon assimilation), thus leading to lower EF values, while the later implies the opposite situation, considered as a more precautionary approach (Monfreda et al., 2004; Venetoulis and Talberth, 2008).

Demand for resource production (material and energy) and waste assimilation are translated into global hectares by dividing the total amount of a resource consumed (or waste generated) by the global average yield of the land type that produces that resource or absorbs that waste. This area is then multiplied by the appropriate equivalence factor to express the total demand in global hectares for each resource. Yields are calculated based on various international statistics, primarily those from the United Nations Food and Agriculture Organization (Kitzes et al., 2007).

Based on the component method, whose results are the most appropriate for the evaluation of production processes and products, the EF basic calculation for each component (V_i), generally applied on a yearly basis, is carried out as shown in Eqn (3.1):

$$\mathrm{EF} = \sum_j \frac{V_i}{\mathrm{NP}_i} F_j + \sum_j \frac{V_i \cdot \mathrm{EV}_i}{\mathrm{EP}_i} F_j \tag{3.1}$$

where NP_i, EV_i, and EP_i are the natural productivity, embodied energy, and energy productivity (associated to fossil fuel) for element i; F_j is the equivalence factor for land type j. Equivalence factors translate a specific land type (i.e., cropland, pasture, forest, fishing ground) into a universal unit of biologically productive area, generally a global hectare, gha (Kitzes et al., 2007), which is the usual unit to express the EF. Thus, in contrast to other composite indicators that sum heterogeneous subcomponents using weights that are based on expert opinion, the EF assigns empirically based weighting coefficients to individual land types based on data on the relative productivity of these different area types (Kitzes and Wackernagel, 2009).

There is a simple formula for the estimation of the EF based on the existence of a relation between EF and LCA identified by Huijbregts et al. (2007) when the ecoindicator (EI) 99 was employed to evaluate a large number of products and services consumed in the Western economy. It was found that the majority of the products have an EF/EI ratio of around 30 ± 5 m^2-eq yr/ecopoint, but deviations occurred when products with high mineral consumption and process-specific metal and dust emissions were evaluated.

The Global Footprint Network is the organization responsible for the launching of the Ecological Footprint Standards (Global Footprint Network, 2009). While working on a new revision, the current standards supposed a series of advances, as for instance the definition of EF assessments for products and organizations or the expansion of the methodology to allow for the use of nonconventional elements in EF analysis, as for example the use of local yields or the incorporation of greenhouse gases other than CO_2.

3.3 Calculating ecological footprint of a textile manufacturing plant

3.3.1 Description of the production process

A general scheme of the production process is shown in Figure 3.1. To manufacture the jackets, the fabric enters the factory to be cut and sewed according to a given pattern. The pieces of the jacket are first drawn on paper and then placed on the fabric,

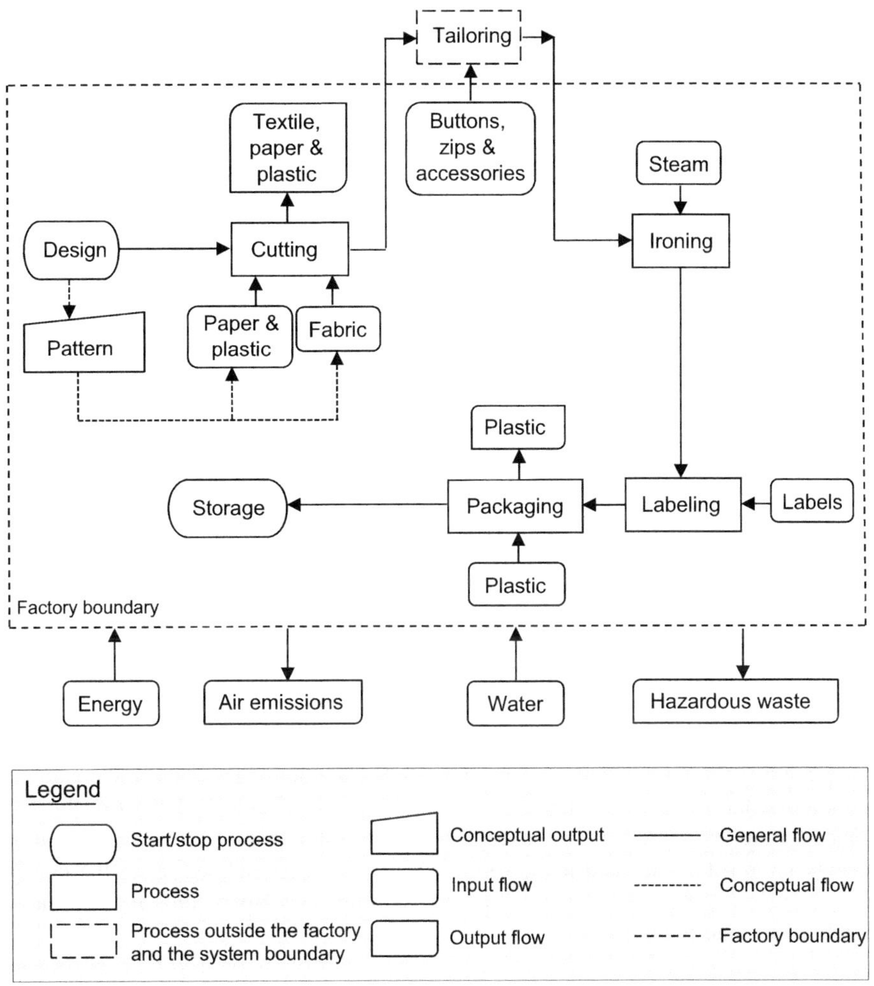

Figure 3.1 Dressmaking factory process flowsheet.
Reprinted from Herva, M., Álvarez, A., Roca, E. Combined application of energy and material flow analysis and ecological footprint for the environmental evaluation of a tailoring factory. Journal of Hazardous Materials 237–238, 231–239. Copyright (2012), with permission from Elsevier.

putting plastic over them to avoid undesired movements during the cutting process. Buttons, zips, and other ornamental elements are added to the item of clothing. This part of the process is carried out externally, although the accessories are provided by the factory, and there are no records of energy consumption during this stage. Back at the plant, the jackets are ironed, labeled, and packed into bags to be stored and later distributed. The sources of energy are electricity, wind power, propane, and gas-oil. The latter two are employed in cogeneration units (common for a series of dressmaking factories belonging to the same company), where air emissions are generated and released. The hazardous wastes are mainly generated in maintenance operations, and water is mainly consumed in the lavatories.

Data collection (Table 3.2) was restricted to the boundaries of the factory. An item of clothing (jacket) was selected as functional unit to facilitate comparisons among production years or factories.

The boundaries established corresponded to a gate-to-gate analysis, focused on burdens derived from the tailoring production process (Figure 3.1). However, it must be noted that, under the EF approach, impacts beyond the inventory data boundaries were included in the form of embodied energy, natural productivity of raw materials, and electricity breakdown into primary sources of energy. The compound method was applied to estimate the EF.

3.3.2 Environmental evaluation

The production process of the dressmaking factory was modeled using the software Umberto® 5.5 (ifu and ifeu, 2005). With this purpose, a project was created and the materials (in Umberto both substances and forms of energy are referred to as materials) involved in the production process were defined and classified into the categories established in the inventory collected in Table 3.2. Then, different scenarios were created for the years studied (2002–2005).

All input materials and products were defined as "good" materials (green triangle), while air emissions and wastes were considered as "bad" materials (red triangle), as can be observed in Figure 3.2.

The basic units are those into which all values are converted when a calculation is conducted (kg for mass-bound substances and kJ for forms of energy). To facilitate the comparability with the inventory data, entry units were defined using a conversion formula and established as display units. In the inventory (Table 3.2) most of the substances were expressed in kg except for water and gas-oil, for which volume display units (m^3 water and m^3 gasoil, respectively) were defined using the density as conversion factor. Similarly, kWh was selected as display unit for electricity, wind power, and natural gas.

A main network showing the input and output flows of the process was constructed, while the energy supply and the different stages of the dressmaking manufacture were modeled in subnets to make the visualization of the process more intuitive. All places connected to the subnet transition must remain the same for both levels. These places are the interface between the upper network level and the subnet level and are called port places (they are marked with a colored square in the symbol for the place).

Table 3.2 **Dressmaking process inventory data for the period 2002–2005**

			2002	**2003**	**2004**	**2005**
INPUT	Raw materials	Cotton fabric (kg)	643,402	651,881	798,199	919,504
		Stitch (kg)	–	–	15,800	35,500
		Lining (kg)	–	–	300,000	350,000
		Paper & cardboard (kg)	[a]5867	[a]5740	6971	7173
		Plastic (kg)	[a]32,153	[a]31,459	24,419	39,313
		Buttons (kg)	28,000	28,000	28,000	31,864
		Zips (kg)	13,500	8100	6300	7164
		Labels (kg)	650	650	650	740
	Energy	Electricity (kWh)	236,193	210,660	322,059	386,621
		Wind power (kWh)	0	8,980	14,711	15,244
		Propane (kg)	0	96.3	123.9	133.9
		Gas-oil (m^3)	61.924	35.470	19.547	34.054
		Natural gas (kWh)	–	485,411	1,045,137	1,105,012
		Water (m^3)	777.5	160.9	110.3	124.6
OUTPUT		Production (number of items)	519,399	508,188	558,078	635,055
	Air emissions	SO_2 (kg)	[a]575	[a]330	[a]182	[a]316
		NO_x (kg)	18,194	3542	3554	6086
		CO (kg)	11,529	11,502	3652	4623
		CO_2 (kg)	261,901	184,975	196,896	262,527

Continued

Table 3.2 Continued

			2002	2003	2004	2005
	Similar to urban waste	Textile (kg)	81,765	83,353	104,632	119,065
		Paper & cardboard (kg)	[a]5867	[a]5740	6971	7173
		Plastic (kg)	[a]605	[a]592	660	740
		Urban waste	–	–	–	–
	Hazardous waste	Paint (kg)	–	–	–	1.185
		Batteries (kg)	1.492	14.967	4.825	2.378
		Fluorescent lamp (kg)	11.114	5.443	13.669	6.817
		Computer waste (kg)	–	3.402	3.586	92.265
		Oil filters (kg)	60.719	11.566	7.706	4.756
		Mineral oil (kg)	104.430	115.658	100.823	–
		Contaminated containers (kg)	0.746	1.565	4.594	3.171

[a]Estimated values.

Reprinted from Herva, M., Álvarez, A., Roca, E. Combined application of energy and material flow analysis and ecological footprint for the environmental evaluation of a tailoring factory. Journal of Hazardous Materials 237–238, 231–239. Copyright (2012), with permission from Elsevier.

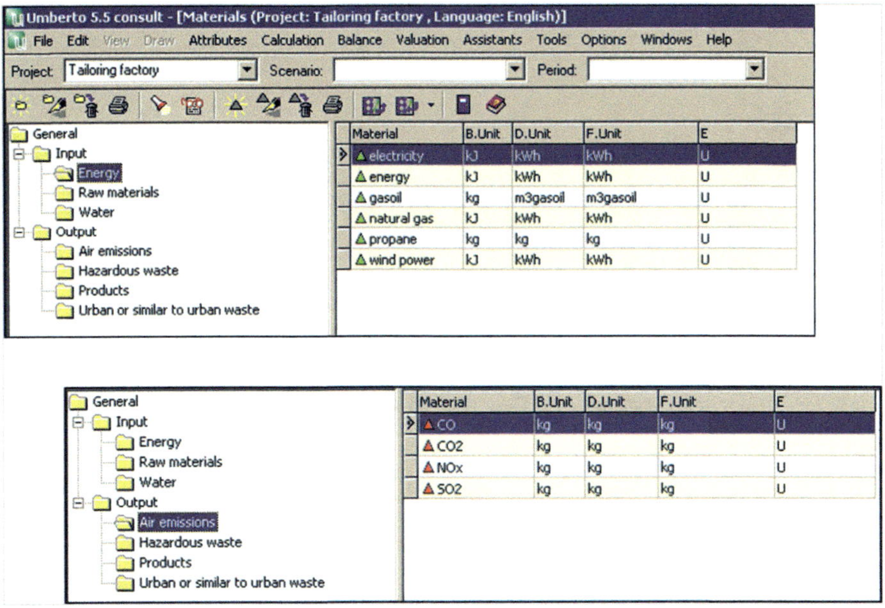

Figure 3.2 Definition of materials and groups of materials for the project, showing the specification for energy and air emissions.

Two methods were employed for the specification of transitions: (1) input/output relations, that is, introducing coefficients that represent the interrelations of the flow quantities; (2) defining the relations between input and output flows by a set of mathematical functions.

Apart from the sources of energy consumed in the factory, an additional material named energy was defined (Figure 3.2) to serve as common flow in the energy conversion subnet once proper conversions were conducted (all sources of energy were finally consumed in the form of electricity). Besides, to be consistent with the functional unit selected, the number of products was selected as reference flow.

The main input and output categories in the production process of jackets (namely energy and resources consumption and wastes generated) were used to model the main network, where the dressmaking process and the energy supply were included as subnets. The Sankey diagrams for these subnets are presented in Figure 3.3.

Since different sources of energy were consumed in the process but all of them were finally used in the form of electricity, a subnet was created to homogenize the energy supply (Figure 3.3(a)). The stages that compound the main process were modeled in the subnet presented in Figure 3.3(b), where an additional transition (T7) was included to represent those general activities implicit in the factory operation that could not be allocated to any of the other stages (e.g., hazardous wastes generation or electricity consumption for lighting and heating). Natural gas was the main supplier of energy to the factory, followed by the electricity from an external power supplier (Figure 3.3(a)). Figure 3.3(b) shows how the total energy consumed in the factory is

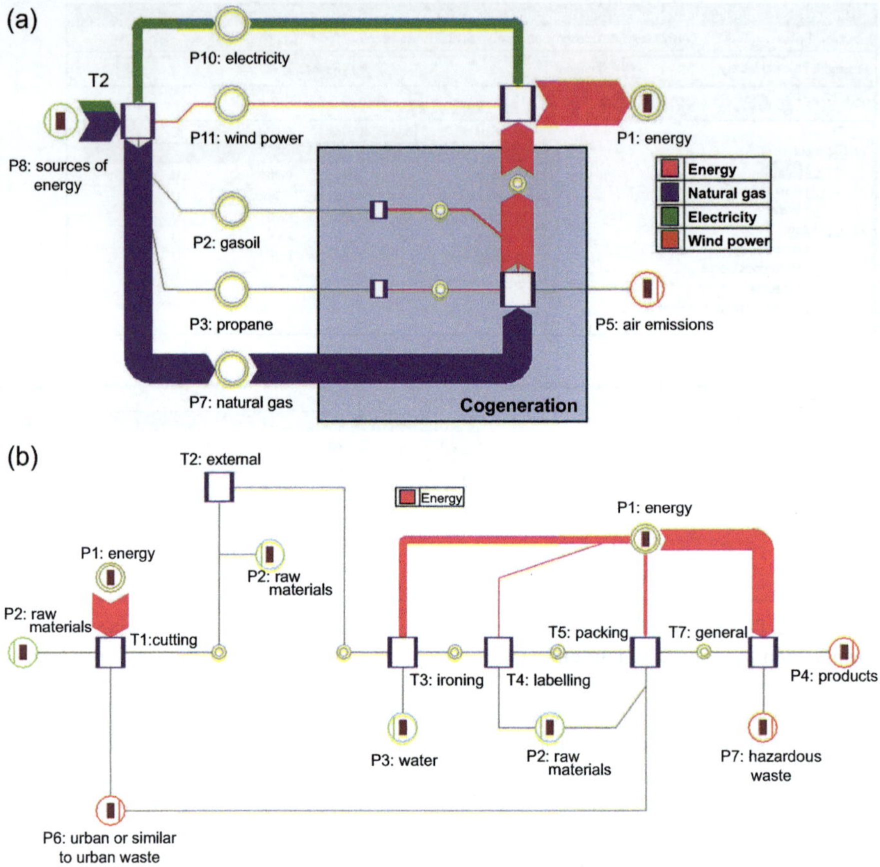

Figure 3.3 Sankey diagram for energy flows (a) in the subnet modeling the energy supply to the dressmaking factory and (b) in the subnet modeling the dressmaking process (year 2005 scenario) using Umberto® 5.5 software.
Reprinted from Herva, M., Álvarez, A., Roca, E. Combined application of energy and material flow analysis and ecological footprint for the environmental evaluation of a tailoring factory. Journal of Hazardous Materials 237–238, 231–239. Copyright (2012), with permission from Elsevier.

distributed among the main stages of the dressmaking process. The cutting stage represented the major consumption of energy, mostly due to the heat sealer used to fix the plastic over the fabric to avoid undesired movements. Precision during the cutting of the fabric was essential to prevent errors that may originate important economic losses and the generation of high amounts of textile waste. This equipment worked continuously because it was proved that the consumption of energy was lower than when it was turned on and off periodically.

The results indicating the contribution to the total EF of the considered categories are shown in Figure 3.4, where the EF is expressed in global hectares (gha). As one of the main contributions to the EF is the amount of fabric employed, an increasing

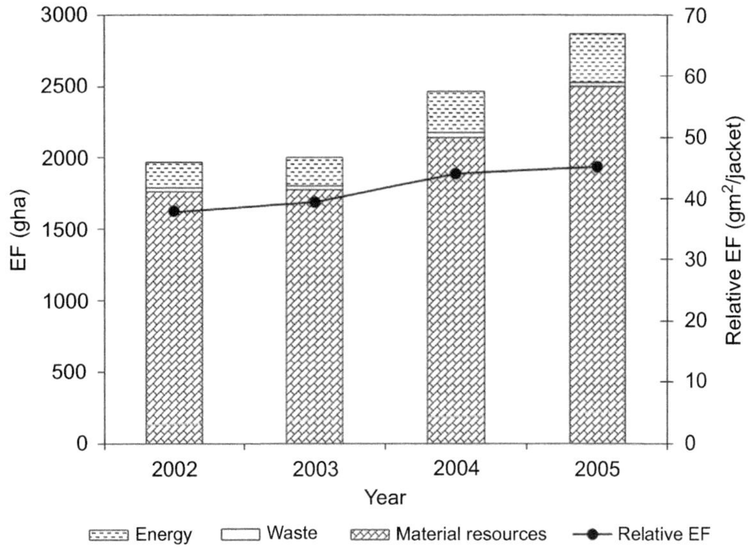

Figure 3.4 Contribution of main categories (energy, material resources, and waste) to the total EF during the period 2002–2005.
Partially modified and reprinted from Herva, M., Álvarez, A., Roca E. Combined application of energy and material flow analysis and ecological footprint for the environmental evaluation of a tailoring factory. Journal of Hazardous Materials 237–238, 231–239. Copyright (2012), with permission from Elsevier.

tendency in the total EF (excluding emissions) was observed (from 1965 gha in year 2002 to 2868 gha in year 2005). The increment in the relative EF was a little bit softer because of the increase in the annual production.

When analyzing a dressmaking factory, the particularity of fashion tendencies dependence has to be considered. Hence, during the design phase, the type and amount of materials to be employed were usually defined according to asthetic rather than environmental criteria. Consequently, it may happen that the jacket design in a certain year employs a major quantity of material—this does not strictly mean that efficiency in the resource use has decreased (in fact, the percentage of material discarded during the cutting phase is maintained) but that fashion preferences have changed. However, a worsening of the environmental performance occurs, a situation reflected in the EF figures obtained, from which the recommendation of moving toward an ecodesign of products approach is derived. The design of the jacket should pursue the fulfillment of a certain function, that is, serving as a piece of clothing rather than following fashion tendencies without considering that these may lead to a worse environmental performance. Therefore, the EF indicator is suitable to inform and guide designers so that environmental criteria are taken into account during the decision-making process.

The increase in the consumption of material resources strongly influenced the overall EF because this category represented more than 80% of the total value of the EF. The main contributor within this category was the cotton textile, which was also the major input material flow to the process. Its EF was composed of two terms, as derived from Eqn (3.1): (1) the arable land required for cotton growing and (2) the fossil area associated with the assimilation of CO_2 emissions of the embodied energy. Hence, the first component represented around 90% of the EF of cotton, corresponding to arable land, meaning that the high amounts of cotton employed required an important extension of land for its growth.

3.4 Calculating ecological footprints of textile products: evaluation of footwear as case study

3.4.1 Description of the case study

In this case study, the EF methodology is applied to evaluate four models of size EU-20 shoes for children: leather red, leather white, synthetic pink, and synthetic white. The former two and the latter two were only differentiated by the color. The production process where these shoes were manufactured, located in Elche (Alicante, SE Spain), was studied (Figure 3.5). On the basis of a pair of shoes as

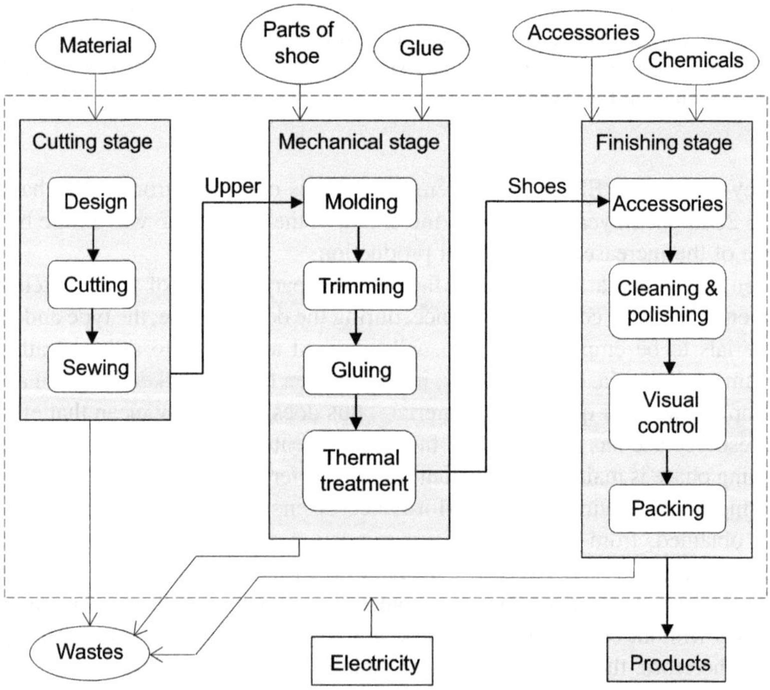

Figure 3.5 Flowsheet of the shoemaking factory.
Reprinted from Herva, M., Álvarez, A., Roca, E. Sustainable and safe design of footwear integrating ecological footprint and risk criteria. Journal of Hazardous Materials 192, 1876–1881. Copyright (2011), with permission from Elsevier.

Table 3.3 **Inventory data to estimate the EF of the synthetic model**

Element	Material	Consumption (ft^2/pair)	Surface density (g/cm^2)[a]	Consumption (g/pair)
Sole	Rubber			[b]29.5208
Upper	Polyurethane	0.70	0.0379	24.6472
Insole	Cotton and polyester	0.18	0.0193	3.2274
Lining A	Leather	0.74	0.0333	22.8932
Lining B	Polyester	0.74	0.0181	12.4434
Velcro	Nylon and polyester	0.02	0.0745	1.3142
Element	**Material**	**Consumption (no. pieces)**	**Mass (g/piece)**	**Consumption (g/pair)**
Rings	Zinc	1	0.4183	0.4183

[a]Experimentally determined.
[b]Directly measured.
Reprinted from Herva, M., Álvarez, A., Roca, E. Sustainable and safe design of footwear integrating ecological footprint and risk criteria. Journal of Hazardous Materials, 192, 1876–1881. Copyright (2011), with permission from Elsevier.

functional unit, inventory data regarding the production process were provided by the factory (Tables 3.3 and 3.4).

A wide number of operations are required for making a pair of shoes, and they are generally performed by a separate machine. After the design is finished, the first stage of the shoemaking process is the cutting of the pieces that will take the form of uppers. This operation needs a high level of skill, especially when the material is leather, to minimize the generation of waste and to avoid the likely defects on the surface that cannot be part of the shoe. Therefore, it is hardly automatized. Next, the component pieces are sewn together to produce the completed upper. The completed uppers are molded into the shape of a foot with the help of a last, a plastic shape that simulates the foot shape, which is later removed from the finished shoe to be used in making other shoes. The surplus of material is trimmed off the seam and the other parts of the shoes (toe cap, stiffener, insole, sole, etc.) are sequentially attached. After gluing steps, a thermal treatment takes places to assure a good adherence. Once the main structure of the shoe is ready, accessories (rings, eyelets, laces, rivets, etc.) are incorporated. At the finishing stage, shoes are cleaned; then, depending on the material and their final use, they could be stained, polished, or waxed to ensure an attractive finish. After visual control, the footwear is packed.

In this case, the component approach based on life cycle inventory was employed (Monfreda et al., 2004). Thus, individual EFs were calculated for each material and

Table 3.4 Inventory data to estimate the EF of the leather model

Element	Material	Consumption (ft^2/pair)	Surface density (g/cm^2)[a]	Consumption (g/pair)
Sole	Rubber			[b]29.5208
Upper	Leather	0.76	0.0714	50.4129
Insole	Cotton	0.18	0.0339	5.6723
Lining A	Leather	0.12	0.0365	4.0692
Lining B	Polyamide	0.12	0.0115	1.2821
Velcro	Nylon and polyester	0.02	0.0692	1.2199
Element	**Material**	**Consumption (no. pieces)**	**Mass (g/piece)**	**Consumption (g/pair)**
Rings	Zinc	1	0.4183	0.4183
Eyelets	Nickel	12	0.1568	1.8816
Rivets	Nickel	4	0.1192	0.4768
Element	**Material**	**Consumption (cm)**	**Mass (g/cm)**	**Consumption (g/pair)**
Lace	Viscose and polyurethane	40	0.0286	1.144

[a]Experimentally determined.
[b]Directly measured.
Reprinted from Herva, M., Álvarez, A., Roca, E. Sustainable and safe design of footwear integrating ecological footprint and risk criteria. Journal of Hazardous Materials, 192, 1876–1881. Copyright (2011), with permission from Elsevier.

energy flow in the inventory data, and then they were aggregated to estimate the total EF of the pair of shoes. This indicator was employed to evaluate the energy and materials consumption (Herva et al., 2008), as well as the solid waste generation according to Herva et al. (2010).

Electricity was the only source of energy used in the factory, and, consequently, air emissions were not released on site. The shoemaking process took place in two different buildings to which electricity was supplied by different companies. Therefore, both electricity mixes were considered to assess the EF.

3.4.2 EF Results

Two EF values were appraised using the inventoried data, yielding the following results: 6.5 gm^2/pair of shoes for the synthetic shoes and 11.1 gm^2/pair of shoes for the leather shoes (Figure 3.6). Given that only an average electricity consumption

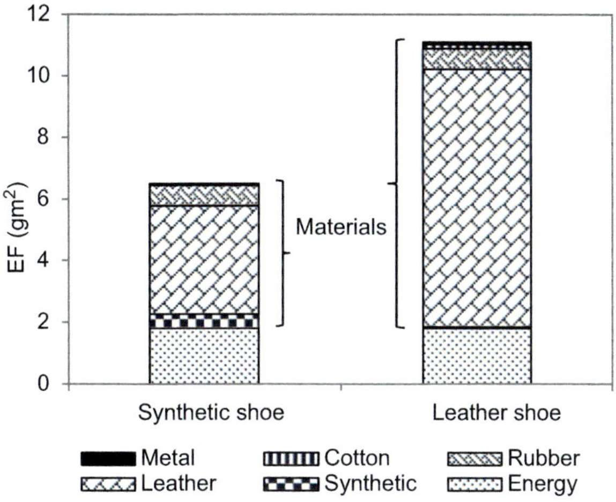

Figure 3.6 EF of the models of shoes evaluated, indicating the contribution from the energy and materials consumption.
Reprinted from Herva, M., Álvarez, A., Roca, E., Sustainable and safe design of footwear integrating ecological footprint and risk criteria. Journal of Hazardous Materials, 192, 1876–1881. Copyright (2011), with permission from Elsevier.

value was available, the EF estimated for the energy category was the same in both cases (1.8 gm^2); nevertheless, this was considered as a good approximation since only shoes for children were manufactured in the factory and similar electricity consumption patterns were expected. For the two models of shoes evaluated, the major contributor to the total EF was the materials category. Among the materials, leather presented the largest EF, thus leading to the higher EF for the leather shoes (contribution from upper and lining).

The reasons why natural materials obtain higher EFs must be sought in the kind of metric that this indicator provides. Cherrett et al. (2005) studied five fiber types, namely conventional cotton, organic cotton, conventional hemp, organic hemp, and polyester, and ranked them with regard to the EF (gha) of producing one ton of spun fiber. The lowest EF figure was 1.5 gha/t for organic hemp. Polyester produced in Europe presented an EF of 1.67 gha/t, in spite of its higher energy requirements. The difference was that polyester did not require the land area for cultivation that cotton and hemp did. Actually, crop cultivation represented the greatest proportion of the EF in the cotton case studies, which ranged from 2.17 gha for organic cotton to 3.57 gha for conventional cotton. Similarly, the main component of the EF of leather was the pastureland for cattle raising; nonetheless, the raw material in the production of leather is a by-product of the meat industry, and allocation of burdens should be considered (Joseph and Nithya, 2009). The consequence was that the synthetic models analyzed obtained a better evaluation from the EF perspective.

3.5 Future trends and challenges in calculating the EF of textile products

In the last decade, the application of the EF has been extended to the environmental evaluation of a wide number of production processes and products and particularly to textile products. However, the effectiveness of the indicator in this field is reduced due to the existence of a series of shortcomings in the methodology that limit its broader application. As sustainability metric, the EF should combine, in a standardized way, different aspects that contribute to quantify how sustainable a product is.

The future trends and challenges in calculating the EF of textile products are precisely some of the most discussed issues regarding the EF methodology. This includes accounting for renewable and nonrenewable material resource consumption (including water use), energy efficiency and renewable energy use, greenhouse gases and other gases emissions, waste minimization (including hazardous wastes) and waste reuse, but keeping its aggregation potential as a single indicator expressed in area units.

The EF methodology cannot differentiate between sustainable and unsustainable use of land (intensive production increases waste, land depletion, and soil degradation); the measure refers to virtual land area and therefore it cannot take into account the quality and quantity of renewable resource use and compare between sustainable and unsustainable agricultural activities. This is a consequence of using world average yields. It is important to have information about the EF of various textile fibers grown in different countries and different finished textile products as well as textile processes (spinning, tailoring, fabric formation, chemical processing, garmenting, etc.). The use of specific yield productivity data is required to differentiate between sustainable and unsustainable agricultural activities, instead of using world average yields.

Similar arguments can be used in the evaluation of energy use, which is focused on emissions and not on the effects of the consumption of resources. The aggregation using physical weights is considered unfair as it refers again to world average technology. The EF accounts for greenhouse gas emissions but only with regard to CO_2; this means that the contribution of other important GHG gases usually emitted in textile processes is not accounted for. The same can be said with respect to other important industrial emissions as acidifying gases. Some research has been done to improve this aspect, however, there is not a standardized method accepted worldwide that would be desirable to make the metric a more powerful tool to appraise textile product sustainability in the whole life cycle, especially when synthetic and natural fibers are compared. Besides, when the sustainability of a production process placed in a specific region is being evaluated, the use of local carbon absorption factors is more adequate than the use of world average values.

Another important issue is the EF evaluation of wastes (especially hazardous ones), which are currently discarded in the EF estimates unless they were considered as recycling material when they have an associated natural productivity, while they may become relevant in most production processes. Herva et al. (2010) proposed a method to evaluate the EF of wastes, including hazardous ones, which could be useful in the evaluation of textile products; however, progress regarding its standardization should be carried out.

The suitability of employing administrative or historical boundaries to calculate the ecological deficit of a country or a region is adequate; nevertheless, it could be irrelevant when considering the EF at a corporate level and specifically in the textile production and supply chain due to its current model of production delocalized in different areas and countries. There is no sense in attributing a biocapacity value to a textile industry. The influence of consumption behavior, functionality and durability of textile products are also important aspects when sustainability is evaluated. These are not properly addressed through the EF methodology and should be studied in the research progress of this indicator.

Last but not least, as in any other methodology of environmental evaluation, the availability of databases and models of EF application to fibers and textile processes and products is another challenge that would help in a wider application and standardization of this methodology in this particular sector.

3.6 Conclusions

The EF indicator determines the space required to support an activity by means of the area needed to provide the resources consumed and to absorb the wastes generated. Its most appealing characteristic is its integrative nature, which allows expressing all environmental aspects in a single score facilitating the understanding and communication of results. A comparison between the EF and the biocapacity shows whether existing natural capital is sufficient to support consumption and production patterns.

In spite of its unquestionable advantages, this indicator presents some acknowledged limitations, that is, no distinction between sustainable and unsustainable use of land, the measure refers to virtual land area and therefore it cannot be compared to real biocapacity, it does not reflect either relative scarcity changes over time or variation over space, the quality and quantity of renewable resource use is missed, the evaluation of energy use is focused on emissions and not on the effects of the consumption of resources, and the aggregation using physical weight is considered unfair as it refers to world average values.

The interest for the application of this methodology to evaluate the level of sustainability in textile processes and products is currently increasing; however, this is still quite scarce, especially for those applications involving textile industries and products. There are two different approaches to determine the EF: compound and component methods. The component method, which is based on life cycle data for each individual component involved in calculations, is the most appropriate for the evaluation of production processes and products including textile products and supply chain.

Two different case studies were used to show the potential interest of the application of EF to textile product (production of jackets and footwear). Through these case studies the reader can see how to apply the methodology in these particular scenarios and what sort of information can be derived with regard to the environmental and sustainability patterns of textile products and processes. In this sense, the contribution to the EF of the main input–output categories (namely energy and resources consumption and wastes generated) and flows within each category involved in the process can be evaluated and thus conclude what is more important with regard to the sustainability of the process or product. This information is very useful in the decision-making process with ecodesign purposes. In this respect, EF can help designers to better choose the materials and energy suppliers considering environmental constraints and functionality of the clothes together with fashion tendencies.

In spite of these very interesting capabilities and usefulness, the EF presents some important drawbacks that should be solved in the future to allow broader applicability of the methodology. These limitations are mainly related to standardization and the unsuitability of the EF to capture most of the impact categories derived from waste and emission flows, which are very important when the EF is applied to evaluate production processes or products.

Acknowledgments

The authors wish to acknowledge Elsevier Ed. for the permissions given to adapt and reproduce some material employed in this chapter. Marta Herva would also like to thank the Spanish Government for the support provided by the Torres Quevedo Subprogram.

References

Best, A., Giljum, S., Simmons, C., Blobel, D., Lewis, K., Hammer, M., Cavalieri, S., Lutter, S., Maguire, C., 2008. Potential of the Ecological Footprint for Monitoring Environmental Impacts from Natural Resource Use: Analysis of the Potential of the Ecological Footprint and Related Assessment Tools for Use in the EU's Thematic Strategy on the Sustainable Use of Natural Resources. Report to the European Commission, DG Environment.

Chambers, N., Simmons, C., Wackernagel, M., 2000. Sharing Nature's Interests. Ecological Footprints as an Indicator of Sustainability. Earthscan, London and Sterling.

Cherrett, N., Barrett, J., Clemett, A., Chadwick, M., Chadwick, M.J., 2005. Ecological Footprint and Water Analysis of Cotton, Hemp and Polyester. Report prepared for and reviewed by BioRegional Development Group and World Wide Fund for Nature—Cymru, Stockholm Environment Institute.

ECOTEC (Research & Consulting Limited), 2001. Ecological Footprinting. A report to the STOA Panel: European Parliament; Birmingham, United Kingdom.

EUROSTAT, European Commission, 2006. Ecological Footprint and Biocapacit: The World's Ability to Regenerate Resources and Absorb Waste in a Limited Time Period. Office for Official Publications of the European Communities, Luxembourg. ISBN 92-79-02943-6.

Ewing, B., Reed, A., Galli, A., Kitzes, J., Wackernagel, M., 2010. Calculation Methodology for the National Footprint Accounts. Global Footprint Network, Oakland.

Ferguson, A., 1999. The logical foundations of ecological footprints. Environment, Development and Sustainability 1, 149−156.

Fiala, N., 2008. Measuring sustainability: why the ecological footprint is bad economics and bad environmental science. Ecological Economics 67, 519−525. http://dx.doi.org/10.1016/j.ecolecon.2008.07.023.

Folke, C., Jansson, Å., Larsson, J., Costanza, R., 1997. Ecosystem appropriation by cities. Ambio 26, 167–172.

Frey, S.D., Harrison, D.J., Billett, E.H., 2006. Ecological footprint analysis applied to mobile phones. Journal of Industrial Ecology 10 (1–2), 199–216. http://dx.doi.org/10.1162/108819806775545330.

Global Footprint Network, 2009. Ecological Footprint Standards 2009. Global Footprint Network, Oakland. Available at: www.footprintstandards.org (accessed 10.08.14.).

Herva, M., Álvarez, A., Roca, E., 2011. Sustainable and safe design of footwear integrating ecological footprint and risk criteria. Journal of Hazardous Materials 192, 1876–1881. http://dx.doi.org/10.1016/j.jhazmat.2011.07.028.

Herva, M., Franco, A., Ferreiro, S., Álvarez, A., Roca, E., 2008. An approach for the application of the ecological footprint as environmental indicator in the textile sector. Journal of Hazardous Materials 156, 478–487. http://dx.doi.org/10.1016/j.jhazmat.2007.12.077.

Herva, M., Álvarez, A., Roca, E., 2012a. Combined application of energy and material flow analysis and ecological footprint for the environmental evaluation of a tailoring factory. Journal of Hazardous Materials 237–238, 231–239. http://dx.doi.org/10.1016/j.jhazmat.2012.08.035.

Herva, M., García-Diéguez, C., Franco-Uría, A., Roca, E., 2012b. New insights on ecological footprinting as environmental indicator for production processes. Ecological Indicators 16, 84–90. http://dx.doi.org/10.1016/j.ecolind.2011.04.029.

Herva, M., Hernando, R., Carrasco, E.F., Roca, E., 2010. Development of a methodology to assess the footprint of wastes. Journal of Hazardous Materials 180, 264–273. http://dx.doi.org/10.1016/j.jhazmat.2010.04.026.

Huijbregts, M., Hellweg, S., Frischknecht, R., Hungerbühler, K., Hendriks, A., 2007. Ecological footprint accounting in the life cycle assessment of products. Ecological Economics 64, 798–807. http://dx.doi.org/10.1016/j.ecolecon.2007.04.017.

ifu (Institute for Environmental Informatics Hamburg GmbH) and ifeu (Institute for Energy and Environmental Research Heidelberg Ltd), Umberto, 2005. A Software Tool for Life Cycle Assessment and Material Flow Analysis. User Manual, version 5. Germany. www.umberto.de.

Joseph, K., Nithya, N., 2009. Material flows in the life cycle of leather. Journal of Cleaner Production 17, 676–682. http://dx.doi.org/10.1016/j.jclepro.2008.11.018.

Kautsky, N., Berg, H., Folke, C., Larsson, J., Troell, M., 1997. Ecological footprint for assessment of resource use and development limitations in shrimp and tilapia aquaculture. Aquaculture Research 28, 753–766. http://dx.doi.org/10.1046/j.1365-2109.1997.00940.x.

Kitzes, J., Wackernagel, M., 2009. Answers to common questions in ecological footprint accounting. Ecological Indicators 9, 812–817. http://dx.doi.org/10.1016/j.ecolind.2008.09.014.

Kitzes, J., Peller, A., Goldfinger, S., Wackernagel, M., 2007. Current methods for calculation national ecological footprint accounts. Science for Environment and Sustainable Society 4 (1), 1–9.

Kitzes, J., Galli, A., Bagliani, M., Barrett, J., Dige, G., Ede, S., Erb, K., Giljum, S., Haberl, H., Hails, C., Jolia-Ferrier, L., Jungwirth, S., Lenzen, M., Lewis, K., Loh, J., Marchettini, N., Messinger, H., Milne, K., Moles, R., Monfreda, C., Moran, D., Nakano, K., Pyhala, A., Rees, W., Simmons, C., Wackernagel, M., Wada, Y., Walsh, C., Wiedmann, T., 2009. A research agenda for improving national ecological footprint accounts. Ecological Economics 68, 1991–2007. http://dx.doi.org/10.1016/j.ecolecon.2008.06.022.

Kratena, K., 2008. From ecological footprint to ecological rent: an economic indicator for resource constraints. Ecological Economics 64, 507–516. http://dx.doi.org/10.1016/j.ecolecon.2007.09.019.

Lenzen, M., Lundie, S., Bransgrove, G., Charet, L., Sack, F., 2003. Assessing the ecological footprint of a large metropolitan water supplier: lessons for water management and planning towards sustainability. Journal of Environmental Planning and Management 46 (1), 113–141.

Mamouni Limnios, E.A.M., Ghadouani, A., Schilizzi, S.G.M., Mazzarol, T., 2009. Giving the consumer the choice: a methodology for product ecological footprint calculation. Ecological Economics 68, 2525–2534. http://dx.doi.org/10.1016/j.ecolecon.2009.04.020.

Monfreda, C., Wackernagel, M., Deumling, D., 2004. Establishing national natural capital accounts based on detailed ecological footprint and biological capacity assessments. Land Use Policy 21, 231–246. http://dx.doi.org/10.1016/j.landusepol.2003.10.009.

Muir, J., 2005. Managing to harvest? Perspectives on the potential of aquaculture. Philosophical Transactions of the Royal Society B 360, 191–218. http://dx.doi.org/10.1098/rstb.2004.1572.

Muthu, S., 2014. Assessing the Environmental Impact of Textiles and the Clothing Supply Chain. Woodhead Publishing Series in Textiles. Ed. Elsevier, Cambridge (UK).

Niccolucci, V., Galli, A., Kitzes, J., Pulselli, R.M., Borsa, S., Marchettini, N., 2008. Ecological footprint analysis applied to the production of two Italian wines. Agriculture, Ecosystems & Environment 128, 162–166. http://dx.doi.org/10.1016/j.agee.2008.05.015.

Saravia-Cortez, A.M., Herva, M., García-Diéguez, C., Roca, E., 2013. Assessing environmental sustainability of particleboard production process by ecological footprint. Journal of Cleaner Production 52, 301–308. http://dx.doi.org/10.1016/j.jclepro.2013.02.006.

Solís-Guzmán, J., Marrero, M., Ramírez-de-Arellano, A., 2013. Methodology for determining the ecological footprint of the construction of residential buildings in Andalusia (Spain). Ecological Indicators 25, 239–249. http://dx.doi.org/10.1016/j.ecolind.2012.10.008.

Stoeglehner, G., Narosdoslawsky, M., 2009. How sustainable are biofuels? Answers and further questions arising from an ecological footprint perspective. Bioresource Technology 100, 3825–3830. http://dx.doi.org/10.1016/j.biortech.2009.01.059.

Van den Bergh, J.C.J.M., Verbruggen, H., 1999. Spatial sustainability, trade and indicators: an evaluation of the 'ecological footprint'. Ecological Economics 29, 61–72. http://dx.doi.org/10.1016/S0921-8009(99)00032-4.

Venetoulis, J., Talberth, J., 2008. Refining the ecological footprint. Environment, Development and Sustainability 10, 441–469.

Wackernagel, M., Rees, W., 1996. Our Ecological Footprint: Reducing Human Impact on Earth, Spanish Edition. (2001): Colección ecología & medio ambiente, LOM edition, 208 pp. (in Spanish).

WCED (World Commission on Environment and Development), 1987. Our Common Future (Brundtland Report). Oxford University Press.

Wiedmann, T., Barrett, J., 2010. A review of the ecological footprint indicator—perceptions and methods. Sustainability 2, 1645–1693. http://dx.doi.org/10.3390/su2061645.

Measuring the reusability of textile products

4

S.S. Muthu
Global Sustainability Services, SGS Hong Kong Limited, Hong Kong

4.1 Introduction

Dealing with waste is one of the important hot-button societal issues of any country, where waste prevention including the capability of reusability has highest priority in the waste hierarchy defined by the waste framework directive (European Union, 2008). Reuse option stands as the second highest priority, next to waste prevention in the order of hierarchy. When we talk about reuse, at times the material under question might need preparation for being reused, which is also included in the reuse option that is discussed here.

The 3Rs of dealing with waste include 'reduce, reuse and recycle.' The first two items are comparatively highly beneficial to the environment. As defined by the U.S. Environmental Protection Agency (EPA), reuse refers to the following: 'Use a product more than once, either for the same purpose or for a different purpose. Reusing, where possible, is preferable to recycling because the item does not need to be reprocessed before it can be used again' (EPA). Reusable means the potential of a product for reuse as defined above and that facilities exist to make such reuse economically feasible (SMART, 2004).

There is a rival school of thought when it comes to reuse. The first one is that a product can be reused for the same function until its functional life permits and the other option is that after a product's functional life ceases, it can be reused for a different purpose (maybe inferior to the previous one). The first type of reuse includes usage of a product by the same user or the product is donated or sold by the first user to the second one, accomplishing the same function in its second life as in its first life. Reusability of a product is defined as the ability of an item to be used repeatedly, unlike a disposable product. A similar definition can easily be adopted for textile products.

Except for disposable products (ones with a very short lifespan such as diapers or sanitary products), in general, textile products are intended to be reusable. The number of times a product can or needs to be reused is determined by being subjected to a number of products, which will be discussed later. Reusable products generally will bring many more advantages in terms of environmental and economic impacts compared to their rivals, but again, this cannot be generalised. Generally, reusable products are very common in medical and health care textiles, and of course with respect to normal textile products for apparel and other applications.

There is scant literature pertaining to this important area of reusability, hence this chapter is included in this handbook to fill some of the gap. To the knowledge of

Handbook of Life Cycle Assessment (LCA) of Textiles and Clothing. http://dx.doi.org/10.1016/B978-0-08-100169-1.00004-6

this author, except for the work on eco-functional assessment reported by the author in other publications (Muthu et al., 2012c, 2013a,b; Muthu, 2011), there is no other work reported so far on measuring the reusability of textile products, particularly with respect to life cycle assessment (LCA). This chapter aims to bring out the complete details of reusability of textile products, its implications for textiles and clothing products, and quantification or measurement of the reusability of textile products.

4.2 Importance of reusability in LCA

Amongst various end-of-life options, it can be commonly understood that reuse is the most beneficial option for textiles, as it will reduce the environmental impacts of the products under question and at the same time, it will provide economic benefits also. With the aid of LCA technique, one can understand the level and magnitude of environmental impacts brought down to a product in its entire life cycle by opting for reuse option at the end of life. Also a LCA study can provide insights on how reuse option can compensate the environmental impacts from the other life cycle stages which may be critical (from cradle to grave). Theoretically, when a product is reused, it avoids or prevents impacts arising from the extraction of natural resources, raw materials and manufacturing process, transportation of a new material, which is the ultimate aim of opting for reuse and the magnitude of the same can be revealed with the aid of an LCA study. LCA studies are not meant for 'apple to apple' comparisons and the crux of LCA is to deal with the functionality/functional aspects of the products to evaluate the environmental impacts arising from a product in its entire life cycle. At this juncture, the longevity of the product under question is very crucial and it is influenced chiefly by the particular product's potential reusability. Many LCA studies have been conducted on this topic to stress the importance of reuse to reduce the environmental impacts of a product from its cradle-to-grave stages (Mata and Costa, 2001; Sadamichi et al., 2003).

4.3 Implications of reusability in textiles and clothing products

Two major issues confronting the clothing sector are:

1. Resource consumption and environmental, social aspects of clothing manufacture;
2. Solid waste management – dealing with the piled-up textile waste in landfills.

Proper waste management that is beneficial to the environment will answer these issues and undoubtedly this means reuse followed by recycling and incineration. Dumping in landfills is not an option, as almost all landfills are full and there is a cost to dispose at landfills (in most countries). Dumping is highly detrimental to the environment, which is obviously known to everyone. Reuse is the most effective and promising way to stop waste at the source and its benefits are twofold:

- It postpones or avoids a textile material being entered into the landfill/disposal stage; (environmental impacts corresponding to dumping at landfills are avoided)

- It prevents/delays the manufacture of a new/virgin material and its associated environmental impacts are prevented. It conserves precious, natural resources and reduces environmental load/pollution.

Apparel in general is reused until the end of its life in the sense that after the first user, it is donated or sold to a second user. In most Asian countries, it is still common that old textile garments are used by siblings (clothes of older ones in the family are then worn by the younger ones until they are torn). This is a scenario of reuse, which normally appears in end-of-life scenarios in LCA. This is not, however, the crux of that which is dealt with in this chapter.

Textile materials, depending on the end use, will have to serve the sufficient life span for which they are intended. Except for disposable products, almost all textile materials, especially the ones for apparel purposes, are expected to have a reasonable lifespan. This chapter is dedicated to deal exclusively about this finest point in LCA. Though the previous sections have dealt with the various aspects of reusability, the crux of this chapter is to deal with this point, which is rarely discussed in the literature.

4.4 Importance of reusability in LCA studies

LCA is one of the widely used and important tools to measure the environmental impacts or environmental performance of a product considering the complete life cycle of a product, encompassing all its life cycle stages from cradle to grave. Environmental metrics of a studied product by LCA are solely dependent upon a term called 'functional unit' (FU), which is the base considered by an LCA tool to map and measure the environmental impacts. Both for single and comparable products (based on the same functional level of course), the definition of a proper FU is pivotal to draw sensible results and to have a meaningful conclusion.

An FU needs to be defined carefully in relation to the defined objectives of any study; in reality it has not received the attention it deserves and this has enabled a huge amount of subjectivity to the LCA practitioner. There is no clearly defined ways to describe a FU in the LCA studies (Cluzel et al., 2013). It is one of the major sources of uncertainty from LCA studies, as indicated by Reap et al. (2008) in their article that surveyed unresolved problems in LCA. This article also stressed the importance of defining a proper FU in LCA studies.

Though there are no clearly specified rules on defining a proper FU for LCA studies, ISO standards earmarked for LCA studies, that is ISO 1404X series emphasises the importance of FUs in comparative LCA assessments. ISO 14040 standard for LCA defines a FU as 'quantified performance of a product system for use as a reference unit' (International Organization for Standardization, 2006a). Additionally, ISO 14044:2006 affirms that the FU must be clearly defined, and it is also stressed that a FU is of a measurable element (International Organization for Standardization, 2006b). However, there is an absence of a properly laid guideline to construct the same, which is one of the major shortfalls of this standard. One of the notable references in LCA is the ILCD handbook, which enables a way to define a FU in LCA by thinking and answering the following four questions, namely what how much how well and for how long? (Joint Research Center – Institute for Environment and Sustainability, 2010).

A proper FU definition is a must and is crucial to understanding the real situation of environmental impacts created by a product across its various life cycle phases. LCA study has many variants such as:

- Cradle-to-gate assessments;
- Gate-to-gate assessments;
- Cradle-to-grave assessments;
- Cradle-to-cradle assessments.

Depending on whether a study consider the entire life cycle phases of a product (full life cycle study; full LCA) or only certain phases of life cycle (partial life cycle study; partial LCA), these variants will be selected. A FU linked to reusability is highly crucial for full LCA studies, covering the usage and disposal stages/phases of life cycle. An example to clarify this point is: Let's say, a 100% cotton garment needs to be assessed for LCA. For partial LCA, it will suffice to define a FU as '1 Kilogram of cotton fibres with primary and secondary packaging from cradle to factory gate' if fibres need to be assessed in case or 1 unit of cotton garment (100% cotton, knitted, solid dyed) weighing 450 g with primary and secondary packaging from cradle to factory gate, in case of assessing the complete product. Of course, these sample FUs can be made much more detailed. But the same FU of 1 unit of garment weighing 450 g with packaging will not be accepted for full LCA study. Here comes the importance of defining a proper FU linked to reusability or functional aspects of the product.

For the same example considered above, a proper FU could be for cradle-to-grave assessment: 100% cotton knitted, solid dyed garment weighing 450 g, used for 1 year with washing thrice in a week in India and disposed in India. In addition, washing, drying conditions and the disposal scenarios can be explained further to make the FU complete. Here, with this FU, the importance of assessing reusability can be understood for the demonstration of an actual scenario of environmental impacts for the product under question.

Assessment of functional aspects of the textile products in the entire product life cycle stages starting from fibre to disposal and linking them to the environmental impacts is crucial and it was first defined as 'eco-functional assessment' by the author. A theoretical framework of eco-functional assessment and an eco-functional model was developed by the author, and the applications of the same were demonstrated for different consumer products by the author in his research work; this work has been published in various research publications (Muthu et al., 2012a,b,c, 2013a,b; Muthu, 2011; Muthu and Li, 2013). Eco-functional assessment is quite new and it has to be applied for various textiles and clothing products, and various eco-functional indices need to be quantified for the clothing supply chain (Muthu, 2014). The aspect of reusability along with the other functional parameters for the entire life cycle stages of clothing products need to be considered to a greater extent in the eco-functional assessment.

4.5 Quantification of reusability of textiles and clothing products

A previous LCA study conducted in 1999 noticed that the LCA studies on textiles do not include the effect of product durability or rate of use, which eventually has to come from the assessment of reusability of textiles and clothing products (Kalliala and Nousiainen, 1999). I still feel that the above gap identified in 1999 is not completely fulfilled by the studies completed by the successors of this LCA study. There are not many contributions to the scientific fraternity of textiles emphasising this point, except the eco-functional assessment methodology related contributions discussed above by the author.

Assessment of reusability and the results from the reusability assessments are highly useful in defining a proper FU and thus defined FUs aid to derive meaningful results from an LCA study. To explain the same with an example, a case example of comparative assessment on shopping bags (by the author) is used here and the FU defined for this study was (Muthu et al., 2011): 'the number of shopping bags used for grocery shopping per year by an average Chinese/Indian/HK resident'. Further, the number of shopping bags in each category used by an average resident of each country under question was calculated based on the reusability potential of each type of shopping bag considered in this study. In this study, it was assumed that the plastic and paper bags were comparable functionally and they were supposed to be equivalent to each other in function. But the major turning point comes when it comes to reusable bags. In the case of nonwoven and woven bags considered in this study, it was assumed that one nonwoven and two woven cotton bags were able to replace 100 single-use plastic and paper bags (this hypothesis was validated with many references). Hence, 1095 plastic and paper bags, 10.95 nonwoven and 21.9 woven bags were calculated to be required to fulfill the FU assumed for this study for average Chinese and HK residents. For Indians, 150 plastic and paper bags, 1.5 nonwoven and three woven bags were calculated to be needed to fulfill this study's FU. This kind of clarity in terms of quantified performance came into the picture when the reusability was included in the consideration of defining the FU; otherwise, if the study would have assumed every bag was equal, then, it would have turned out to be senseless. In this study, the reusability parameters were assumed in terms of functionality. Understanding the importance of this parameter in LCA studies, the very first of its kind, an eco-functional tester was designed by the author to quantify various parameters of different types of shopping bags, which will be used to calculate the useful lifetime of a particular type of shopping bag. Many trials were conducted for various types of shopping bags, and the results were reported (Muthu et al., 2013b). Further, the reusability results from this tester were used in LCA results (for eco-functional assessment) and the usefulness of these values was demonstrated (Muthu et al., 2012c, 2013a; Muthu, 2011).

For the assessment of reusability of textile materials, there is a long list of parameters that need to be tested to draw conclusions about the reusability of a specific material. Lists of tests that need to be carried out to assess the functionality chiefly depend on the product type (woven or knitted or nonwoven) and its end use (apparel/technical textiles/surgical use/disposable item), apart from other considerations.

For example, in the shopping bags case, what we discussed earlier needs to have the following parameters principally in terms of reusability (the below parameters most appropriately can be termed as eco-functional properties rather than functional properties (Muthu et al., 2013b)):

- Impact resistance;
- Number of times a bag can be reused – reusability;
- Load-bearing capacity.

If one looks at these properties along with their definitions, it can be clearly understood that these three parameters chiefly decide the functional life of a shopping bag. Results from the tests pertaining to these three parameters can directly be utilised to assess the functional life of shopping bags and can be used in LCA calculations. Apart from these three parameters, there are many other physical, mechanical and chemical (safety related) test parameters that can also determine the functional life of such products. These are tabulated in Table 4.1.

Another example from the LCA study conducted in 1999 was on hotel textiles made out of cotton and polyester fibres, and its blends had the following parameters tested to assess their durability to include in LCA (Kalliala and Nousiainen, 1999):

- Abrasion resistance
- Changes in warp and weft strength
- Number of rubbing cycles
- Loss in weight
- Shrinkage

These were all tested after 100 laundering cycles for different types of hotel textiles to decide upon their durability and hence used accordingly in LCA calculations.

Most of the time, the textiles are discarded after they lose their functional value, and of course a slight intervention comes in terms of fashion too. Therefore, when deciding the reusability, one has to be careful in testing all the relevant parameters that decide the fate of the particular textile product in terms of its functional value. Some of the basic parameters to be tested to decide the functional life of the textiles and clothing products are listed below (of course, it is not limited to the following ones):

1. Areal density
2. Thickness
3. Tensile strength
4. Tear strength
5. Bursting strength (for knitted ones)
6. Seam strength and slippage
7. Dimensional stability and appearance after cleaning
8. Fiber content analysis

Table 4.1 Functional parameters of shopping bags (Muthu et al., 2013a; Muthu, 2011)

Test parameter	Test standard followed
Material composition	ISO 1833-1/FTIR/HPLC
Tensile strength and elongation	ASTM D 5034 grab test
Tear strength	Elmendorf tear test ASTM D 5734
Thickness	ISO 5084
Weight	ISO 9073-1:1989
Bursting strength	ISO 13938-2
Colour fastness to friction	ISO 105-X12
Colour fastness to washing	ISO 105-C10:2006
Colour fastness to water	ISO 105-E01:2010
Colour fastness to perspiration	ISO 105-E04:2008
Colour fastness to light	ISO 105- B 02
Impact resistance and toughness	Eco-functional tester
Load carrying capacity	Eco-functional tester
Ph	ISO 3071
Formaldehyde	ISO 14184-1
Air permeability	ISO 9237
Water proof	AATCC 127
Water vapour permeability	ASTM E 96

9. Air permeability
10. Oil proof
11. Water vapour permeability
12. Water repellence and resistance
13. Colour fastness to light, water, washing, rubbing, perspiration, etc.
14. pH
15. Formaldehyde
16. Flammability
17. Toxicity
18. Anti-static
19. Heavy metals
20. Toxicity tests
21. Other eco-parameters such as azo dyes, carcinogenic dyes, and allergenic dyes.

As emphasised earlier, depending upon the type of product, fibre content, end use, the parameters to be tested need to be decided reflecting an idea of the functional life of the products. This functional life has to be taken into consideration when deciding the useful lifetime of the products, and it especially has to be used in LCA calculations. Integration of the functional parameters into the LCA models and calculations have already been described in the author's previous publications (Muthu et al., 2012c, 2013a; Muthu, 2011).

4.6 Future trends and remarks

It is very important to decide upon the functional parameters of a textile/clothing product before the start of LCA study. This functional parameter has to be considered to judge the actual lifetime of a product under question and this has to be included in the LCA study when deciding the FU and further issues pertaining to LCA calculations. Functional parameters must be included in the LCA study (such as eco-functional assessments, which includes the entire cradle-to-grave concept or at least the crucial aspects of conventional LCA approaches). If the FUs of LCA studies on textiles and clothing are decided based upon this, it will be worthwhile to relate these potential environmental impacts arising from LCA studies to the actual functionality/functional limits of the studied products. There must be a worldwide guideline/procedure to help perform the calculations pertaining to decide the functional life of the products from the different test parameters and integration of such functional values/parameters and functional limit of the products to LCA calculations.

4.7 Conclusions

'Product longevity of life' is crucial when it comes to the assessment of environmental impact of a product. With this characteristic only, reusable products score higher over the disposable products in LCA studies (again, this cannot be generalised, as it heavily depends on the resource consumption needed in the use phase of the reusable product and the impacts from the other phases pertaining to the reusable product vis-à-vis the disposable one). Reusability of a product is the crux to decide the ultimate lifetime of a product (to decide the functional limit of the product at which the useful lifetime of a product seizes) and hence the environmental impacts across the entire life cycle phases of a product. This general phenomenon is quite applicable to textiles and clothing products as well, where reusability is a major concern. Measurement/quantification of reusability of textiles and clothing products and accordingly utilising the same in LCA calculations is crucial, as LCA studies are not meant truly for 'apple-to-apple' comparisons. This chapter highlighted this finest part with various LCA examples pertaining to textiles and clothing sector. This chapter also discussed the basic parameters needed to be tested to judge the functionality of textiles and clothing products.

References

Cluzel, Francois, Leroy, Yann, Yannou, Bernard, August 2013. Toward a structured functional unit definition framework to limit LCA results variability. In: 6th International Conference on Life Cycle Management – LCM 2013. Gothenburg, Sweden. Paper no. 484.

http://www.epa.gov/epaoswer/non-hw/muncpl/reduce.htm#reuse.

European Union, 2008. Directive 2008/98/EC of the European Parliament and of the Council of 19 November 2008 on Waste and Repealing Certain Directives.

International Organization for Standardization, 2006a. ISO 14040:2006–Environmental Management – Life Cycle Assessment – Principles and Framework (ISO Standard).

International Organization for Standardization, 2006b. ISO 14044:2006–Environmental Management – Life Cycle Assessment – Requirements and Guidelines (ISO Standard).

Joint Research Center - Institute for Environment and Sustainability, 2010. ILCD Handbook – General Guide for Life Cycle Assessment – Detailed Guidance v1.0 (No. EUR 24708 EN). European Commission.

Kalliala, E.M., Nousiainen, P., 1999. Life cycle assessment. environmental profile of cotton and polyester-cotton fabrics. Autex Research Journal 1 (1), 8–20.

Mata, Teresa M., Costa, Carlos A.V., 2001. Life cycle assessment of different reuse percentages for glass beer bottles. International Journal of Life Cycle Assessment 6 (5), 307–319.

Muthu, Subramanian Senthilkannan, Li, Y., September 2013. Assessment of Environmental Impact by Grocery Shopping Bags: An Eco-functional Approach. Springer, Singapore.

Muthu, S.S., Li, Y., Hu, J.Y., Mok, P.Y., January 2011. Carbon footprint of shopping (grocery) bags in China, Hong Kong and India. Atmospheric Environment 45 (2), 469–475.

Muthu, S.S., Li, Y., Hu, J.Y., Mok, P.Y., 2012a. Quantification of environmental impact and ecological sustainability of textile fibres. Ecological Indicators 13, 66–74.

Muthu, S.S., Li, Y., Hu, J.Y., Mok, P.Y., 2012b. Quantification of recyclability potential Index (RPI) for textile fibers. Ecological Indicators 18, 58–62.

Muthu, S.S., Li, Y., Hu, J.Y., Mok, P.Y., 2012c. Eco-functional assessment combined with life cycle analysis: concept and applications. Energy Education Science & Technology Part A 29 (1), 435–450.

Muthu, S.S., Li, Y., Hu, J.Y., Mok, P.Y., 2013a. Modelling and quantification of eco-functional index: the concept and applications of eco-functional assessment. Ecological Indicators 26, 33–43.

Muthu, S.S., Li, Y., Hu, J.Y., Mok, P.Y., Mao, Y.F., Wu, X.X., Li, Q.H., 2013b. Assessment of eco-functional properties of shopping bags: development of a novel eco-functional tester. International Journal of Clothing Science & Technology 25 (3), 208–225.

Muthu, S.S., 2011. Eco-functional Assessment of Grocery Shopping Bags (Ph.D. thesis), The Hong Kong Polytechnic University, Hong Kong.

Muthu, Subramanian Senthilkannan, March 2014. Assessing the Environmental Impacts of Textiles & the Clothing Supply Chain. Woodhead Publishing, UK.

Reap, J., Roman, F., Duncan, S., Bras, B., 2008. A survey of unresolved problems in life cycle assessment -Part 1: goal and scope and inventory analysis. The International Journal of Life Cycle Assessment 13 (4), 290–300. http://dx.doi.org/10.1007/s11367-008-0008-x.

Sadamichi, Y., Kimura, Y., Widiyanto, A., Kato, S., Maruyama, N., Nishimura, A., 2003. LCA evaluation of reuse/recycle impact for environmental conscious industrial products. In: Environmentally Conscious Design and Inverse Manufacturing, 2003. EcoDesign '03. 2003 3rd International Symposium, 8−11 December 2003, pp. 339−343. Tokyo, Japan.

SMART, December 15, 2004. Sustainable Textile Standard 2.0 -Promoting Sustainable Textile Achievement, the Institute for Market Transformation to Sustainability, Final Approved Consensus Ballot Version.

Further reading

Cooper, J., 2003. Specifying functional units and reference flows for comparable alternatives. The International Journal of Life Cycle Assessment 8 (6), 337−349. http://dx.doi.org/10.1007/BF02978507.

Collado-Ruiz, D., Ostad-Ahmad-Ghorabi, H., 2010a. Fuon theory: standardizing functional units for product design. Resources, Conservation and Recycling 54 (10), 683−691. http://dx.doi.org/10.1016/j.resconrec.2009.11.009.

Collado-Ruiz, D., Ostad-Ahmad-Ghorabi, H., 2010b. Comparing LCA results out of competing products: developing reference ranges from a product family approach. Journal of Cleaner Production 18 (4), 355−364. http://dx.doi.org/10.1016/j.jclepro.2009.11.003.

Esterman, M., Fumagalli, M.E., Thorn, B., Babbitt, C., 2012. Towards parametric environmental profiles of complex industrial systems in preliminary design stage. In: Proceedings of the ASME 2012 International Design Engineering Technical Conferences & Computers and Information in Engineering Conference, IDETC/CIE 2012. Chicago, Illinois, USA.

Kato, S., et al., July 10−12, 2001. LCA of industrial products (ecological improvement by reuse and recycle of vending machines). In: Proceedings of the 11th Symposium on Environmental Engineering 2001. Kawasaki, Japan, pp. 425−428.

Zaraket, T., Yannou, B., Leroy, Y., Minel, S., et al.Chapotot, E., 2012. An experimental approach to assess the disparities in the usage trends of domestic electric lighting. In: Proceedings of International Design Engineering Technical Conferences (IDETC), Chicago, United States. IDETC2012−70835.

Environmental impacts of the use phase of the clothing life cycle

5

S.S. Muthu
Global Sustainability Services, SGS Hong Kong Limited, Hong Kong

5.1 Introduction: the use phase in the life cycle of clothing

Life cycle assessment (LCA) of a product involves quantification of potential environmental impacts of a product by considering the different life cycle phases of a product including raw material extraction and processing, process of manufacturing, transportation and distribution, and use and disposal phases. LCA helps in quantifying the potential environmental impacts of a product and very importantly helps in identifying the critical phase of the life cycle phase of a product that creates the maximum impacts, allowing one to devise a suitable plan to work on reducing those impacts arising from the identified critical phase. Depending upon the scope of assessment (cradle to gate or gate to gate or cradle to grave), a product category might have more impact at a particular life cycle phase. LCA of textiles and clothing products is slowly gaining its importance and again, depending upon the boundary of assessment, a particular life cycle phase will contribute to a major amount of impact arising from the LCA of the entire product life cycle.

Although textiles and clothing products are meant to be used for a longer span, there is a particular category of textiles, disposable textiles, that are meant to be disposed of right after the first use. This class of textiles is a perceived symbol of a throwaway society and common examples falling under this category are napkins, diapers, hygiene products, etc. For such products, phases apart from use stage such as raw material extraction or manufacturing phase or disposal stage at times might be detrimental in terms of creating maximum impacts across the life cycle phases. However, in the case of reusable products, the use phase is mostly the one that causes the maximum amount of impact across the various life cycle phases and it is found to be the most detrimental phase of all (hotspot). The use phase of clothing products involves actions such as washing, drying and ironing in general and at times this might also include maintenance for some product categories.

5.2 Quantifying the use phase of the clothing life cycle

As discussed above, in general, the use phase is one of the major hotspots in the life cycle of a clothing product; however, it depends on various factors such as the product

Handbook of Life Cycle Assessment (LCA) of Textiles and Clothing. http://dx.doi.org/10.1016/B978-0-08-100169-1.00005-8

type and its characteristics, end use, fibre type of the product, consumer behaviour, washing frequency demanded by the product and governed by consumer behaviour, etc. This section will deal with the use phase implications in the life cycle of different types of textile products, namely cotton textiles (a T-shirt and jeans, since the washing frequencies, product characteristics are different for both of these products), linen textiles, woollen textiles and synthetic textiles (again, a jacket and a blouse). This section deals with the LCA results of the above-mentioned products by highlighting the implications/proportion of use phase impacts.

5.2.1 Cotton textiles

5.2.1.1 T-shirts

A 2009 LCA study on 100% Australian-cotton T-shirt reported that consumer use phase dominates the life cycle entirely and the use phase alone was reported to have 96% of total life cycle impacts (as far as GHG emissions were considered). In this study, the garment under study was assumed to be washed and dried 75 times, and this was found to be responsible for creating 277 kg CO_2e/T-shirt. Out of different operations in the consumer use phase, tumble drying caused maximum emissions with 182.9 kg CO_2e/T-shirt, followed by domestic washing, which caused 93.6 kg CO_2e/T-shirt and ironing caused the least impacts (1.2 kg CO_2e/T-shirt) (Life Cycle Assessment, 2009).

Another 2009 case study reported by Continental Clothing on a cotton T-shirt reported that the total life cycle footprint of one unit of a women's short-sleeve T-shirt in size small, colour charcoal, with one colour print was 2.34 kg CO_2e. Hotspot analysis by breaking down this total figure into different life cycle phases revealed that the use phase (including automatic washing, tumble drying and ironing 25 times during the product's lifetime) contributed a major share, almost 48%, of the total carbon footprint of the T-shirt studied. Again, similar to the previous study, consumer use phase was bifurcated in terms of contribution to the global warming of use phase by different elements, and it was noticed that automatic washing caused 24%, tumble drying was responsible for 45% and ironing constituted to 31% of the use phase impacts (Carbon Footprint, 2009).

A third study reported in 2009 was mainly based on a study of life cycle inventory (Quantification of Energy Use and Four Major Pollutants) of a cotton T-shirt produced in India and consumed in Germany. It was a cradle-to-grave assessment focusing on direct and indirect energy uses, use of chemicals and water use in various life cycle phases of a cotton T-shirt. Electricity consumption, consumption of detergents and water treatment were included in the assessment while evaluating the consumer use phase. The functional unit defined in this study was, '100 days of a garment being worn'. A cotton T-shirt weighing 250 g, estimated to be washed once every two wearings, that is, 50 times total during its entire life period, was chosen for this study. The life cycle inventory analysis conducted in this study primarily revolved around quantifying four important pollutants, namely carbon dioxide (CO_2), nitrogen oxides (NO_x), particulates (PMs) and sulphur dioxide (SO_2) as well as the primary nonrenewable energy (Steinberger et al., 2009). As revealed from this study, a cotton T-shirt in its entire life cycle with the assumed conditions depleted 240.11 MJ, where drying and washing consumed a majority of it. Drying for 50 times alone consumed 114.23 MJ, followed by washing for 50 times that consumed 62.16 MJ, and which were found to be the major consumers of energy in the total life cycle of the studied T-shirt. As far as the four major pollutants are considered, during the lifetime of 100 wearings, one cotton T-shirt was found to emit 14 kg CO_2, 48.98 g NO_x, 32.70 g PM and 32.03 g SO_2. Consumer usage in the assumed life span of the product under question (which predominantly involved washing and drying) was found to be the major threat for emissions of CO_2 and PMs. Drying and washing during consumer usage were also found to be the major contributors of SO_2 emissions. Many scenarios were analysed in terms of use phase, and based on the analysis, the following were recommended to reduce the use phase impacts:

- Over 60% reduction in the use phase impacts can be achieved if a consumer chooses air drying instead of machine drying.
- When it comes to the washing temperatures, 30% reduction of the washing phase emissions can be achieved at 40 °C compared to washing at 60 °C.
- In terms of appliance efficiency rating, 20% reduction in terms of choosing A-rated appliances compared to the C-rated ones, if other parameters are held constant (Steinberger et al., 2009).

The carbon footprint of a white long shirt was conducted in 2010; the carbon footprint of this product for its entire life cycle was reported to be 10.75 kg CO_2e, which is 50 times the net weight of the studied product. In this study, 31% of the emissions were from use phase (3.30 kg CO_2e); again, in this study use phase was found to be the major hotspot (Jungmichel, 2010).

5.2.1.2 Cotton denim

An LCA study of cotton denim was undertaken by Levi Strauss & Co. in 2006. A medium stone-washed product, produced from a shrink-to-fit fabric, to cater to the needs of the American market was selected for this full LCA (cradle-to-grave assessment) study. This study was conducted based on ISO 14040 standard, and it covered three major impacts, namely climate change, energy use and water consumption. Functional unit details pertaining to this study were not clearly obtained from any of the respective references, even after repeated efforts were spent in searching, as mentioned in my previous publication; I assume that it must have been conducted for a unit of jeans (Muthu, 2014). From the full LCA results, it was noticed that the consumer use phase alone was responsible for 58% of the total climate change and energy use impacts in the entire life cycle impacts; in the case of water consumption, again, consumer use was found to be one of the major hotspots, however, it was the second major contributor, followed by the cotton production. This study recommends that consumers consider reducing the number of washes in the lifetime of a pair of jeans, since the consumer use phase was found to be the key to reducing the major impacts studied. It was recommended that customers change the washing frequency to once every two weeks from once per week and perhaps even to once per month; the percentages of savings in climate change impacts, energy use and water consumption were showcased for the various washing frequencies. The results from this study confirmed that switching over to washing once every two weeks or once a month could reduce 32% and 48% of climate change impacts, respectively; 20% and 40% of energy use, respectively; and 23% and 35% of water consumption, respectively (Product Life Cycle, 2009).

5.2.2 Woollen textiles

A full LCA study focusing on carbon footprint as a single issue conducted on a woollen sweater (without buttons, laces, zippers or other accessories) was reported in 2012. An average sweater, made out of 100% merino wool with four colours, picked up from the 2009 winter collection with a weight of 264.85 g (without accessories) was considered as a functional unit. However, the functional unit was defined properly in this study (Muthu, 2014). A medium life of 5 years was assumed for the chosen product for this study and for the use phase it was assumed that the product was washed 15 times per year at 30 °C of washing temperature. For each cycle of wash, it was estimated that 10 L of water and 130 mL of chemicals, namely soap and conditioner, were required. From the results, it was noticed that the total life cycle of the selected garment emitted 1947 kg CO_2 from the cradle-to-grave stages, and from this, the use phase was responsible for only 0.280 kg CO_2 (Bevilacqua et al., 2012). Differing from the previous studies mentioned on cotton textiles, this study did not cherry pick consumer phase as the major hotspot, which might be due to the nature of the product and the frequency of the washing demanded by this product.

5.2.3 Linen textiles – linen shirt

To discuss the use phase impacts of linen textiles, a comparative study of linen and cotton shirts conducted in 2008 is discussed here. Twelve indicators were included in this study; however, the results of only five indicators were presented in the eco-profile: primary energy consumption, water consumption, global warming potential, eutrophication and freshwater aquatic ecotoxicity potential. This cradle-to-grave study defined the functional unit of the study as ‘Wearing a shirt for a day’. This study compared the environmental impacts of a linen shirt (cultivated in France, manufactured in China, transported to France for use and disposal) and a cotton shirt (cotton was cultivated and the shirt was manufactured in China, and then transported to France for use and disposal). The chosen shirts of linen and cotton were of the same weight, 253 g, including 240 g of linen or cotton fabric, 10 g of polyester yarn and 3 g of plastic buttons. It was assumed that both of these shirts were worn on average 60 times before disposal by the primary user. The disposal scenario was assumed to be 30% of the disposal with the household waste and 70% of reuse. With this scenario, the average total lifetime of a shirt worked out to 100 times of use. As far as the use phase was concerned, both were assumed to be washed at 40 °C and ironed after each use with different ironing periods (linen shirt needed to be ironed for 9 min and the cotton shirt for 7 min). From the results of LCA, it was revealed that wearing a linen shirt for 1 day consumed 6 MJ of energy, 6.4 L of water, emits 130 g of CO_2 and is responsible for 0.10 g phosphate units of eutrophication and 11 g 1,4 dichlorobenzene of freshwater aquatic ecotoxicity potential. Results further revealed that the use phase was found to be the major hotspot for energy and water consumption and freshwater aquatic ecotoxicity potential, whereas the manufacturing phase was found to the major contributor to global warming potential and eutrophication impacts (Linen Shirt Eco-Profile, 2008).

5.2.4 Synthetic clothing

5.2.4.1 Polyester garments – blouse and jacket

A Life Cycle Inventory (LCI) study conducted on a polyester blouse was reported in 1995. This was a cradle-to-grave study performed by Franklin Associates, Ltd for the American Fiber Manufacturers Association on a women's blouse made out of 100% polyester knit fabric. This study focused mainly on quantifying the energy requirements, atmospheric emissions, waterborne wastes and solid wastes from industrial and post-consumer operations of a polyester blouse. This study was conducted on a short-sleeve, pullover women's blouse made out of 100% knitted polyester filament weighing 0.12 lbs without any accessories such as buttons or zippers, and the study included the primary and secondary packaging associated with this product. The functional unit was assumed to be one million wearings of the selected polyester blouse. Load size of washing during consumer use phase was assumed to be 20 blouses, and the frequency of washing was assumed to be once after every two wearings. From the results of energy use, it was understood that approximately 82% of the total energy use was at consumer use phase, where two-thirds of this total energy was needed for washing and the remaining one-third for drying. As far as solid wastes are concerned, 66% of the total results were created at consumer use phase. When it comes to atmospheric emissions (air- and waterborne emissions), most of the air emissions emerged from the electricity production demanded by the laundering process, and the wastewater from the laundering process was the major contributor for large quantities of waterborne emissions quantified in this study (Smith and Barker, 1995).

Using the example of another polyester garment, let us discuss the major findings from a comparatively recent study, conducted in 2009. This study revolved around the development of LCI of a polyester jacket, produced in China and consumed in Germany and focused on all life cycle stages from cradle to grave. The functional unit defined in this study was, '100 days of a polyester jacket, weighing 0.5 kg being worn', every 2 days for two seasons. It was assumed that the selected jacket was washed three times a season or six times during its entire lifetime period. This study quantified four important pollutants, namely carbon dioxide (CO_2), nitrogen oxides (NO_x), PMs and sulphur dioxide (SO_2) as well as the primary nonrenewable energy in the entire life cycle of the selected polyester jacket (Steinberger et al., 2009). The results of this study revealed that in the total life cycle, a polyester jacket consumed 156.39 MJ of energy for the assumed functional unit. Washing the garment six times in the entire lifespan consumed 14.92 MJ and drying it six times consumed 27.41 MJ of energy (Steinberger et al., 2009). In terms of all the main elements quantified in the study (energy, four pollutants), it was noticed that the use phase was not the major hotspot, rather the production phase was found to be more detrimental than the consumer phase, in contrast to the cotton and linen apparel discussed in Sections 5.2.1 and 5.2.3. This is in line with the findings of the woollen sweater discussed in Section 5.2.2.

From the above case studies, it is evident that consumer use phase is the major hotspot for apparel for normal wear/use and, of course, the degrees of impact vary subject to various factors discussed herein.

5.3 Consumer use phase in LCA studies of textiles and clothing products – implications for LCA

The previous section included some sample case studies to highlight the use phase impacts of textiles and clothing products. Washing, drying and ironing are the major processes involved in the consumer use phase, and the impacts from each process in the use phase vary for different product categories. Apart from the previously discussed case studies, several other studies also reported use phase as the most detrimental phase in terms of different environmental impacts (Jungmichel, 2010; Collins and Aumônier, 2002; BSR, 2009; Cotton Incorporated, 2011). A comparative look at the literature references pertaining to consumer use phase is tabulated in Table 5.1.

Most of the studies have clearly demonstrated the change in various environmental impacts with different use phase parameters such as

- washing temperatures, class of washing machines, washing cycles and washing frequencies;
- drying timings, temperatures, frequencies, drying methods; and
- ironing timings and temperatures.

Based on the various scenarios analysed using the above-mentioned parameters, most of the studies where use phase was found to be the major hotspot have recommended the best option to be followed while using a particular clothing type (washing, drying and ironing parameters).

Understanding the importance of use phase impacts and looking at ways to reduce them, a recent study demonstrated a new idea of designing clean clothes by altering the structure and composition of garments with a focus on three main approaches of clothing design to promote low-impact laundering by working around washing temperature, washing frequency and size of load (Hu, 2012). However, the implications of this low-impact laundering on the LCA results were not evaluated and presented.

Another dimension of the argument was presented in other research published in 2010. Focusing on clothing design for sustainable use, this research chiefly discussed how to minimise the life cycle impacts of clothing in its use phase by means of improved design. Demanding complete systems thinking, this article pointed out that the use phase should not be considered only as the consumers' responsibility; rather

Table 5.1 Literature results of use phase impacts of clothing products

Product	Consumer use phase	Share of primary energy use in the total results (%)
Knit cotton golf shirt (study conducted by Cotton Incorporated, 2011)	Average conditions – 54% cold wash/46% heated wash; load size, medium; washer efficiency, 70% conventional/30% Energy Star; water heater type, 50% elec./50% nat. gas. Drying, 84% dryer/16% air dry; dryer efficiency, 70% conventional/30% Energy Star	84% (approx. 1000 MJ/kg)
Knit cotton golf shirt (study conducted by Cotton Incorporated, 2011)	Best conditions – 100% cold wash; load size, extra large; washer efficiency, 100% Energy Star; water heater type, 100% nat. gas; drying, 100% air dry; dryer efficiency, n/a	51% (approx. 200 MJ/kg)
Knit cotton golf shirt (study conducted by Cotton Incorporated, 2011)	Worst conditions – 100% heated wash; load size, small; washer efficiency, 100% conventional; water heater type, 100% elec.; drying, 100% electric dryer; dryer efficiency, 100% conventional	93% (approx. 2500 MJ/kg)
Cotton T-shirt (Steinberger et al., 2009)	50 washes	73%
Polyester blouse (Smith and Barker, 1995)	40 washes/94 °F + drying	82%
Polyester blouse (Smith and Barker, 1995)	40 washes/cold no drying	2%
Polyester trousers (Collins and Aumônier, 2002)	92 washes	47.5%
Men's cotton briefs (Collins and Aumônier, 2002)	104 washes	84%
Denim jeans (BSR, 2009)	104 washes	57%
All clothing types (BSR, 2009)	Aggregated	39%

Adapted from van der Velden et al. (2014).

this responsibility must be shared and should have been taken into account during the design and production phases. As mentioned in this research, improved design offers a way to have more sustainable use from the garment through producing a better quality garment or product.

For instance, an improved design in a textile product may fetch a longer shelf life (use periods) and broader areas of use. However, despite all the technical developments implemented in the clothing production, use phase elements such as washing machines and detergents are significant in terms of life cycle impacts, and these elements can add to the designer's solution space as well. Hence, it is very important to consider consumer behaviour in future research study (Laitala and Boks, 2010).

5.4 Challenges and conclusions

It is clear that use phase impacts are very significant in a clothing product's life cycle, and for the clothing products that are intended for normal/casual wear (apparel purposes), use phase is the most detrimental phase compared to the other life cycle phases. However, this statement is made with caution, as we do not have many studies reported in the literature to confirm this. There is a dearth of LCA studies on various textile and clothing products with a special emphasis on use phase of clothing. No literature data or empirical studies on wearing and laundry behaviour of garments could be found as well.

It is very complicated to analyse consumer behaviour, especially in terms of how a consumer wears a particular garment and maintains it. Hence, most of the studies do not include use and disposal behaviour. However, this must be addressed as early as possible, as this is very crucial when it comes to LCA of textiles. Product category rules (PCRs) may be an option to address this; however, we do not have very many PCRs developed and available right now for various clothing products. Development of PCRs for clothing products is still in its infancy.

If we need to address the issue of consumer use phase impacts to reduce them, it is pertinent to quantify the impacts at use stage for various clothing products. With the basis of those scientific results only, one must be able to address and advise customers on sustainable use behaviour. Many research studies on LCA of textiles and clothing products with a chief focus on use phase and development of PCRs for various clothing products would enable a way to ensure sustainable consumer behaviour, which is of course, the need of the hour.

References

Bevilacqua, M., et al., 2012. Design for Environment as a Tool for the Development of a Sustainable Supply Chain, Chapter 10: Case Study: A Carbon Footprint Analysis in Textile Supply Chain. Springer-Verlag London Limited. http://dx.doi.org/10.1007/978-1-4471-2461-0_10, pp. 219–239.

BSR, 2009. Apparel Industry Life Cycle Carbon Mapping. Business for Social Responsibility.

Collins, M., Aumônier, S., 2002. Streamlined Life Cycle Assessment of Two Marks & Spencer Plc Apparel Products. Environmental Resources Management, Oxford.

The Carbon Footprint of a Cotton T-shirt, March 2009. Continental Clothing Co. Ltd. Executive Summary.

Cotton Incorporated, 2011. Life Cycle Assessment of Cotton Fibre and Fabric Executive Summary, 15 pp.

Hu, Y., September 2012. A study on the sustainable fashion design in the process of use. International Journal of Arts and Commerce 1 (4), 54–59.

Jungmichel, N., 2010. The Carbon Footprint of Textiles. Systain Consulting. Available at: http://www.ci-romero.de/fileadmin/media/informieren-themen/gruene_mode/Jungmichel._Systain.pdf.

Laitala, K., Boks, C., 2010. Clothing design for sustainable use: social and technical durability, knowledge collaboration & learning for sustainable innovation. In: Erscp-emsu Conference, Delft, The Netherlands, October 25–29, 2010.

Life Cycle Assessment of a 100% Australian-Cotton T-shirt, 2009. Institute for Sustainable Resources, Queensland University of Technology. Available at:http://cottonaustralia.com.au/uploads/publications/Case_Study_-_Life_Cycle_Assessment,_Peter_Grace.pdf.

The Linen Shirt Eco-Profile, Press Release, Masters of Linen; Eco-profile of a Linen Shirt and Comparative Analysis of Linen and Cotton Shirts, February 2008. Bio Intelligence Service. Available at: http://www.saneco.com/IMG/pdf/linen_shirt_eco-profile.pdf.

Muthu, S.S., April 2014. Assessing the Environmental Impacts of the Textiles and Clothing Supply Chain. Woodhead Publishing, Cambridge, UK.

A Product Life Cycle Approach to Sustainability, March 2009. Levi Strauss & Co., San Francisco, CA. Available at: http://www.levistrauss.com/sites/default/files/librarydocument/2010/4/Product_Lifecyle_Assessment.pdf.

Smith, G.G., Barker, R.H., 1995. Life cycle analysis of a polyester garment, resources. Conservation and Recycling 14, 233–249.

Steinberger, J.K., Friot, D., Jolliet, O., Erkman, S., 2009. A spatially explicit life cycle inventory of the global textile chain. International Journal of Life Cycle Assessment 14, 443–455.

van der Velden, N.M., Patel, M.K., Vogtländer, J.G., 2014. LCA benchmarking study on textiles made of cotton, polyester, nylon, acryl, or elastane. International Journal of Life Cycle Assessment 19, 331–356.

Further reading

Allwood, J.M., Laursen, S.E., de Rodríquez, C.M., Bocken, N.M.P., 2006. Well Dressed? The Present and Future Sustainability of Clothing and Textiles in the United Kingdom. University of Cambridge, Institute for Manufacturing, Cambridge.

Bristwistle, G., Moore, C.M., 2007. Fashion clothing – where does it all end up? International Journal of Retail & Distribution Management 35 (3), 210–216.

Domina, T., Koch, K., 1999. Consumer reuse and recycling of post-consumer textile waste. Journal of Fashion Marketing and Management 3 (4), 346–359.

Ha-Brookshire, J.E., Hodges, N.N., 2009. Socially responsible consumer behaviour? Exploring used clothing donation behaviour. Clothing and Textiles Research Journal 27 (3), 179–196.

Lilley, D., Lofthouse, V., Bhamra, T., 2005. Towards instinctive sustainable product use. In: 2nd International Conference: Sustainability Creating the Culture. Aberdeen.

Madsen, J., Hartlin, B., Perumalpillai, S., Selby, S., Aumônier, S., 2007. Mapping of Evidence on Sustainable Development Impacts that Occur in Life Cycles of Clothing: A Report to the Department for Environment, Food and Rural Affairs. Environmental Resources Management Ltd., London.

Tham, M., 2010. Languaging fashion and sustainability – towards synergistic modes of thinking, wording, visualising and doing fashion and sustainability. The Nordic Textile Journal 1, 15–23.

Open- and closed-loop recycling of textile and apparel products

6

A. Payne
Queensland University of Technology, Brisbane, Queensland, Australia

6.1 Introduction

Manufacturing of textiles requires extraction of natural resources, energy consumption and frequently results in toxic emissions to air, water and soil. Recycling textiles can play a role in reducing solid waste to landfill and reducing the need for virgin fibre with its associated water, chemical and energy consumption. Yet in 2010 in the United States, 13.2 million tonnes of textiles were sent to municipal solid waste (MSW) (United States Environmental Protection Agency, 2010) even though the majority of textiles are recyclable in some form. This chapter examines the open-loop recycling (OLR) and closed-loop recycling (CLR) of textile products with particular reference to the fashion and apparel context. It describes the fibres used within apparel, the current mechanical and chemical methods for textile recycling, and applications within apparel for each. These are discussed within the qualitative framework of life cycle thinking (LCT): considering the cradle-to-grave impacts of a product encompassing production, use and disposal. This chapter also discusses quantitative life cycle assessment (LCA) findings for several textile recycling approaches. Barriers for more effective recycling include ease of integration into existing textile and apparel design methods as well as coordinated collection of post-consumer waste. Despite these barriers, there are a number of innovations that hold promise for more effective OLR and CLR within the apparel industry in the future.

6.2 Recycling textiles and apparel

6.2.1 LCA and LCT

Before examining the approaches to recycling textiles, it is important to place this discussion in the context of the product life cycle. The methodology of LCA is one approach to quantitatively assess the environmental advantages of recycling fibre. An LCA typically considers the energy, water and chemical impacts of a product system from cradle (raw materials) through to production, distribution, use by the consumer and disposal. Formal LCA follows ISO 14040 and ISO 14044 standards. It requires determination of the functional unit of the product or raw material in question to be assessed, for example '1 kg of cotton fibre', or 'one viscose blouse'. Importantly, LCA is conducted under defined system boundaries. For example, the LCA may be

Handbook of Life Cycle Assessment (LCA) of Textiles and Clothing. http://dx.doi.org/10.1016/B978-0-08-100169-1.00006-X

bounded as cradle to gate (final product before market) or cradle to grave, which includes the use phase and the disposal phase. It also analyses a product system rather than a discrete product (Klöpffer and Grahl, 2014).

Aside from the quantitative, LCA is also useful as a qualitative framework for tracing material flows through complex systems such as the interconnected fibre, textile and apparel industries. Life cycle thinking may be applied in the analysis of the social context into which apparel products enter. Given apparel is such a complex product composed of many inputs that each could be subject to an LCA, the use of LCT helps conceptualise the waste across the garment life cycles as a whole. Life cycle thinking approaches the topic from a qualitative framework, exploring the role of waste, the role of the designer, the use of the product as well as disposal.

There are challenges to analysing the environmental benefits of recycling textiles using LCA methodology. These include how to set the boundaries or cutoff points for the LCA, or how to allocate energy savings from recycling between different product systems. The following sections describe both the quantitative LCA of recycling methods as well as a qualitative approach of LCT in discussing the system of apparel production, reuse and recycling as a whole. The combination serves to illuminate the social and behavioural implications of recycling as well as the measurable benefits of further recycling of textile products.

6.2.2 Fibre and the apparel industry

Fibres and textiles have a wide range of applications from industrial textiles used in medical and construction industries to domestic apparel and soft furnishings. In 2012, world fibre consumption was approximately 83 million metric tonnes (International Cotton Advisory Committee, 2013). In the apparel context, growing global population and rising standards of living mean that demand for fibres and textiles will continue to rise. Although affected by the global economic downturn in 2009, apparel consumption continues to grow (Ruiz, 2014). US households spend on average $1700 per household per year on apparel (United States Bureau of Labour Statistics, 2012), and in Australia over one billion units of apparel are imported yearly (Wells, 2011).

The fashion industry is highly globalised, characterised by rapidly changing garment styles in response to consumer trends. In contrast to the rapid changes in design, apparel manufacturing has a low level of technological change. Production relies heavily on human manual handling with a low level of automation, and low wages in developing countries means there is little incentive for companies to automate the industry (Ruiz, 2014). With low innovation in manufacturing technologies, the industry is reliant on traditional approaches to garment assembly that may hinder attempts to recycle garments at end of life.

6.2.3 Characteristics of textiles used in the apparel industry

Fibres used in apparel are chosen with consideration of characteristics such as softness, breathability, ease of dyeing and care. Textiles used are developed with consideration of the comfort, appearance, drape, price, handle and colour fastness required for the design's appearance and functionality.

Textiles for apparel are commonly woven or knitted from natural or manufactured fibres. Natural fibres include cellulosic fibres such as cotton, linen and hemp, or protein fibres such as silk and wool. Growing of natural fibres frequently involves significant water, land and chemical use. Natural fibres are biodegradable, however, because of the chemicals used in the finishing and dyeing processes, these fibres can still negatively impact soil and groundwater upon disposal.

Manufactured fibres include those from a petrochemical base such as polyesters, the most common being polyethylene terephthalate (PET). Other manufactured fibres include nylon and acrylic as well as regenerated cellulosic fibres such as viscose, cuprammonium, acetate (a semisynthetic fibre), rayon and bamboo. Put simply, manufactured fibres are produced by extruding the liquefied feedstock (whether from a petrochemical or cellulosic base) through a spinneret to form a fibre that can then be spun into yarn for weaving or knitting into textiles. Manufactured fibres from a petrochemical origin (synthetic fibres) are produced from a nonrenewable resource and require significant energy inputs in their production. Additionally, they do not biodegrade but rather photodegrade and as such will persist in the environment for centuries.

Synthetic fibres, followed by cotton, are the most common in apparel production. Although cotton consumption has risen steadily in the past two decades, synthetic consumption has grown much faster and now dominates global fibre production. Cotton accounts for 32.9% of global textile production, synthetic fibres including polyester, acrylic, nylon (polyamide) and polypropylene for 60.1%, wool 2.1%, flax (linen) 1.0% and other cellulosic 3.9% (Shui and Plastina, 2013). In the apparel context, manufactured fibres can be engineered to mimic natural fibres in handle, function and aesthetic, which makes them attractive for both apparel manufacturers and end consumers.

Importantly, fibres are commonly blended together in order to give fabric more desirable qualities appropriate for apparel applications. For example, cotton and polyester blends have the breathability and wearability of cotton but will crease less and so require less pressing. A small percentage of petrochemical-based fibre elastane is often blended with other fibres for added stretch in both wovens and knits. This wide variety of fibres and fibre blends used in textiles for apparel makes effective recycling difficult due to the complexity of separating and sorting constituent fibres from textile blends.

6.2.4 Recycling and reuse

Recycling refers to the breakdown of a product into its raw materials in order for the raw material to be reclaimed and used in new products. In contrast, reuse refers to an existing product being used again within the same production chain. Textile recycling may involve reclaiming pre-consumer waste or post-consumer waste. There are a number of ways to identify the kinds of recycling possible within the textile and apparel context.

Wang (2010) identifies four recycling approaches as follows:

1. Primary: recycling industrial scrap;
2. Secondary: processing a post-consumer product into raw materials;
3. Tertiary: converting plastic wastes into basic chemicals, monomers or fuels;
4. Quaternary: incinerating waste as a way of reclaiming the embedded energy.

In apparel production, primary recycling is the collection of pre-consumer offcuts of fabric from production. The nature of cut-and-sew manufacture of apparel means that there is significant textile waste resulting from the leftover fabric between individual pattern pieces. Rissanen (2008) estimates this as up to 15% of the yield required for each garment.

Secondary recycling involves collection and recycling of post-consumer textile wastes such as garments and other household textiles. Depending on consumer action, this apparel may be sent to MSW or donated to charities or collection services. Collection agencies will then sort the clothing into higher quality clothing suitable for resale and lower quality clothing suitable for recycling.

Tertiary recycling involves chemical breakdown of nylon or PET for repolymerisation. This requires either clean, sorted pre-consumer waste or post-consumer waste, sorted by fibre type, as a feedstock. In quaternary recycling, the embedded energy of textiles can be reclaimed through incineration (Wang, 2010). This final approach is not addressed in this chapter.

After the raw material is recycled, it can be further classified according to the new product stream it enters. These two classifications are OLR and CLR. In OLR, a garment reaches the end of its useful life and the fibre is reclaimed for use in different products, therefore entering into a new product life cycle (Curran, 2012). This may also be termed 'downcycling' if the new product is of lesser economic value. Similarly, pre-consumer textile offcuts from apparel production may be collected and recycled into fibre for use in other products.

In CLR, collected textiles are reprocessed into new fibre to use for new garments, re-entering the same production system that the textile originally came from. In more sophisticated approaches to CLR, the recycled fibre can be repeatedly recycled, while OLR may only temporarily divert waste from landfill or incineration. Both systems are relevant and important. While OLR has had greater uptake within the fashion sector, CLR holds greater potential for systemic change towards a more sustainable fashion industry. Sections 6.3 and 6.4 examine these two approaches to textile recycling in greater depth.

6.3 Open-loop recycling

Open-loop recycling refers to a system in which a product's raw material is broken down to be used in a second, often unrelated product system (see Figure 6.1). Generally, the second product will not be recycled and instead be disposed of at the end of its life (Curran, 2012). For this reason, OLR is a worthy reuse of the first product's virgin materials and reduces the need to consume virgin materials in the manufacture of the

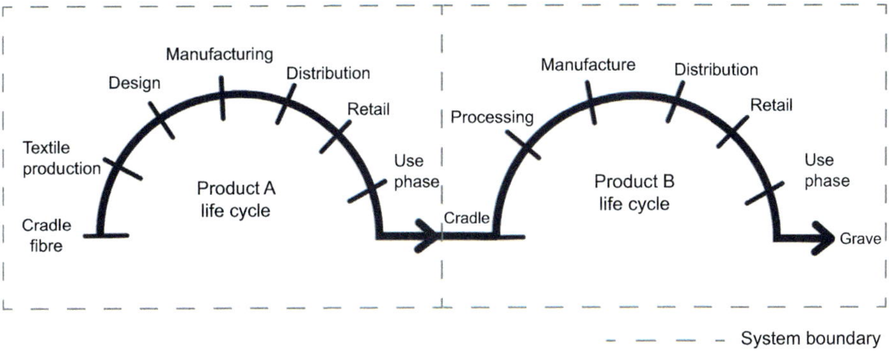

Figure 6.1 Open-loop recycling.

second product. However, OLR typically only delays a material's ultimate passage to MSW, as there are limits to how often a material can be recycled without quality being degraded.

Typically, OLR recycled items in textiles and apparel include

1. pre-consumer textile waste such as offcuts from the cutting process;
2. post-consumer textile waste in the form of whole garments; and
3. post-consumer PET bottles that may be manufactured into recycled PET (RPET) fibre.

The recycling of the waste in examples 1 and 2 above can be viewed as OLR when the textile waste is used as material in other product systems. When examples 1 and 2 are recycled for use within the apparel industry, this is CLR addressed in greater detail in Section 6.4.

6.3.1 Discussion of methods for mechanically recycling textiles

Textiles can be mechanically or chemically broken down to the fibre for recycling. Additionally, other forms of post-consumer waste such as PET bottles may be recycled into fibre for apparel. For all forms of recycling, considerations include the blend of fibres or polymers, the length of these fibres, to what degree the textile or product can be broken down and the cleanliness or degree of impurities of the feedstock to be recycled (Gulich, 2006b).

6.3.1.1 Textile to fibre

Mechanical methods for converting fabrics back to fibre have been in use since the Industrial Revolution. Carded yarn spinning and mixed shoddy spinning are traditional practices for converting wool into useable yarn to be re-woven or knitted (Gulich, 2006b). Newer processes to develop reclaimed fibre include 'the rotor spinning process, the yarn covering process, and the fraction spinning process DREF which allows economical production' (Gulich, 2006b).

Mechanical recycling machines cut fabric into smaller pieces and then progressively shred the fabric until it is in a fibrous state suitable for other processes such as re-spinning into yarn or for manufacturing into nonwoven textiles. A typical process is as follows:

- Fabrics are sorted, metal components are removed and fabrics are then baled.
- Fabrics are cut with a rotary blade into small pieces, for example 1 × 8 cm strips.
- Fibres are then separated through a process known as 'picking', 'pulling' or 'tearing' as fabrics are rolled on progressively smaller spiked surfaces to break them apart to remove the fibres (Langley and Kim, 2006).

This process is used to recycle fabrics made from natural fibres such as cotton and wool as well as synthetic fibres including polyesters, nylons and blended fibres. Hawley (2006) describes the mechanical processing technique used in facilities in Prato, Italy, where acrylic textiles are shredded down to fibre. In her example, acrylic garments were sorted and cut up, mechanically shredded to fibre, and then re-spun into acrylic yarn for weaving into blanketing (Hawley, 2006).

6.3.1.2 Bottle to fibre

PET bottles to fibre is an OLR method in which PET bottles are recycled into PET flakes, re-spun into fibre and then woven or knitted into textiles. Unlike the method described above, where textile waste enters a second product life cycle, this OLR approach sees waste from other product cycles (i.e. bottles from the food and beverage industries) being utilised in textile and apparel production.

As PET is nonbiodegradable, recycling is an appropriate use of the resource. PET bottles have been recycled since the 1970s (Awajaa and Pavelb, 2005). However, the quality of the resulting recycled polyethylene terephthalate (RPET) depends on whether the bottles to be recycled contain contaminants including dust, colouring, acids and water (Awajaa and Pavelb, 2005).

The process for mechanically recycling PET bottles to fibre is as follows:

1. Bottles are collected, cleaned and sorted by colour.
2. Labels are removed.
3. Bottles are processed into PET flakes.
4. The PET flakes are liquefied and extruded through the spinneret into new fibres.

The quality of RPET in mechanical processing is generally not high when compared to virgin PET as purity depends heavily on impurities or contaminants within the bottles (Terazono et al., 2004). As well as the mechanical process, PET can be recycled using a chemical process. Steps 1–3 are identical to the above; however, the PET flakes are chemically transformed and returned to an oligomer or monomer (Shen et al., 2010).

6.3.2 Applications for recycled fibres

The range of applications for OLR yarns varies considerably depending on the quality of the feedstock. It is important to distinguish between pre-consumer textile waste collected from factories and post-consumer textile waste collected from used garments and other household textiles. For example, when many fibre types from post-consumer textile waste are recycled mechanically, the resulting fibre is generally of a low quality in terms of fibre length and colour (generally black or grey), making it inappropriate for reuse within apparel (Gulich, 2006b).

As apparel textiles are often comprised of fibre blends that are not easily separated, the most efficient approach to recycling is shredding the textiles to fibre and using the shredded fibre for applications such as nonwoven textiles or insulation. Mechanically recycling textiles into fibre that can then be spun into a good quality yarn is a more difficult proposition, and rather than using post-consumer waste, best results come from using clean pre-consumer textile waste of the same colour and fibre type. When pre-consumer textile waste is mechanically recycled (e.g. denim offcuts) being from the same fabric allows a yarn to be produced with a consistent staple length and thus of a quality suited to textiles for apparel. However, even with clean pre-consumer textile waste, the mechanical recycling process will still result in shorter fibre lengths. For this reason, recycled fibres are frequently blended with virgin fibres for apparel applications.

Depending on the quality of the recycled yarn, applications include

- Carpet underlay
- Building materials such as insulation
- Wipers and rags for industrial use
- Nonwoven textiles
- Fibre for the paper industry.

Chen and Burns (2006) estimate that up to 75% of pre-consumer textile waste is collected by industry. It is then sold as a raw material to other industries and thus diverted from landfill. In contrast, collection and recycling of post-consumer textile waste is a more difficult proposition, and recycling rates vary considerably from country to country.

6.3.3 Discussion of LCA findings in the context of OLR

Open-loop recycling of waste from one product system into feedstock for another brings benefits in the form of reducing the amount of solid waste to landfill. Life cycle assessments conducted on the OLR of PET bottles confirm that there are energy benefits to utilising recycled PET bottles in fibre for apparel. A series of LCAs conducted by Shen et al. (2010) examined mechanical and chemical recycling of PET and found in all instances that recycled PET has greater environmental benefits than virgin PET across seven to eight of the nine categories examined including energy consumption, greenhouse gas potential, eutrophication and acidification.

Using RPET and recycled yarns in industries requiring fibre presents clear environmental benefits. When compared with their virgin counterpart, recycled fibres may require less energy, reduce MSW and reduce the need to consume virgin materials.

OLR of pre-consumer waste also has environmental benefits. An LCA conducted by Muthu et al. (2012a) concluded that the recycling of pre-consumer textile waste, or process waste, would directly lower the carbon footprint of the product. However, not all textiles have the same environmental benefits when recycled. In developing their Recyclability Potential Index assessing both environmental and economic benefits to recycling, Muthu et al. (2012b) found that polyester and polypropylene are ranked highest in recyclability with nylon 6,6 ranked last.

It is important to note that there are challenges in calculating the benefits of OLR using LCA methodology. Chief of these are how to determine the parameters of the assessment. For example, when pre-consumer textile waste ('Product System A') is collected for use in housing insulation ('Product System B'), the LCA must account for how the environmental benefits of recycling are allocated to both Product System A and Product System B (Klöpffer and Grahl, 2014).

As Muthu et al. (2012a) describe, more work is needed to accurately assess environmental benefits of integrating recycled materials into new products. The sheer diversity in textile types leads to difficulty in separating fibres for efficient recycling, with the energy expended to collect, sort and manufacture the shoddy into a new product complicating the analysis.

6.3.4 Examples of apparel companies utilising open-loop recycled yarns

Given that OLR is the recycling of waste from one product system into another, strictly speaking the use of RPET is the clearest case of OLR yarns being utilised within the apparel industry. The use of OLR yarns in apparel has been well established since the 1990s. A number of apparel companies advertise their use of recycled yarns, but it is difficult to gauge what percentage of their ranges this may apply to.

The use of PET bottles for feedstock for RPET production is well advanced and has become increasingly utilised within the apparel industry since the 1990s, with the outdoors brand Patagonia being one of the earliest adopters (Brown and Wilmanns, 1997). BIONIC® yarn is comprised of 40–45% recycled content from plastic bottles (Bionic, 2014). Owned by hip-hop artist Pharrell Williams, the Bionic Yarn company has collaborated with Topshop and active wear label Moncler to develop apparel collections using recycled yarns in conjunction with cotton or virgin PET. In 2014, Bionic developed a denim line for G-Star Raw that used harvested ocean plastic as the feedstock.

The surf wear label Billabong has been using recycled PET fabric since 2005 for some ranges of board shorts. Called *ecosuede*, this material is comprised of 100% recycled PET bottles, with between 10 and 20 bottles utilised for each pair of board shorts (Billabong, 2012).

Many other retailers have utilised RPET in apparel, including Max Mara and H&M. There has also been niche designer interest, with Italian label Armani creating a gown of 100% recycled bottles for celebrity Livia Firth in 2012.

6.3.5 Implications of OLR

OLR has proved feasible in the fashion context, both in collection of pre- and post-consumer textile waste for use in other products, and in collection of used bottles for recycling into textiles. Garments are suitable as a feedstock for products of lesser value, for example carpet underlay or insulation, with many applications possible. However, the varieties of fibre types and colours mean that the resulting shoddy is in unattractive greys or blacks that are unsuitable for spinning into apparel-quality yarn. The wide variety of products that utilise pre-consumer textile waste and even post-consumer textile waste demonstrates that the recycling of textiles works when entering product streams of lesser value.

The recovery and use of plastic bottles into RPET yarns has had strong uptake from a number of apparel companies. Currently, OLR of PET bottles to fibre has had the greatest success for reuse as a material in the fashion sector, with an open loop of waste from the first product (PET bottles) used as feedstock for the second product system (polyester fabric to garment). A common approach is to blend the recycled yarns with virgin fibres to create textiles that are of apparel quality.

6.4 Closed-loop recycling

6.4.1 Definition of CLR in the textiles and apparel context

There are a number of ways to define CLR practices in the apparel industry. This section describes three: recycling apparel textile waste (pre- or post-consumer) in order for it to re-enter the apparel supply chain; cradle-to-cradle (C2C) streams of biological and technical materials; and reuse of existing garments.

In the first definition, CLR refers to recycling methods whereby the material being recycled is the same material being produced: ‘a product enters the production chain of the same product again after use’ (Klöpffer and Grahl, 2014). Under this definition, provided that the waste textile or fibre re-enters a garment production chain, both pre- and post-consumer mechanically recycled textiles may be considered closed-loop recycled. Two approaches are illustrated in Figures 6.2–6.4.

6.4.1.1 Cradle to cradle

The cradle-to-cradle methodology is a radical approach to CLR in which a CLR fibre will be recyclable as well as recycled into the same production chain (see Figure 6.5). In McDonough and Braungart’s (2002) C2C closed-loop system, waste is reclaimed and used again in the production of products of the same or higher value. Waste is diverted into either biological or technical streams. Biological waste can be composted, while technical waste can be reused within industry to create the same products again.

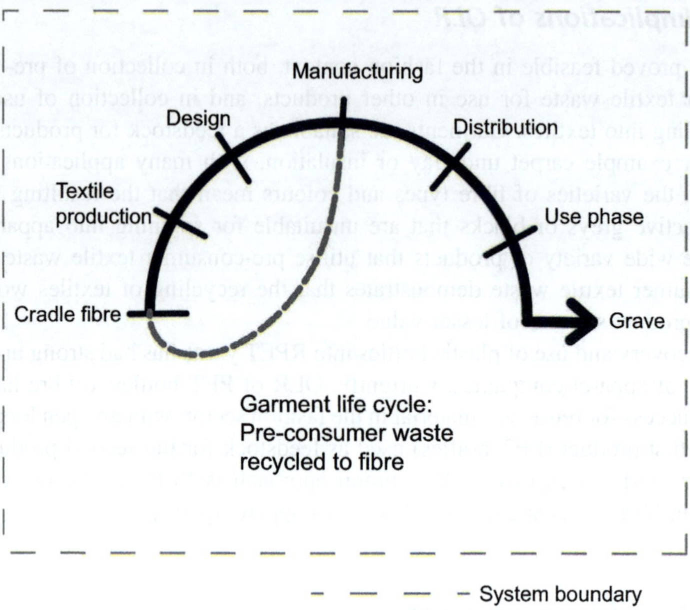

Figure 6.2 Pre-consumer waste from the manufacturing process is collected for mechanical processing for use in apparel.

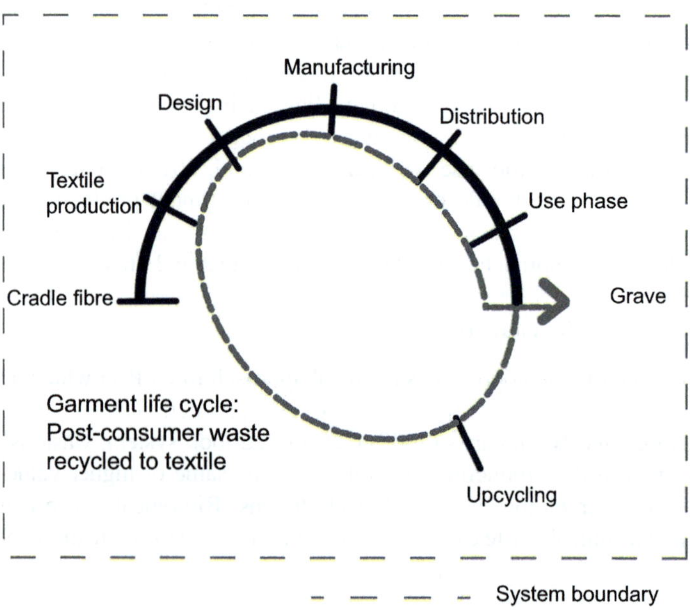

Figure 6.3 Post-consumer waste is disassembled in order to reuse the textile.

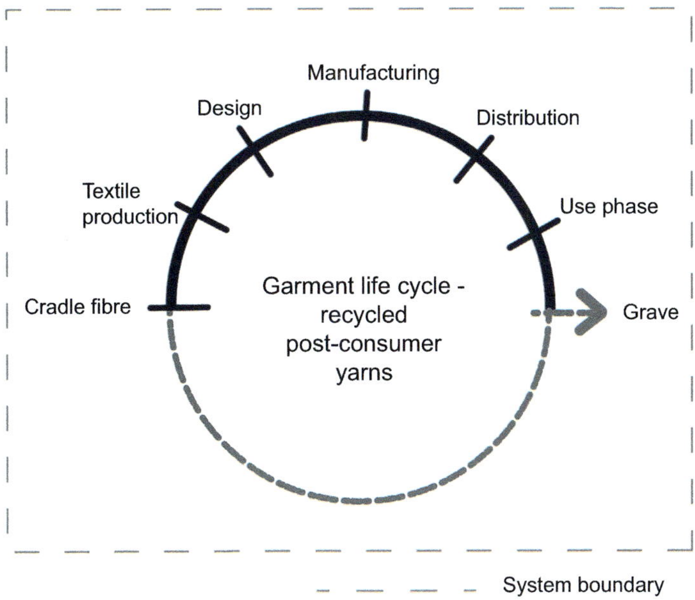

Figure 6.4 Post-consumer waste is shredded to fibre for reprocessing back to yarn.

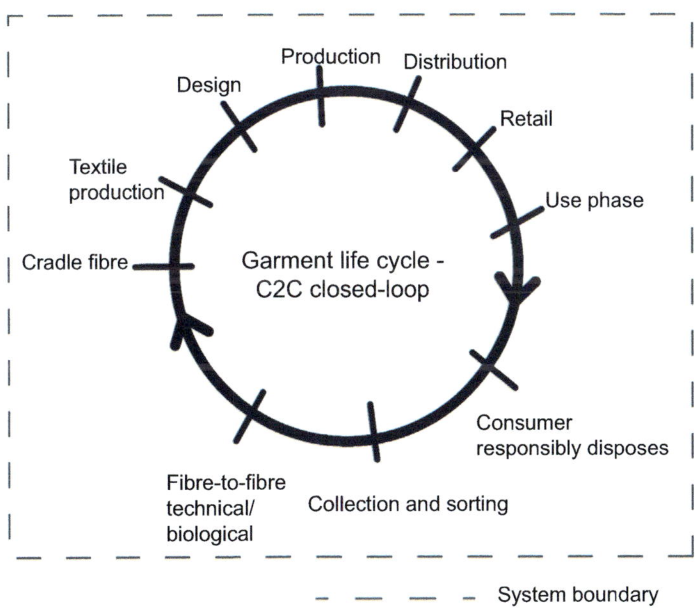

Figure 6.5 Cradle-to-cradle closed-loop recycling.

Biological CLR refers to fibres that can be safely composted at end of life to return nutrients to the soil. Technical CLR refers to the synthetic products that are not biodegradable. In textiles, this is frequently the synthetic polymer-based fibres such as polyester, acrylic and nylon. Blending of the two kinds of streams is referred to by McDonough and Braungart (2002) as a 'monstrous hybrid', meaning that the two kinds of waste streams cannot be effectively separated for ease of recycling. In the apparel context, monstrous hybrids abound in the form of cotton/polyester, or viscose/polyester, or cotton and spandex blends.

6.4.1.2 Closed-loop reuse

A related closed-loop approach is in the reuse of garments. Both reusing of clothing and also upcycling of clothing are related to CLR. Closed-loop reuse implies clothing can have multiple useful lives on the second-hand market. Although reuse of garments is not recycling in the sense of breaking down a product into its raw materials, it equates to CLR in that the product may enter a new life cycle within the same production chain. Researcher Jana Hawley (2006) has examined the global market in used clothing extensively. She notes that the used clothing sold around the world varies considerably from lightly used clothing suitable for resale through to the 'diamonds' or sought-after vintage or designer pieces.

6.4.2 Methods for chemical recycling of polymer-based fibres

There are several methods for CLR of synthetic polymer-based fibres such as PET or nylon. This process includes chemically breaking down the fibre to a polymer and then repolymerising it before passing through the spinneret to generate new fibre to be spun into yarn, ready for weaving or knitting into fabric. This process is related to the chemical recycling of polymer-based synthetic fibres discussed in Section 6.2.3; however, it results in a higher quality fibre that can have high equivalence to virgin fibre. Typically, polymeric recycling is a challenge as the polymer degrades with each subsequent attempt at recycling. Teijin Textiles in Japan has developed a method in which the resulting RPET is of as high quality as virgin PET (Patagonia, 2009).

6.4.3 Manufacture and application of biodegradable textiles in the apparel context

All textiles from natural sources will biodegrade in time, yet many are not safely biodegradable due to the chemicals used in textile finishing and dyeing processes. Therefore, to be effectively closed-loop biodegradable textiles, toxicity of all chemicals used in the dyeing and finishing processes needs to be considered. One example is the Climatex fabric (McDonough and Braungart, 2002). Lyocell is a biodegradable fabric that contains no harmful chemicals in its production. If dyed using natural dyes and nontoxic mordants, Lyocell fabrics would be safely biodegradable. Biopolymers such as PLA are another example.

6.4.4 Discussion of LCA findings in the context of CLR

The environmental gains through taking a C2C approach may be difficult to measure, partly because C2C is a qualitative approach whereas LCA is a quantitative approach. The two systems may complement each other in theory but may not always be complementary in practice. C2C refers more to innovation and looking ahead to system-level changes rather than to quantifiable approaches. As Bakker et al. (2010) identify, depending on the use of the energy and water, the C2C approach to CLR may not be appropriate. The use phase of the garment life cycle is typically the most energy and water intensive (Smith and Barker, 1995). Therefore, energy gains and avoiding waste through development of C2C processes may not make an appreciable difference to key areas measured by an LCA such as water, energy and chemical use.

However, reuse as a CLR strategy has considerable environmental advantages. An LCA of used and donated clothing found energy savings of 65 kWh when second-hand clothing displaces a kilogram of virgin cotton or 90 kWh when a kilogram of polyester is displaced (Woolridge et al., 2006). Given that the use phase of the garment life cycle is the most energy intensive, the reuse of garments is thus a viable and important step towards reducing the amount of post-consumer waste prior to recycling.

Primary recycling of pre-consumer waste has met with greater success for CLR in apparel production than the recycling of post-consumer waste. As the pre-consumer waste can be relatively clean and homogeneous in fibre and colour, for example, in the case of denim offcuts, then the resulting recycled fibre will be of higher quality for re-entering the apparel supply chain. Similar to the method described earlier, textile offcuts may be mechanically recycled through shredding and then re-spinning into yarn.

This approach has been used with cotton pre-consumer waste. Typically, the fibre will be of lower quality than virgin cotton; however the recycled cotton fibre can be blended with other fibres to improve the quality. Recycling cotton offcuts in this way has some benefits when compared with using virgin cotton. In an LCA of recycled cotton T-shirts, conducted for the retailer Esprit, it was found that use of recycled cotton instead of virgin fibre has water savings of up to 75% (Esprit, 2014). Although this claim has not been independently verified, it does follow that through utilising waste cotton, there would be water savings when compared with virgin cotton.

6.4.5 Examples of apparel companies utilising closed-loop recycled yarns

6.4.5.1 Pre-consumer waste

As described above, collection of pre-consumer waste is advanced in apparel manufacturing, with an estimated 75% of offcuts collected. While this waste may be open-loop recycled into other products, it is less common to find apparel companies utilising the recycled waste in apparel. The clothing brand Esprit, stocked in

40 countries worldwide, has several ranges of garments that utilise recycled yarns, including cotton, polyester and nylon. Their pre-consumer textile waste is collected and mechanically recycled to be spun into yarn for knitting into T-shirts (Esprit, 2014). Esprit also utilises recycled wool in their collections, as well as OLR PET.

British high street chain store Topshop has also used pre-consumer waste in a capsule collection through a partnership with Reclaim to Wear. This project used ends of rolls and fabric offcuts to construct new garments. Unlike Esprit, this process did not involve mechanically recycling back to fibre but instead kept the offcuts of textile whole. This approach to reusing cloth (whether pre- or post-consumer) is known as upcycling and has wide uptake in niche fashion production by independent designers and labels.

Closed-loop recycled nylon is a new innovation that has been used by several large apparel companies. In 2012, Esprit developed a range of swimwear of 70% recycled nylon blended with elastane, and in 2013 released a swimwear range of 100% recycled nylon blended with nylon (Esprit, 2014). Patagonia has also used CLR nylon. Patagonia has been a leader in both pre- and post-consumer waste utilisation. Recycling of nylon 6,6 has been in use in recycling carpeting since the 1990s, but its adaptation into CLR within apparel is recent.

6.4.5.2 Post-consumer waste

The Swedish fashion brand Hennes & Mauritz (H&M) has made advances in recycling post-consumer waste for use in their product lines. Since January 2013, the company has collected used clothing in store and mechanically recycled it through a partner firm. In 2014, H&M released a recycled denim collection containing 20% recycled yarns from the collected clothing (Jobling, 2014). Given the low-quality yarn that emerges from the mechanical recycling process, currently the denim blend cannot have a higher percentage of recycled yarn; however, H&M is working with recyclers to further develop the technology.

6.4.5.3 Biodegradable textiles

A number of fashion companies have utilised materials that are safely biodegradable, hence 'closing the loop' on production. One example is the Oat shoe that can be composted. These commonly use biodegradable plastics such as corn-based polylactic acid (PLA). Additionally, where the dyes and mordants are nontoxic and the garments may be safely biodegradable.

6.4.6 Implications of CLR for the fashion sector

As described above, for the fashion sector, effective CLR has proved more difficult at scale than OLR. However, Patagonia, PUMA and H&M have been leaders in attempts to 'close the loop'. As discussed before, weighing up the environmental benefits of CLR for apparel, when taking into account the high energy and water consumption in the use phase of the garment, the environmental savings may not be great. However, from the qualitative perspective of life cycle thinking, the implied systemic shift towards better management of waste streams is important.

6.5 Barriers to effective OLR and CLR

Effective recycling within apparel production is a difficult proposition given the fragmented, globalised nature of textile and apparel supply chains, as well as the many fibres, whether natural, synthetic or blended in use. Effective recycling involves action from stakeholders throughout the supply chain, including fibre and textile producers, apparel companies, manufacturers, consumers, local and national governments and nongovernment organisations.

6.5.1 Collection of pre- and post-consumer textile waste

In textile waste, there are production challenges, consumption challenges and logistical challenges. From a production perspective, use of recycled yarns is limited to availability, handle and use. Post-consumer waste comes with another range of challenges. Chief of these is organising efficient collection of garments from individual consumers. Often this role is filled by nonprofit organisations such as Oxfam or the Red Cross. As Muthu et al. (2012a) note, given that the consumer is ultimately responsible for disposal or recycling textiles at end of life, consumer behaviour and government policy are crucial concerns in any effective recycling scheme.

6.5.2 Ease of integration into existing production processes

Once the consumer or post-consumer waste has been collected, the sorting stage also presents challenges. In the case of post-consumer waste, garments need to be carefully sorted and partially disassembled before they can be recycled. But disassembly in readiness for recycling is a challenge given that garments are generally comprised of so many different components. This may include but is not limited to

- One or more base fabrics, each of which may be a blend of fibres
- Metallic or PET zippers
- Acrylic, wood, shell or PET buttons
- Metallic snap fastenings
- Cotton/polyester sewing threads
- Nonwoven or knitted interfacings and interlinings
- Polyester brand and care labels

To disassemble these garments requires considerable manual handling with each garment being assessed case by case for disassembly. Some processing plants use magnets to remove metallic elements; however, in the first instance most rely on human sorting.

Gam et al. (2011) conducted a study in which they developed a tailored jacket that could be disassembled and streamed into the technical and biological elements of the

C2C model. The study revealed considerable barriers to taking this design into the mainstream design process due to the level of technical challenge and the difficulty of resolving these challenges by garment design in an industry setting.

Aside from the challenges of deconstructing the garments, the large amount of fibre blends in use within the apparel industry is a further barrier to effective recycling. While garments of a single fibre can be sorted and processed together, contemporary apparel is typically highly diverse in design, fibre content and trims. The need to process all these fibres together generally results in a lower quality fibre of mixed colour and fibre length.

In the case of pre-consumer textile waste, fibre blends such as cotton/elastane (also known by as spandex or Lycra) are a problem for effective recycling as the highly elastic fibres have very different characteristics from other fibres. As such, firms may send textiles with spandex to the landfill rather than attempt recycling (Langley and Kim, 2006).

Pre-consumer waste such as offcuts from the cutting process results in relatively clean and sortable fabrics, as one type of fabric can be reclaimed over a production run and will be of a uniform colour and fibre type/blend. For this reason, recycled yarns that are then returned into the apparel product cycle are more likely to be manufactured using pre-consumer textile waste. This also applies in the case of 'upcycling' whole fabric pieces into new apparel.

6.5.3 Barriers to use of closed-loop recycled yarns within fashion

Closed-loop recycling brings additional challenges, given that in addition to recycling waste textiles, to close-loop recycle a firm must integrate the recycled yarns back into its apparel production chain. Significant barriers exist towards effective CLR in the apparel context. As Johnsen et al. (2014) note, from a supply chain perspective, firms are required to not only consider traditional supply chain activities but also the reverse logistics of returns and product collection, testing, sorting, remanufacturing and remarketing. Aside from these considerable logistical challenges, strategic issues such as whether the approach is economically viable or aligns with the firm's values are also difficult to weigh up (Johnsen et al., 2014).

6.6 Future trends in the apparel industry

6.6.1 Innovations in stakeholder engagement

Many stakeholders in the process need to be engaged to ensure effective recycling. These stakeholders include upstream and downstream suppliers, apparel brands and manufacturers, consumers, governments and nongovernment organisations. All are needed to create an environment in which recycling, whether open loop or closed loop, is efficient and economically viable.

A number of large apparel brands have taken a lead role in stakeholder engagement in recent years. Patagonia, H&M, Marks & Spencer and Gap have all instigated clothing collection programmes in various forms. Marks & Spencer and Gap have each partnered with charities to manage the collected clothing; however, H&M has taken a different approach by selling the collected garments to garment recyclers (Stock, 2013). All schemes offer a discount to incentivise consumer garment returns. Most retailers only reward recycling of their own garments, but H&M developed the first global project to accept donated clothing from any brand. H&M reported recycling over 5500 tonnes of textiles in the 15 months between February 2013 and May 2014 (Jobling, 2014).

Individual apparel retailers have also branched out to inform the public of recycled content through labelling. R-Cert is an approach to labelling and tracking of recycled content developed by Esprit. Although a new initiative, it demonstrates a way to communicate with the public regarding the recycled content in a garment.

6.6.2 Design-led approaches for effective recycling

Opportunities exist for designers to selectively choose items from the waste stream to remodel. This is a niche upcycling approach within fashion design practice and was popularised in a high-fashion context by labels such as Junky Styling and designers Orsola De Castro and Christopher Raeburn. Sherbourne (2009) refers to this method as 'design-centricity' in the approach to recycling. Importantly, before moving to recycling, designers can consider extending the lifespan of garments and also encouraging repair and reuse amongst consumers. These kinds of design strategies point to emerging attitudes in the sector; however, is no replacement for reduced consumption (Niinimäki and Hassi, 2011).

Other approaches to design include developing a product with recycling in mind. This may be design for ease of later disassembly or designing with monomaterials. In Gulich's (2006a) analysis, use of single materials is the most effective way to ensure ease of recycling, with multiple materials that can be detached next in effectiveness. Most difficult to recycle are multi-materials with fixed connections, which characterises the majority of apparel products. Given the low rate of technological innovation in most apparel manufacturing, a challenge for effective recycling is how to develop new manufacturing methods that allow for garments to be more easily disassembled at end of life.

6.6.3 Future of recycled textiles

Technological innovations continue to be explored in the field of textile recycling. A key barrier to effective recycling is the proliferation of textiles in different fibre blends that are difficult to separate for recycling, such as cotton and polyester. Researchers have examined how to separate the cotton from the polyester using environmentally sound approaches to dissolve the cotton in order to reclaim the polyester for recycling (De Silva et al., 2014).

Another barrier to effective recycling is the resulting poor colour quality of shoddy. The dark grey of recycled fibre is unsuited to most apparel applications or indeed for most general products. A Japanese study proposes a colour coding to enable the resulting shoddy to be more easily processed into usable yarns (Motoko et al., 2013). As both mechanical and chemical recycling technologies develop, more opportunities may appear for CLR of textiles within apparel, as well as continued OLR of textiles into other products.

Trends towards environmental concerns see consumers more likely to expect firms to offer sustainable textiles and consider the ecological impacts of their products. Large retailers such as H&M have the reach to collect used clothing at scale, and by taking a lead they demonstrate that they see the social value as well as economic value in pursuing recycling options.

6.7 Conclusion

Many opportunities exist for improved textile recycling. Analysis of the energy and water usage advantages in utilising recycled textiles is evident across LCAs of both OLR and CLR fibres. However, considerable barriers exist towards greater uptake of recycled fibre on the part of companies. Additionally, although there have been advances in the past 10 years, more efficient recycling and collection of textile waste is required from diverse stakeholders such as apparel manufacturers, individuals and governments.

There has been considerable success with OLR in utilising the textile waste from apparel in other product systems, such as into building insulation. Similarly, in the specific case of PET bottles, the apparel industry has had success in integrating waste from another product system, the beverage industry, into textiles.

In contrast, CLR, or recycling waste from apparel (whether pre- or post-consumer) back into the apparel production system is a far more difficult proposition. There are several reasons for this. First, the sheer diversity of garments in style, trims, textiles and fibre types means that sorting and disassembling garments for recycling is highly labour intensive. Although this is also a problem for OLR of textile waste into other products, for recycling fibre to use in apparel, the quality standards of the recycled yarn in colour and purity need to be far higher than for a product such as insulation.

The advantages for using recycled fibres within garments are measurable in LCA across water, energy and land use indicators. Many stakeholders can play a role in improving apparel recycling rates and recyclability. Designers and manufacturers can play a crucial role in backward and forward engineering garments to be disassembled for ease of recycling and also in selecting textiles that contain a percentage of recycled fibre. Consumer trends towards more ecologically considered products are already prompting the apparel sector to successfully utilise recycled yarns.

List of abbreviations

C2C	Cradle to cradle
CLR	Closed-loop recycling
LCA	Life cycle assessment
LCT	Life cycle thinking
MSW	Municipal solid waste
OLR	Open-loop recycling
PET	Polyethylene terephthalate
RPET	Recycled polyethylene terephthalate

References

Awajaa, F., Pavelb, D., 2005. Recycling of PET. European Polymer Journal 41 (7), 1453–1477.

Bakker, C.A., Wever, R., Teoh, Ch., De Clercq, S., 2010. Designing cradle-to-cradle products: a reality check. International Journal of Sustainable Engineering 3 (1). http://dx.doi.org/10.1080/19397030903395166.

Billabong, 2012. Billabong Environment. http://www.billabongbiz.com/phoenix.zhtml?c=154279&p=environment (accessed 11.04.12.).

Bionic, 2014. Material Science: Manufacturing Process. http://www.bionicyarn.com/manufacturing-process/ (accessed 12.07.14.).

Brown, M.S., Wilmanns, E., 1997. Quick and dirty environmental analyses for garments: what do we need to know? The Journal of Sustainable Product Design 1, 28–35.

Chen, H.-L., Burns, L.D., 2006. Environmental analysis of textile products. Clothing and Textiles Research Journal 24 (248). http://dx.doi.org/10.1177/0887302X06293065.

Curran, M.A., 2012. Life Cycle Assessment Handbook: A Guide for Environmentally Sustainable Products. Wiley, Hoboken.

De Silva, R., Wang, X., Byrne, N., 2014. Recycling textiles: the use of ionic liquids in the separation of cotton polyester blends. RSC Advances 4, 29094–29098. http://dx.doi.org/10.1039/c4ra04306e.

Esprit, 2014. Sustainability in Practice. http://www.esprit.com/company/sustainability/sustainability_in_practice/ (accessed 12.7.14.).

Gam, H.J., Cao, H., Bennett, J., Helmkamp, C., 2011. Application of design for disassembly in men's jacket. International Journal of Clothing Science and Technology 23 (2/3), 83–94.

Gulich, B., 2006a. Designing textile products that are easy to recycle. In: Wang, Y. (Ed.), Recycling in Textiles. Woodhead, London, pp. 25–37.

Gulich, B., 2006b. Development of products made of reclaimed fibres. In: Wang, Y. (Ed.), Recycling in Textiles. Woodhead, London, pp. 117–136.

Hawley, J.M., 2006. Digging for diamonds: a conceptual framework for understanding reclaimed textile products. Clothing and Textiles Research Journal 24 (3), 262–275.

International Cotton Advisory Committee, 2013. Overview of world fiber production. In: Paper Presented at the 1st International Conference on Natural Fibres Guimarães, Portugal, June 9–11, 2013. http://dnfi.org/wp-content/uploads/2013/08/Portugal-Paper.pdf (accessed 29.11.14.).

Jobling, A., 2014. H&M: Where Next for Sustainability?. http://www.wgsn.com.ezp01.library.qut.edu.au/content/board_viewer/-/145030/page/1.

Johnsen, T.E., Howard, M., Miemczyk, J., 2014. Purchasing and Supply Chain Management: A Sustainability Perspective. Routledge, Oxon, UK and New York.

Klöpffer, W., Grahl, B., 2014. Life Cycle Assessment (LCA). Wiley, Hoboken.

Langley, K.D., Kim, Y.K., 2006. Manufacturing nonwovens and other products using recycled fibers containing spandex. In: Wang, Y. (Ed.), Recycling in Textiles. Woodhead, London, pp. 137–164.

McDonough, W., Braungart, M., 2002. Cradle to Cradle: Remaking the Way We Make Things. North Point Press, New York.

Motoko, U., Teruo, K., Tetsuya, S., 2013. Study on recycling system of waste textiles based on colour. Journal of Textile Engineering 59 (6), 159–164.

Muthu, S.S.K., Li, Y., Hu, J.-Y., Ze, L., 2012a. Carbon footprint reduction in the textile process chain: recycling of textile materials. Fibers and Polymers 13 (8), 1065–1070.

Muthu, S.S., Li, Y., Hu, J.-Y., Mok, P.-Y., 2012b. Recyclability Potential Index (RPI): the concept and quantification of RPI for textile fibres. Ecological Indicators 18, 58–62. http://dx.doi.org/10.1016/j.ecolind.2011.10.003.

Niinimäki, K., Hassi, L., 2011. Emerging design strategies in sustainable production and consumption of textiles and clothing. Journal of Cleaner Production 19, 1876–1883.

Patagonia, 2009. Closing the Loop - a Report on Patagonia's Common Threads Garment Recycling Program. http://www.thecleanestline.com/2009/03/closing-the-loop-a-report-on-patagonias-common-threads-garment-recycling-program.html (accessed 22.01.13., last modified 4.03.09.).

Rissanen, T., 2008. Creating fashion without the creation of fabric waste. In: Hethorn, J., Ulasewicz, C. (Eds.), Sustainable Fashion - Why Now: A Conversation Exploring Issues, Practices, and Possibilities. Fairchild, New York, pp. 184–206.

Ruiz, B., 2014. Global Apparel Manufacturing. http://clients1.ibisworld.com.au.ezp01.library.qut.edu.au/reports/gl/industry/default.aspx?entid=470 (accessed 10.08.14.).

Shen, L., Worrell, E., Patel, M.K., 2010. Open-loop recycling: a LCA case study of PET bottle-to-fibre recycling. Resources, Conservation and Recycling, 34–52.

Sherbourne, A., 2009. Achieving sustainable textiles: a designer's perspective. In: Blackburn, R.S. (Ed.), Sustainable Textiles: Life Cycle and Environmental Impact. Woodhead, London, pp. 3–32.

Shui, S., Plastina, A., 2013. World Apparel Fiber Consumption Survey 2013. Food and Agriculture Organization of the United Nations and International Cotton Advisory Committee, Washington, DC. https://http://www.icac.org/cotton_info/publications/statistics/world-apparel-survey/FAO-ICAC-Survey-2013-Update-and-2011-Text.pdf (accessed 11.05.14.).

Smith, G.G., Barker, R.H., 1995. Life-cycle analysis of a polyester garment. Resources, Conservation and Recycling 14 (3–4), 233–249.

Stock, K., 2013. The Brilliant Business Model behind H&M's Clothes Recycling Plan. http://www.businessweek.com/articles/2013-06-24/the-brilliant-business-model-behind-h-and-ms-clothes-recycling-plan (accessed 23.07.14., last modified 24.06.13.).

Terazono, A., Yoshida, A., Yang, J., Moriguchi, Y., Sakai, S-ichi, 2004. Material cycles in Asia: especially the recycling loop between Japan and China. Journal of Material Cycles and Waste Management 6 (2), 82–96.

United States Bureau of Labour Statistics, 2012. Spotlight on Statistics: Fashion: United States Bureau of Labour Statistics. http://www.bls.gov/spotlight/2012/fashion/pdf/fashion_bls_spotlight.pdf (accessed 23.06.14.).

United States Environmental Protection Agency, 2010. Municipal solid waste generation, recycling, and disposal in the United States: facts and figures for 2010. EPA. http://www.epa.gov/osw/nonhaz/municipal/pubs/msw_2010_rev_factsheet.pdf (accessed 20.06.14.).

Wang, Y., 2010. Fiber and textile waste utilization. Waste Biomass Valorization 1 (1), 135–143.

Wells, K., March 18, 2011. Jo Kellock, Textile and Fashion Industries of Australia, Interview. Craft Australia. http://www.craftaustralia.org.au/pdf/generate.php?sd=interview&id=jo_kellock_textile_fashion_industry (accessed 20.12.11.).

Woolridge, A.C., Ward, G.D., Phillips, P.S., Collins, M., Gandy, S., 2006. Life cycle assessment for reuse/recycling of donated waste textiles compared to use of virgin material: an UK energy saving perspective. Resources, Conservation and Recycling 46, 94–103.

Life cycle assessment method for environmental impact evaluation and certification systems for textiles and clothing

7

S.H. Eryuruk
Istanbul Technical University, Istanbul, Turkey

7.1 Introduction

Environmental sustainability is an important subject today. Some of the environmental unsustainability problems include greenhouse effect and climate change (emissions of CO_2, N_2O, CH_4), ozone depletion, acidification, toxic contamination, depletion of some native species, deforestation, land degradation, depletion of fossil fuels and minerals, discharge of polluting chemicals and other materials into the environment (Ekins, 1999). Today, recycling and reusing goods, buying recycled products, use of renewable resources and redesigning the production processes according to the environmental issues have gained importance.

The textile and apparel industry has been causing a number of different environmental problems like polluted air, land and water through the release of both toxic and hazardous waste. Moreover, a large amount of water, energy and other valuable resources are consumed during the production process. For conventional cotton cultivation, about 18 g pesticides are used per kg of cotton. A study showed that 1 kg of polyester fibres consume around 97.4 MJ of energy and 17.2 kg of water with 2.31 kg of CO_2, 18.2 g of carbon monoxide, 39.5 g of CH emissions to air and 3.2 g of water emissions. Furthermore, to produce 1 kg of polyamide 66, it is required to use 138.62 MJ of energy and 663 kg of water. In addition, 1 kg of Nylon 6 production demands 120.47 MJ of energy and 185 kg of water. Polypropylene (PP) production needs 73.37 MJ of energy and 43 kg of water for 1 kg of production and emits 1.7 kg of CO_2. Also, 1 kg of acrylic requires 175 MJ of energy, 210 L of water and emits approximately 5 kg of CO_2. During other textile production processes like spinning, fabric manufacturing, dyeing, finishing and clothing manufacturing various sources like water, energy, dyes, chemicals and auxiliaries are used. Clothing processes consume 2.472 MJ of energy of which 49.8% is used in sewing, 29.6% in cutting and 20.6% in packaging. Moreover, a large amount of waste (such as chemical, lubricants, packaging materials, fibre, yarn, fabric and apparel waste) create other environmental problems. Regarding the product life cycle of textile and clothing industry, many key inputs and outputs need to be evaluated in detail to

Handbook of Life Cycle Assessment (LCA) of Textiles and Clothing. http://dx.doi.org/10.1016/B978-0-08-100169-1.00007-1

identify their environmental effects and provide actions to mitigate these effects (Horrocks et al., 1997; Kalliala and Nousiainen, 1999; Laursen et al.,2007; Cepolina, 2012; Muthu, 2014).

Today, consumers are becoming more aware of protecting the environment and demanding cleaner and more efficient production methods. Textile and clothing companies have started to use some terms to promote their goods or services with eco-labels. Eco-friendly, environmentally friendly, nature-friendly and green are some of these terms used for the goods, services, laws, guidelines and policies that claim minimal or no harm to the environment.

Like many other industries, there is an urgent need for the textile and apparel sectors to follow environmental standards and regulations for preventing environmental damage. This chapter will first discuss the current environmental situation and life cycle analysis of the textile and clothing industry. Then environmental standards and environmental management systems will be discussed briefly. The chapter concludes with a discussion of eco-labelling and certification of textiles and apparel.

7.2 Life cycle assessment and environmental impacts of textiles and clothing

Life cycle assessment (LCA) is a method to measure total environmental effects of a product 'from cradle to grave'. Every product has a life cycle and a supply chain starting with the product design, then continues with raw material supply, production processes, logistics, use and ends with the product's reuse, recycling and disposal. Products passing through all these stages have an impact on the environment such as energy, water usage and CO_2 and waste production (Duda and Shaw, 1997; Guinée et al., 2002; Nowack et al., 2012; Muthu, 2014).

LCA method has four stages. The first step includes defining the goals of the study, describing the product to be assessed. In the second step, called inventory, data from all the processes of the product's life cycle are collected and processed, from cradle to grave. In this step, energy and raw material requirements, environmental emissions and discharges of the product, and other processes are computed and presented for all stages of production. This information is then used to calculate consumption and discharges from all processes of the product's life cycle. The third stage is called the impact assessment stage, in which the inventory data are translated into effects on human health, ecological health and resource depletion. In the final step, recommendations are made based on the results of the inventory and impact stages. This improvement analysis step may require choosing different raw materials, modifying production processes or choosing one product over another (Duda and Shaw, 1997; Werf and Turunen, 2008; Laursen et al., 2007). LCA is a very important procedure to assess the environmental effects and identify the most polluting stages in order to improve new methods and processes; there are many studies in the literature that have researched this subject (Søndergard et al., 2004; Werf, 2004; Nieminen et al., 2007; Herva et al., 2008; Thai et al., 2011).

The life cycle assessment of textile products is common, for example, extraction of raw materials, manufacturing processes (yarn and fabric), chemical finishing and treatment stage, apparel manufacturing, packaging, transportation, use phase and end of life including reusing, recycling and disposal of waste. A general model of a textile product's life cycle is presented in Figure 7.1. During the life cycle of a textile product, there are many inputs like energy, water and raw materials and outputs like air

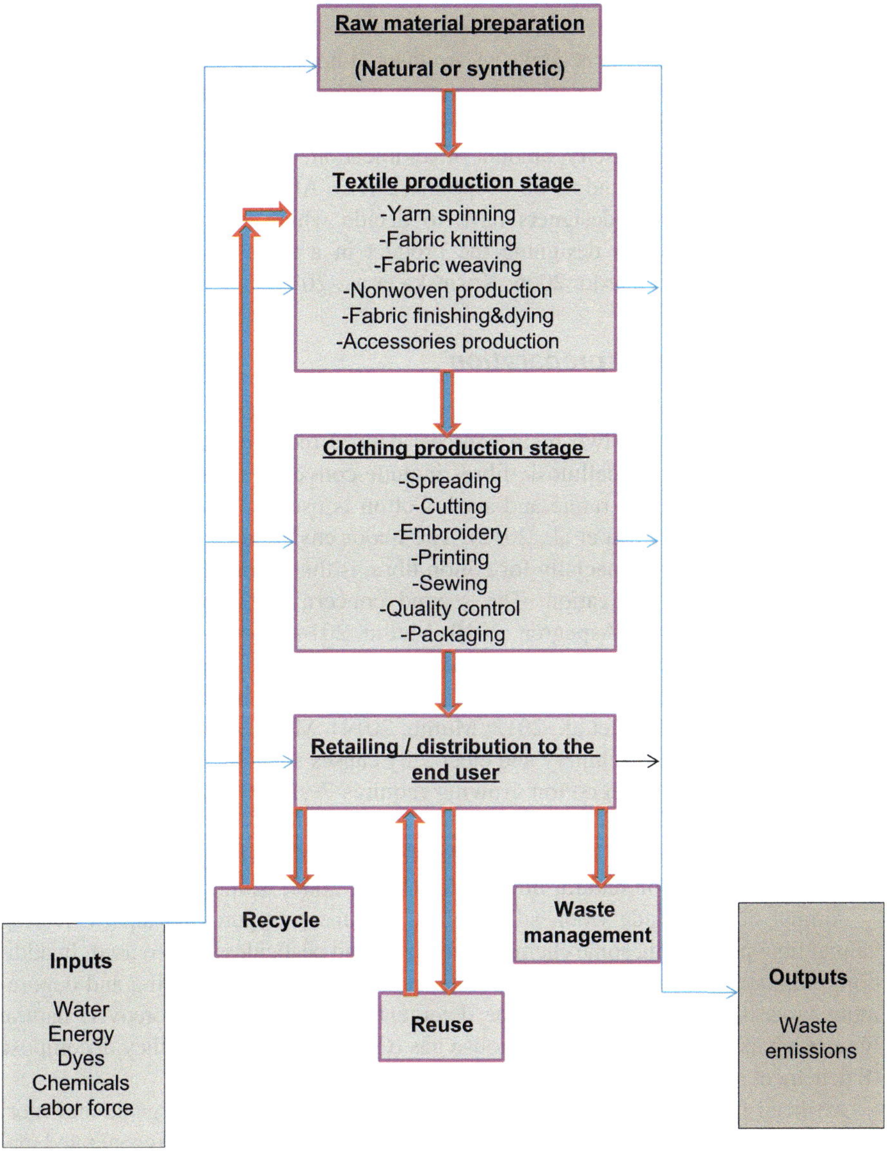

Figure 7.1 Textile and clothing production life cycle.

emissions, discharges into water and air, solid wastes and other environmental outputs that are harmful to the environment. This part of the chapter investigates the entire life cycle of textile and the clothing products in terms of various processes from fibre production to finished products – in other words from cradle to grave – and their environmental impacts in the different stages.

7.2.1 Environmentally sustainable design

Today, environmentally sustainable design or environmentally conscious design is very important to eliminate negative environmental impacts of products. For environmentally friendly products, designers must consider different environmental inputs and outputs for sustainable life cycle products design and manufacturing such as energy use, water use, toxicity, amount of acidification, CO_2 emissions, ozone depletion, resource depletion, and so on (Cepolina, 2012). After analysing and comparing different product types, designers have to decide which environmental hazard(s) they will focus on when designing the product in a more environmental-friendly manner (Bovea and Gallardo, 2006; Kosinska et al., 2014).

7.2.2 Raw material preparation

The starting point in the textile supply chain is the raw material preparation. Textile fibres are obtained from two main sources: natural (cellulose or animal) fibres or synthetic fibres. Natural cellulosic fibres include conventional and organic cottons, rayon, linen, hemp, jute, ramie and sisal. Cotton is used to produce 40% of world textile products (Saicheua et al., 2012). The major environmental concern in cellulosic fibre production, especially for cotton fibre, is the chemical fertilizers and pesticides used during cultivation. The second concern is the high level of water consumption (Dave and Aspegren, 2010; Muthu, 2014). Cotton is one of the most popular natural fibres used in the world. Three percent of the world's cultivated land is used for cotton production and 16% of the world's insecticides are used on this crop alone (Saicheua et al., 2012; Muthu, 2014). Moreover, the use of chemical fertilizers, pesticides, machinery and electricity causes some human health and environmental problems. Also cotton growing requires 7–29 tonnes of water per kg of raw cotton fibres (Kalliala and Nousiainen, 1999). Other types of cellulosic fibres are hemp and flax, which can be considered to be the most significant sustainable fibres in the non cotton natural fibre sector (Werf, 2004; Muthu, 2014).

Animal fibres include wool, silk, mohair, cashmere, angora and alpaca. During animal fibre production, some chemical fertilizers and pesticides are also used. In addition, pollution generated by the wastewater released during the cleaning and degumming processes always contain some detergents and chemicals. Moreover, natural fibres such as wool create the greenhouse gas (GHG) methane when they decompose (Saicheua et al., 2012).

Artificial fibre products are made from two types of materials, organic and inorganic. Synthetic fibres can be produced using true synthetic polymers, regenerated materials and modified natural ones. Dry spinning, wet spinning and melt spinning technologies also can be used to produce synthetic fibres (Slater, 2003). The true synthetics are polyester, nylon, acrylic, modacrylic and polyolefin fibres that are produced using synthetic polymers. The semisynthetics are viscose rayon, acetate, modal and cellulose acetate that are produced from natural polymers (Dayioglu and Karakas, 2007). Synthetic fibres are used to produce more than half of the world's garment needs. In 2011, 79.1 million tonnes of textile fibres were produced, 61.3% of which were synthetic, around 31.2% of them were cotton, 6% were man-made cellulosic fibres and 1.5% were wool fibre (Laursen et al., 2007). Polyester fibre is the most used synthetic fibre worldwide. Less water and land are used during the manufacturing process of synthetic fibres but there are more effects on the environment. Synthetic fibres are mostly obtained from nonrenewable resources (e.g. fossil fuels) and in the production stage a high level of energy is needed. Moreover, higher GHG emissions are released and the production requires a high level of chemicals. Environmental problems in waste management such as nonbiodegradability pose various health and toxicological threats that are also negative impacts of the production of synthetic fibres (Saicheua et al., 2012; Muthu, 2014).

Today renewable resource-based raw materials and organic clothing made from organic fibres has started to be considered important (Shishoo, 2012, Chapter 4). Key factors of sustainable clothing are new fibre sources and renewability without the use of agro-chemicals, pesticides during the entire manufacturing process starting from raw fibre production to the textile product manufacturing (Url-1).

7.2.3 Textile and clothing production stage

Yarn spinning is usually carried out using one of two methods, ring spinning or open-end spinning. Fibres are produced in many different places, thus the first step of the yarn manufacturing is the transportation of fibre from the place of production to the spinning factory. Spinning production includes opening, carding, combing, drawing, roving, spinning and winding processes. A high level of energy usage, dust, fibre and yarn waste production are the main issues related to the environmental protection during spinning (Martínez, 2010).

During the yarn spinning production, different combinations of fibres are mixed to obtain required blends such as natural/natural or synthetic/natural fibre combinations. Moreover, some sources such as lubricants, chemicals, water, packaging materials and other disposals are used in the various processes during blowing, carding, drawing, combing, roving, ring spinning, open-end spinning and other types of spinning processes. One of the largest environmental risks in these processes is inhalation of dust that may cause a dangerous disease called byssinosis (commonly called 'Brown Lung') (Laursen et al., 2007).

Fabric structure can be obtained generally using three methods: knitting, weaving and nonwoven production. Knitting is a simple process to produce fabric and generally some oils or lubricants are used during this process. In woven fabric production, after beaming and before weaving, warp yarns are processed with sizing agents that can be natural or synthetic, such as polyvinyl alcohol or carboxymethyl cellulose. All knitted and woven raw fabrics are washed in order to remove impurities and machine oils using water and some chemicals. Moreover, in all types of fabric manufacturing processes high levels of energy consumption; solid waste production; various chemicals, water, packaging materials use; disposals and excessive noise can be issues (Kalliala and Talvenmaa, 2000; Kalliala, 2003).

Fabric wet processing can be classified into three stages: pretreatment, colouring and finishing. Moreover, fabric finishing can be divided into two processes: physical and chemical. Wet processes include bleaching, carbonizing, neutralizing, scouring, sizing, desizing, mercerizing, dyeing, softening, printing and other functional finishing processes like shrink resistance, antistatic finishing, easy-care, wrinkle reduction, water repellent finishes, stain and odour resistance, fireproofing and mothproofing, disinfectants and mildew prevention finishes (Kalliala and Talvenmaa, 2000). Environmental effects of textile chemicals in textile wet processes must be focused and cleaner production systems must be developed to reduce the aquatic toxicity of chemicals (Moore and Ausley, 2004; Fransson and Molander, 2013). Although the amount of water, energy and chemical consumption depends on the type of fibre being processed, on average the textile industry uses 50–150 L of water per kg of textile material (You et al., 2009). To dye 1 kg of fibre, nearly 80–100 L of water and various amounts of chemicals are needed. During the dyeing process, formaldehyde and aromatic amines from azo dyes may be used, and these materials may cause skin cancer and allergies (Saicheua et al., 2012).

The final stage in clothing or garment production is the textile production supply chain. Clothing production includes operations like spreading, cutting, sewing, heat pressing, ironing and packaging. The important considerations here in the environmental impact assessment are high level of energy usage; fabric waste, accessories waste and rejected garments; lubricants, water and chemical usage; use of packaging and hanging materials, production of high carbon emissions during transportation operations and use of child labour force during manufacturing. Accessories, linings, pockets, seam bindings, ribbons, cords, labels and lace must be 100% natural fibre; buttons, zippers and clasps must be nickel and chrome-free; packing materials may not contain PVC. Also, fastness values of fabrics, like colour fastness to rubbing, sweat, light and washing must be high enough, especially for children's wear (Muthu, 2014).

The various production stages from raw materials to finished textile goods are generally located in different places around the world, which makes logistical activities an important component along the textile supply chain. After the production stage, textile and garment products are sent to wholesalers, retailers, distribution centres and shops by using some combinations of land, maritime, rail or air transportation that have different amounts of energy consumption values and CO_2 emission levels. It is estimated that freight transport accounts for nearly 8% of energy-related CO_2 emissions worldwide. Considering the warehousing and goods handling, this value increases around 2–3% in total. Green logistics aim to reduce freight transport externalities, city logistics, reverse logistics and to develop corporate environmental strategies towards green logistics and green supply chain management (Sarkis, 2003; Seuring, 2004; Mckinnon et al., 2010; Toke et al., 2012). McKinnon and Woodburn characterized four levels for green logistics. The first step is the strategic decision relating to the numbers, locations and capacities of factories, warehouses, shops and terminals. The second step includes commercial decisions on product sourcing, subcontracting of production processes and distribution of finished goods. Then operational decisions on the scheduling must be made about production and distribution. In addition, the rate of inventory rotation in warehouses must be specified. The last step is to make the functional decisions relating to the management of logistical resources (McKinnon and Woodburn, 1996).

After the product is obtained by the end user, the clothing usage phase starts. End users have to be aware of the environmental impacts of their caring attitudes or behaviour (of their garments) during the end use phase of their textiles and garments. Washing frequency, amount of water used, amount of detergents, softener usage, selected washing temperature, tumble drying and ironing directly influence the amount of energy and water consumption and chemical disposal to the environment.

In all stages of textile production, energy consumption is a very important consideration and there are several studies in the literature about energy consumption (Martínez, 2010; Kemal, 2004; Werf and Turunen, 2008; Brand et al., 1996; Bruggen et al., 2004; Hubenbach et al., 2006). Electricity, natural gas or fuel can be used as the energy source. Generally, air pollutants are produced from fossil fuels. In order to prevent air pollution, direct emissions of hazardous gases must be prohibited. A limited source of water is used at every step of the finishing processes and water becomes full of chemicals with dyes, defoamers, bleaches, detergents, optical brighteners, equalizers and many other chemicals. Furthermore, these chemicals evaporate into the air or are absorbed through our skin from the fabrics that contain residuals of chemicals used during their manufacture. To protect the environment and conserve available water resources, wastewater reclamation and reuse has become an important subject of consideration (Lu et al., 2010; Url-2).

7.2.4 Reducing, recycling, reusing and textile waste management

Textile and clothing waste can be classified into two categories: pre-consumer and post-consumer waste. Pre-consumer textile waste includes by-products obtained from the manufacturing facilities and they are mainly used to produce new raw materials for the automotive, furniture, mattress, coarse yarn, home furnishings, paper and other industries. Post-consumer textile waste includes any type of garments or household articles, and it is beneficial to reuse and recycle these textiles to reduce water and energy consumption, dyes and fixing agents (Woolridge et al., 2006).

During the textile supply chain, *reducing* is a very important technique to achieve a minimized consumption rate of scarce materials and/or energy. *Recycling* refers to activities performed to recover materials from products. *Reusing* is the concept of using intact parts of used products for manufacturing activities. *Product/material recovery* includes the activities carried out to regain the product value at the end of its life cycle, and such activities include repair/refurbish and disassembly (Farahani et al., 2009; Eryuruk, 2012).

There are five different ways to reuse or recycle for the post-consumer waste (Url-3);

1. Manufacturing of a textile or clothing product from recycled consumer waste such as plastic bottles, waste polyester yarns or fabrics.
2. The reuse of waste textile and clothing products in a way that avoids throwing the items away, such as shredding the products into fibres for sound insulation.
3. Reusing the items in the form of second-hand clothing via charity shops or textile merchants (also known as rag collectors).
4. Convert the waste to heat, ash and flue gas.
5. Disposal to landfill.

Today, nearly most of the collected textiles are reused and recycled. In many African countries, over 80% of the population buy their clothing from second-hand clothing shops. The University of Copenhagen published a study in 2008 that analysed the environmental advantages obtained from the collection of used clothing. By collecting 1 kg of used clothing, 3.6 kg of CO_2 emissions, 6000 L of water consumption, 0.3 kg of the use of fertilizers and 0.2 kg of the use of pesticides can be reduced (Url-4). Reuse and recycling are very important life cycle phases to decrease water, energy and raw material consumption (Cepolina, 2012).

7.2.5 Relationship between LCA, standards and eco-labelling certifications

In order to measure and reduce the environmental impacts of textile products, some tools such as LCA methodology can be used as an objective process to define key impact areas, evaluate potential environmental threats, make environmental decisions and ensure eco-labelling and certification of textiles. Life cycle assessment is a scientific base for environmental declaration and labelling. The most polluting stages of textile processing can be identified with the help of LCA methodology to develop production technologies correctly.

Environmental examinations of textiles and clothing during the manufacturing processes can be conducted starting from raw material production to waste materials by using LCA. These data are then compared with the criteria of the predetermined environmental labels and standards (e.g. EU eco-labelling and the Öko-Tex standard). Today, there are many different types of certification systems and a wide range of requirements have been introduced by various environmental labels for textile and clothing products. Some of the environmental labels are only based on LCA of the products and consider environmental effects across the product chain from the production of raw materials to waste management like ISO, EU eco-label and Nordic environmental labels. On the other hand, some environmental labels only focus on user safety like the Öko-Tex standard that considers the possible harmful effects of finished products. Moreover, some labels may certify only the quality of raw material. Environmental data on textile products can be collected using LCA method or any other required procedure and these data are compared to the criteria of selected eco-labelling and standardization systems to determine whether or not the product delivers the required conditions (Kalliala, 2003; Nieminen et al., 2007).

7.3 Standards and associations for environmental management systems

The eco-certification rules were created with the cooperation of international organizations. In this section, the United Nations Environment Programme (UNEP), International Organisation for Standardisation, ASTM Environmental Standards, Global Organic Textile Standard and NATURTEXTIL BEST are introduced.

7.3.1 *United Nations Environment Programme*

UNEP, founded in 1972, is the designated authority in the United Nations system. UNEP coordinates and supports the environmental activities of the United Nations for the sustainable development of the global environment. UNEP's mission is (Url-5):

> *To provide leadership and encourage partnership in caring for the environment by inspiring, informing, and enabling nations and peoples to improve their quality of life without compromising that of future generations.*

One of the missions of UNEP is to determine global, regional and national environmental conditions and trends. Then it develops international and national environmental instruments and strengthens institutions for the wise management of the environment (Url-5).

7.3.2 *ISO 14000 family of environmental management standards*

ISO is the International Organization for Standardization. It has a membership of 160 national standard institutes worldwide. ISO has developed many standards for helping organizations to take action to manage environmental issues, and ISO 14000 is a family of environmental management standards. ISO has standards for GHG accounting, verification and emissions trading, and for measuring the carbon footprint of products. ISO develops documents to encourage the inclusion of environmental aspects in product design. For sampling and test methods related to specific environmental challenges, ISO has developed a wide range of standards. To observe the air, water and soil quality, noise and radiation levels and also the transport of dangerous goods, ISO also developed 570 international standards. They also serve as the technical basis for environmental regulations in many countries (Url-6).

ISO technical committee ISO/TC 207, environmental management, is responsible for developing and maintaining the ISO 14000 family of standards. The aims of ISO/TC 207 are:

- Environmental management systems
- Environmental auditing and related environmental investigations
- Environmental performance evaluation
- Environmental labelling

- Life cycle assessment
- Environmental communication
- Environmental aspects of product design and development
- Environmental aspects in product standards
- Terms and definitions
- GHG management and related activities
- Measuring the carbon footprint of products.

The ISO 14000 family has a Plan-Do-Check-Act (PDCA) cycle as in all ISO management systems standards:

PLAN: This stage includes the environmental management system implementation and addresses environmental aspects in products and product standards.

DO: This stage consists of standards for conducting life cycle assessment and managing environmental aspects and GHGs.

CHECK: This stage has standards related to conducting audits and evaluating environmental performance. It also evaluates GHG performance.

ACT: This is the last stage and includes standards about communication and use of environmental declarations and claims.

ISO standards related to the environmental management systems are the world's most recognized framework for environmental management systems. ISO 14001: 2004 guides the organizations to manage better the impact of their activities on the environment and to prove environmental management. ISO 14004:2004 complements ISO 14001 by presenting additional guidance and useful explanations. ISO 14005: 2010 includes the use of environmental performance evaluation. ISO 14020:2000 is a series of standards including approaches about environmental labels and declarations, eco-labels, self-declared environmental claims, and environmental information about products and services. ISO 14031:2013 standard guides the organizations on how to evaluate their environmental performance and to select suitable performance indicators. Then the performance can be evaluated according to the criteria set by management and reported as internal and external environmental performance. ISO 14040: 2006 standards present guidelines on the principles of LCA and give information to reduce the overall environmental impacts of products and services.

Other standards under the title environmental management also include the following standards:

- ISO/TS 14033:2012 Environmental management – Quantitative environmental information – Guidelines and examples
- ISO 14006:2011 Environmental management systems – Guidelines for incorporating ecodesign
- ISO 14063:2006 Environmental management – Environmental communication – Guidelines and examples
- ISO 14045:2012 Environmental management – Eco-efficiency assessment of product systems – Principles, requirements and guidelines
- ISO 14051:2011 Environmental management – Material flow cost accounting – General framework

There are also ISO standards related to the GHGs. ISO 14064-1-2-3:2006 series are the international GHG accounting and verification standards. These standards present requirements and support organizations and proponents to reduce GHG emission levels. ISO 14063 includes environmental communication guidelines and examples, to show important links to external stakeholders. ISO 14065 specifies requirements to accredit organizational bodies that commit GHG validation or verification using ISO 14064 or other relevant standards. ISO 14065:2013 introduces the requirements for GHG validation and verification bodies for use in accreditation or other forms of recognition. ISO 14066:2011 gives the competence requirements for GHG validation teams and verification teams. ISO/TS 14067:2013 presents the requirements and guidelines for quantification and communication of carbon footprint of products.

ISO standards related to the life cycle assessment are listed below:

- ISO 14044:2006 Environmental management — Life cycle assessment — Requirements and guidelines
- ISO/TR 14047:2012 Environmental management — Life cycle assessment — Illustrative examples on how to apply ISO 14044 to impact assessment situations
- ISO/TR 14049:2012 Environmental management — Life cycle assessment — Illustrative examples on how to apply ISO 14044 to goal and scope definition and inventory analysis
- ISO/TS 14071:2014 Environmental management — Life cycle assessment — Critical review processes and reviewer competencies: Additional requirements and guidelines to ISO 14044: 2006
- ISO/TS 14067:2013 GHGs — Carbon footprint of products — Requirements and guidelines for quantification and communication

7.3.3 ASTM environmental standards

ASTM technical committees have published hundreds of standards about environmental safety. The aim of these efforts is to obtain cleaner air and water, eco-friendly homes and office buildings; enhanced waste management and recycling programmes; new innovations in oil spill response, and improved environmental assessment processes (Url-7).

List of environmental standards developed by ASTM:

- Atmospheric Analysis Standards
- Environmental Assessment Standards and Risk Management Standards
- Environmental Toxicology Standards
- Waste Management Standards
- Water Testing Standards

7.3.4 Global Organic Textile Standard

The Global Organic Textile Standard International Working Group has four member organizations: OTA (USA), IVN (Germany), Soil Association (UK) and JOCA (Japan). They contribute to the GOTS together with other international stakeholder organizations for their expertise in organic farming and environmentally and socially responsible textile processing. GOTS was first announced in Düsseldorf (Germany) at the Intercot Conference in 2002.

The Global Organic Textile Standard (GOTS) is a textile processing standard for organic fibres and considers all parts of the textile supply chain including ecological and social criteria. The main aim of the standard is to specify requirements for certifying organic textiles starting from harvesting through environmentally sensitive manufacturing and to give guarantees to the end user. With this certification manufacturers also prove that their fabrics, garments or products are organic. Version 4.0 was published March 1, 2014. The standard includes processing, manufacturing, packaging, labelling, trading and distribution of all textiles made from organic natural fibres.

The standard presents a two label-grade choices. The first is label-grade 1, which is 'organic'. If a textile product has the GOTS label grade 'organic', it must have a minimum of 95% certified organic fibres. Label-grade 2 is 'made with X% organic'. If a product has the label-grade 'made with organic', it must contain greater than 70% certified organic fibres, less than 30% nonorganic fibres. Also, the nonorganic fibre part must have a maximum 10% synthetic fibres (25% for socks and sportswear) (Url-8).

7.3.5 NATURTEXTIL BEST

Internationale Verband der Naturtextilwirtschaft (IVN) was founded in 1989, and today it has members from all levels of textile production. Since 2000, the IVN awards the two-step quality seal "NATURTEXTIL" that addresses all levels of textile production and includes social standards.

The aim of IVN is to increase the awareness for eco-friendly textiles among consumers, press and the retail trade. It defines and implements specific criteria to set ecological and social accountability in production processes and also high quality standards in the finished product. Products labelled with ' NATURTEXTIL BEST' must be produced using 100% certified organic fibres and restricted fibre processing methods (bleaching, chlorination, mercerization, etc.). The use of hazardous dyes, auxiliaries and substances (e.g. formaldehyde, heavy metals and many more) are forbidden, and accessories (buttons, pockets, etc.) must be made using high-quality natural raw materials. The association has also presented a new quality seal NATURLEDER (Url-9).

7.4 Eco-labelling and certification of textiles and clothing

Since the early 1980s, consumers seeking environmental-friendly products and organizations have increasingly become aware of environmental management systems, socially responsible behaviour, sustainable growth and development around the world (Phau and Ong, 2007; Saicheua et al., 2012).

The eco-label is an 'ecological label' that supports minimizing adverse environmental effects starting from the design through installation, usage and disposal stages of a product and highlights 'green products' for customers. Moreover, in order to compete in the global market, more and more companies have tried to get eco-labels and certifications to show their customers that their products and services fulfil the international regulations and environmental standards and that they have reduced environmental impacts along the life cycles of their products. Today there are 460 eco-labels worldwide and 109 of them are related to textile products (Url-10).

Many studies have been conducted by researchers about eco-labelling, applications, effects of eco-labelling on consumer buying behaviour and eco-labelling schemes. Eco-labels can be classified as voluntary or mandatory. A third-party independent body is required to attest to the required standards for mandatory programmes. Eco-labels in textiles and clothing are not enforced by mandatory rules. Voluntary programmes may be established by firms, business associations as well as third parties. After a product meets the required standards or criteria, an eco-label is issued either by government or private enterprises. Some government-based eco-labels are Blue Angel (Germany), Eco Mark (Japan), White Swan (Nordic Countries) and Green Label (Singapore). Some private labels are Eco-tex and Oeko-Tex (textiles and clothing) (Germany) (Mccarthy and Burdett, 1998; Nimon and Beghin, 1999; Jackson, 1999; Atilgan, 2007; Ruževičius and Waginger, 2007; Mehta and Sridhar, 2009; Pammi and Rohit, 2010; Koszweska, 2011, 2013; Goel, 2012; Dekhili and Achabou, 2014).

ISO defines the voluntary environmental labelling in three classes (Mehta and Sridhar, 2009; Koszewska, 2011).

ISO Type I Environmental Labelling: Most of the eco-labelling schemes can be included in this type. This programme is a voluntary, multiple criteria-based, third-party programme. It awards a license for a product type considering the whole life cycle of this product category. Blue Angel, EU Ecolabel and Nordic Swan can be included in this class.

Type II Self-Declaration Claims: These are environmental claims of products declared by their manufacturers, importers and distributors. There are no accepted criteria that need to be verified independently, so this is the least informative type of label.

Type III Environmental Declarations: This is a voluntary programme. Predetermined criteria for different categories of products are determined by a qualified third party and verified by another qualified third party.

Eco-labelling supports environmentally sound innovation and leadership and increases consumer awareness for protecting the environment (Mehta and Sridhar, 2009). The following section discusses the most frequently used eco-labels.

7.4.1 EU Ecolabel

The Environment Directorate General of the European Commission (DG Environment) was established in 1973 to protect and improve Europe's environment for present and future generations. The main aim of the commission is to suggest policies and legislation for protecting natural habitats, preserve air and clean water, assure the safe disposal of waste, raise awareness about toxic chemicals and guide businesses for operating close to a sustainable economy (Url-11).

The EU Ecolabel includes a wide range of product groups, from manufacturing to tourist accommodation services. There are key experts who develop the criteria for each product group to lower the environmental impacts over the entire life cycle, and every four years these criteria are revised. Revision is needed to consider revised technical innovations of materials, production processes, emission reduction and market changes.

The Commission's decisions for textile product groups, which establish the ecological criteria for the award of the EU Ecolabel for textile products, were accepted on June 5, 2014 with an end date of June 5, 2018. The aim of these criteria is to source materials using more sustainable agriculture forms; to use resources and energy more efficiently; to use cleaner, less polluting processes; to use less hazardous substances; and to design and specify high quality and durable products (Url-12).

7.4.2 Eco-text (Oeko-Tex) standards

An eco-label specific to textiles was founded in 1993 by the Austrian Textile Research Institute, called the Öeko-Tex label. Öeko-Tex certificate is the most widely recognized textile environmental standard in the world (El-Nagar, 2012; Url-13).

According to the Oeko-Tex criteria, textile ecology includes four sectors: production ecology, human ecology, performance ecology and disposal ecology.

- Production ecology examines the impact of production processes on people and environment, for example, occupational health and safety, material, water and energy consumption, wastewater and waste treatment as well as generation of dust and noise.
- Human ecology deals with the impact of textiles and their chemical ingredients on the health and well-being of humans.
- Performance ecology comes in at the usage phase of textile products. It examines the environmental impact of washing, cleaning and caring for textiles.
- Disposal ecology analyses the problems connected with disposal, reuse, recycling and removal (thermal recycling or landfill) of textiles.

7.4.2.1 OEKO-TEX® Standard 100

The OEKO-TEX® Standard 100 is an independent testing and certification system for textile raw materials, intermediate and end products at all stages of production. Products are analysed for harmful chemicals, illegal substances such as carcinogenic colourants; legally regulated substances such as formaldehyde, plasticizers, heavy metals or pentachlorophenol; known harmful (but not legally regulated) chemicals such as pesticides, allergenic dyes or tin-organic compounds, as well as parameters for health care. The laboratory tests currently contain around 100 test parameters based on international test standards and other recognized testing procedures such as colourfastness and a skin-friendly pH-value.

Moreover, if the skin contact of a product is more intensive, stricter human ecological requirements must be met. Accordingly, there are four product classes (Url-14):

Product class I: Textile items for babies and toddlers up to 3 years (clothing, toys, bed linens, terry cloth items, etc.)

Product class II: Textiles used close to the skin (underwear, bed linens, T-shirts, etc.)

Product class III: Textiles used away from the skin (jackets, coats, etc.)

Product class IV: Furnishing materials (curtains, tablecloths, upholstery materials, etc.)

In order to get certification according to OEKO-TEX® Standard 100, all components of a textile product have to conform to the required criteria without exception, meaning that in addition to the outer material also sewing threads, linings, prints, etc. as well as nontextile accessories such as buttons, zip fasteners, rivets, etc. have to comply with the requirements. The OEKO-TEX® certificate is valid for a duration of 12 months.

7.4.2.2 Sustainable Textile Production

Sustainable Textile Production (STeP) is the new OEKO-TEX® certification system. This new system replaced the previous certification of production sites according to OEKO-TEX® Standard 1000 from July 2013.

The aim of STeP certification is to achieve environmental-friendly production processes, optimum health and safety and socially acceptable working conditions. It also allows certified companies to have the best possible competitive position with the improvement of their environmental protection achievements, their social responsibilities as well as their efficiencies.

Certification is possible for all processing stages from fibre production, spinning, weaving and knitting to finishing facilities and clothing manufacturers. Certificate is valid for three years and can be extended any number of times.

7.4.2.3 OEKO-TEX® Standard 100 plus

The OEKO-TEX® Standard 100 plus is a product label. With this label, manufacturers certify to their end users that their products have been optimized for human ecology and also their production conditions are environmentally friendly. Products with this label are tested for harmful substances according to OEKO-TEX® Standard 100 and produced at environmentally friendly production sites according to Sustainable Textile Production (STeP) (Url-15).

7.4.3 Eco-labels in Europe

7.4.3.1 The Blue Angel label

Blue Angel is a German eco-label established in 1978 and is the oldest eco-label in the world. The criteria are established by the German Federal Environment Agency. The label has four different protection goals: health, climate, water and resources. Blue Angel contributes both environmental and consumer protection (Url-16; Url-17; Url-18).

7.4.3.2 eco-INSTITUT

eco-INSTITUT is a German eco-label that tests emissions and residues of hazardous chemicals in many different materials such as construction products, floor coverings, mattresses and bedding, furniture and furnishings, textiles and leather and other types of consumer goods. The label tests emissions measurement capabilities according to ISO 16000 and EN 717-1 and conducts dynamic environmental chamber testing, content analysis for phthalates, organotin compounds, formaldehyde, chromium VI, triclosan, ortho-phenylphenol, external and quality inspection in the context of Ü and CE marking, toxicological and ecological evaluations (Url-19).

7.4.3.3 SG Label (Schadstoffgeprüft-Zeichen)

SG Label (Schadstoffgeprüft-Zeichen) certifies especially leather goods, leather products, leather garments and shoes. The SG® symbol demonstrates that goods are tested for harmful substances, comply with certain emission limits and are safe for health. The criteria refer only to the finished product; requirements for the production process along the textile chain are not provided (Url-20).

7.4.3.4 Ecoproof

This label is especially for textiles made from cotton. Certain criteria have to be fulfilled by the product throughout the entire life cycle, like environmental fulfilment, and also decrease the negative effects on human health. This label provides some benefits like promoting conformity for the requirements such as pollutants and quality. Also consumers will be certain about the environmental safety of their products. Tests and analysis include chemical tests, testing for harmful substances (forbidden azo dyes, formaldehyde, nickel, allergenic disperse dyes, etc.), textile physical exams and quality audits extract, and certification (Url-21).

7.4.3.5 Green Shape

VAUDE is a modern, environmentally friendly family run company that has an outdoor outfitter with three divisions: Mountain Sports, Bike Sports and Packs'n Bags. Green Shape is VAUDE's label for products that have been ecologically manufactured. Products made of at least 90% organic cotton or recycled materials, coloured using the VAUDE ecolour dyeing process or made to conform to the textile standard bluesign® can use the Green Shape quality label (Url-22).

7.4.3.6 NF Environnement label

The NF Environnement mark was created in 1991; it is a French ecological certification for Association française de normalization (AFNOR). Products having this label have a reduced effect on the environment and also equivalent performance properties. It is a collective certification mark that certifies the quality and safety of the products and services (Url-24).

7.4.3.7 The Skal label

The Skal label is from an organization in the Netherlands and is an eco-label assigned by the Ministry of Economic Affairs. It focusses on a reliable organic product and production methods. Its main goal is to offer the consumer an organic product that originates from an organic production process (Url-25).

7.4.3.8 Netherland (Stichting Milieukeur)

Milieukeur was developed and is managed by the Stichting Milieukeur (the Ecolabel Foundation), an independent organization, which was founded in 1992 on the initiative of the Ministry of Housing, Spatial Planning and the Environment and the Ministry of Economic Affairs (Goel, 2012).

7.4.3.9 The Nordic Ecolabel – Nordic Swan

The Nordic Ecolabel is the official Ecolabel of the Nordic countries; it was established in 1989 by the Nordic Council of Ministers. Its main aim is to provide an environmental labelling scheme to offer a sustainable consumption. It is an ISO 14024 Type 1 Ecolabelling system and conformity with Nordic Ecolabel or 'Swan' standard is verified by an independent organization (third party) following ISO 17011 Accreditation, ISO 17021 Management system certification, ISO 17025 Testing and Calibration Laboratories. Companies that produce and market their products in Sweden, Norway, Finland, Iceland or Denmark apply in their own country. Other companies must apply in the country in which the Ecolabelled products primarily will be sold. The Swan makes sure that products conform to certain criteria by controlling samples from independent laboratories and making control visits (Url-26; Url-27).

7.4.3.10 The KRAV label

KRAV was established in Sweden in 1985 and is a label for the organic market. The KRAV label is well-known among Swedish consumers. The KRAV standards also fulfill the EU standards for organic production in the regulation (EC) No 834/2007. The KRAV standards cover a variety of areas, such as certification of restaurants and industrial kitchens, fishing and textiles (Url-28).

7.4.3.11 bluesign® standard

The bluesign® is a government organization and system that was founded in Switzerland in 2000. The bluesign® system considers all of the textile manufacturing chain for sustainable textile production. It prevents the use of harmful substances starting from the beginning of the manufacturing process. It also determines the standards for environmentally friendly production. The label ensures that consumers get sustainable products conforming to safety requirements worldwide (Url-29).

7.4.3.12 Coop Naturaline: Switzerland

Coop Naturaline is a Swiss label established in 1993. Products for textiles and natural cosmetics made from cotton are certified and controlled for biological cultivation according to the guidelines of BIO Suisse or the European Union. Moreover, Coop Naturaline covers the entire textile chain and undertakes additional pollution testing by external labs. To produce the textiles labelled with COOP Naturaline, companies must fulfill the basic standards of the International Labour Organization (ILO), at least according to the criteria of the Business Social Compliance Initiative (BSCI), and in the medium term, they must meet the SA 8000 certification (Url-30).

7.4.3.13 IMO certified

Institute for Marketecology (IMO) was founded in 1989, and IMOgroup AG was founded in 2006 in Switzerland. The Institute for Marketecology is an international agency for inspection, certification and quality assurance of eco-friendly products. For more than 20 years, IMO has been active in the field of organic certification, but it is also active in the sectors of natural textiles, sustainable forestry and social accountability monitoring (Url-31).

7.4.3.14 Global Recycled Standard

The Global Recycled Standard (GRS) is used by companies that are making and/or selling products with recycled content. The standard applies to the full supply chain and addresses traceability, environmental principles, social requirements and labelling. Although GRS developed with the textile industry, it can be applied to products from any industry (Url-32).

7.4.4 Other eco-labels in the world

7.4.4.1 Japan EcoMark

Japan EcoMark was established in 1989 and is managed by the Japan Environment Association (JEA). A committee including academics, governments, consumer groups and experts from different industries sets standards and carries out the certification. Products labelled with this mark include those with relatively less environmental impact compared to similar products, starting from product extraction, collecting the product materials, to the manufacturing, distribution, use, consumption, disposal and recycling. The EcoMark programme is directed in accordance with the standards and principles of the ISO (ISO 14020 and ISO 14024). The EcoMark committee for product certification conducts an audit for product certification of the EcoMark based on certification criteria for the corresponding product category. After getting the certification, an Eco Mark Utilization Contract is signed between the applicant and the JEA for every product type. The EcoMark may be used until the expiry date indicated in the certification. The programme offers a choice of products with a lower environmental impact; however, excess use of EcoMark products does not lead to an environmentally sound society. Also, the programme is not intended to protect consumers' direct benefits such as product quality and safety (Url-33).

7.4.4.2 Singapore Green Label Scheme

Singapore Green Label Scheme (SGLS) was announced in May 1992 by the Ministry of Environment. It was handed over to the Singapore Environment Council (SEC) on

June 5, 1999, and is currently under the authority of the SEC. The main aim of the SGLS is to help the public identify environmentally friendly products that meet certain eco-standards. It also aims to identify the growing demand for greener products in the market and increase awareness of the level of eco-consumerism in Singapore (Url-34; Url-35).

7.4.4.3 BMP certified cotton

BMP is the Australian cotton industry's guide for growing cotton in accordance with the natural environment. The Australian BMP Cotton trademark guarantees that the branded textile products are made of Australian cotton, which is grown under best management practices by producers considering the environment (Url-36).

7.4.4.4 Good Environmental Choice Australia

Good Environmental Choice Australia has certified facilities since 2000. It conducts Australia's only independent, nonprofit, multisector eco-labelling programme and also is the Australian member of the Global Ecolabelling Network (GEN). This certification programme has standards that are developed following ISO 14024 principles for global best practice in eco-labelling. Products with this label have a lower impact on the environment, human health and address important social considerations (Url-37).

7.4.4.5 India Eco-Mark

India Eco-Mark was launched by the Government of India in 1991. The label is awarded to consumer goods that meet the requirements of specified environmental criteria and the quality elements of Indian Standards. The criteria follow a cradle-to-grave approach, from raw material extraction to manufacturing and disposal. The logo: an earthen pot means the usage of renewable resources, producing nonhazardous waste and consuming little energy in production (Goel, 2012).

7.4.4.6 Thai Green label

This label was established by the Thailand Environment Institute in association with the Ministry of Industry in August 1994. Products made from cloths such as daily clothes, baby clothes, home textiles and also footwear are the main textile product categories, and these products are evaluated under their life cycle consideration. The label also has high priority national goals such as waste reduction and water and energy conservation (Url-38).

7.4.4.7 China Environmental Labelling

The China Environmental Label was established in 1993 and provides environmental standards for construction materials, textiles, vehicles, cosmetics, electronics, packaging and more (Url-39).

7.5 Conclusions

Textile production has a complex structure and different material types, fibre production methodologies and yarn spinning processes; fabric and garment production technologies can be used for various product types. The textile manufacturing process consumes resources like water, fuel and different types of chemicals along its life cycle and generates a significant amount of waste. The main environmental problems can be listed as air pollution, high toxicity level, wastewater and solid waste production.

Today consumers and producers have started to be more aware of protecting the environment by preferring more environmentally friendly products and production methods. Many textile and clothing brands have started to promote their goods or services with eco-labels around the world. LCA methodology can be used as one of the objective processes to define key impact areas, evaluate potential environmental threats, make environmental decisions and ensure eco-labelling and certification of textiles.

Although the certification process is very hard and costly, demand for eco-labelling and certification of textile products has been growing worldwide. Consumers want to purchase textiles made from sustainable materials and produced using sustainable production methods that comply with environmental standards. For the textile and clothing industry, this demand will increase worldwide and will shape the position of the products in the international markets.

References

Atilgan, Turan, April/June 2007. Eco-labelling applications in the textile & apparel sector in Turkey. Fibres & Textiles in Eastern Europe 15 (2), 14–23 (61).

Brand, J., Drechsler, K., Arendts, J., 1996. Mechanical performance of composites based on various three-dimensional woven-fiber performs. Composite Science Technology 56, 381–386.

Bruggen, B., Curcio, E., Drioli, E., 2004. Process intensification in the textile industry: the role of membrane technology. Journal of Environmental Management 73, 267–274.

Bovea, M.D., Gallardo, A., 2006. The influence of impact assessment methods on materials selection for eco-design. Materials and Design 27, 209–215.

Cepolina, S., February 2012. Textile and clothing industry: an approach towards sustainable life cycle production. International Journal of Trade, Economics and Finance 3 (1).

Dayioglu, H., Karakas, H., 2007. Elyaf Bilgisi. Ajans Plaza, İstanbul, ISBN 978 975 01610 1 8.

Dave, G., Aspegren, P., 2010. Comparative toxicity of leachates from 52 textiles to *Daphnia magna*. Ecotoxicology and Environmental Safety 73, 1629–1632.

Dekhili, S., Achabou, M.A., 2014. Eco-labelling brand strategy independent certification versus self-declaration. European Business Review 26 (4), 305–329.

Duda, M., Shaw, J.S., November/December 1997. Life cycle assessment. Social Science and Public Policy 38–43.

El-Nagar, K., 2012. Eco-tex in Relation to World Textile Market, State of the Arts. Lap Lambert Academic Publishing.

Ekins, P., 1999. Economic Growth and Environmental Sustainability: The Prospects for Green Growth. Routhledge, p. 7.

Eryuruk, S.H., 2012. The greening of textile and clothing industry. Fibres & Textiles in Eastern Europe 20 (6A,(95)), 22–27.

Farahani, R.Z., Asgari, N., Davarzani, H., 2009. Supply Chain and Logistics in National, International and Governmental Environment, Concepts and Models. Physica-Verlag.

Fransson, K., Molander, S., 2013. Handling chemical risk information in international textile supply chains. Journal of Environmental Planning and Management 56 (3), 345–361.

Goel, B., 2012. Eco-Labels and standards for textile products to save environment. Man-made Fibres in India 337–345.

Guinée, J.B., Gorrée, M., Heijungs, R., Huppes, G., Kleijn, R., de Koning, A., van Oers, L., Wegener Sleeswijk, A., Suh, S., Udo de Haes, H.A., de Bruijn, H., van Duin, R., Huijbregts, M.A.J., 2002. Handbook on Life Cycle Assessment. An Operational Guide to the ISO Standards. Kluwer Academic Publishers, Dordrecht, The Netherlands.

Herva, M., Franco, A., Ferreiro, S., Alvarez, A., Roca, E., 2008. An approach for the application of the ecological footprint as environmental indicator in the textile sector. Journal of Hazardous Materials 156, 478–487.

Horrocks, A.R., Hall, M.E., Roberts, D., 1997. Environmental consequences of using flame-retardant textiles – a simple life cycle analytical model. Fire and Materials 21 (5), 229–234.

Hubenbach, W., Kroll, L., Böhm, R., Langkamp, A., Czulak, A., 2006. Development of piping elements from textile reinforced composite materials for chemical apparatus construction. Journal of Materials Processing Technology 175, 231–240.

Jackson, P., 1999. The European Union eco-labelling scheme. Surface Coatings International (4), 172–175.

Kalliala, E.M., Nousiainen, P., 1999. Life cycle assessment environmental profile of cotton and polyester-cotton fabrics. AUTEX Research Journal 1 (1), 8–20.

Kalliala, E., Talvenmaa, P., 2000. Environmental profile of textile wet processing in Finland. Journal of Cleaner Production 8, 143–154.

Kalliala, E.N., 2003. Environmental indicators of textile products for ISO (Type III) environmental product declaration. AUTEX Research Journal 3 (4), 206–218. December.

Kemal, H., 2004. Energy usage and cost in textile industry: a case study for Turkey. Energy 30, 24–46.

Kosinska, B., Czerwiński, K., Struszczyk, M.H., 2014. Safety and labelling requirements for textile products - design and use aspects. Fibres & Textiles in Eastern Europe 22, 2 (104), 19–24.

Koszewska, M., 2011. Social and eco-labelling of textile and clothing goods as means of communication and product differentiation. Fibres & Textiles in Eastern Europe 19 (4(87)), 20–26.

Koszewska, M., 2013. Role of ecological and social labels in consumer evaluation of textile and clothing products. Polish Journal of Commodity Science 47–61.

Laursen, S.E., Hansen, J., Knudsen, H.H., Wenzel, H., Larsen, H.F., Kristensen, F.M., 2007. EDIPTEX-environmental Assessment of Textiles. Danish Environmental Protection Agency. Working Report No. 24.

Lu, X., Liu, L., Liu, R., Chen, J., 2010. Textile wastewater reuse as an alternative water source for dyeing and finishing processes: a case study. Desalination 258, 229–232.

Mehta, S., Sridhar, R., 2009. Influence of eco-labelling on consumer buying behaviour – a study of FMCG & apparel sector at Ludhiana. In: The 21st Century Consumers a Behavioural Perspective, Publish by Global Vision Publishing House, 81-102.

Martínez, C.I.P., 2010. Energy use and energy efficiency development in the German and Colombian textile industries. Energy for Sustainable Development 14, 94–103.

Mccarthy, B.J., Burdett, B.C., 1998. Eco-labelling and Textile eco-labelling. Review of Progress in Coloration and Related Topics 28, 61–70.

McKinnon, A.C., Woodburn, A., 1996. Logistical restructuring and freight traffic growth: an emprical assessment. Transportation 23 (2), 141–161.

Mckinnon, A., Cullinane, S., Browne, M., Whiteing, A., 2010. Green Logistics: Improving the Environmental Sustainability of Logistics. Kogan Page.

Moore, S.B., Ausley, L.W., 2004. Systems thinking and green chemistry in the textile industry: concepts, technologies and benefits. Journal of Cleaner Production 12, 585–601.

Muthu, S.S., 2014. Assessing the Environmental Impact of Textiles and the Clothing Supply Chain. Woodhead Publishing, Cambridge, U.K.; Waltham, Mass.

Nieminen, E., Linke, M., Tobler, M., Beke, B.V., 2007. EU COST Action 628: life cycle assessment (LCA) of textile products, eco-efficiency and definition of best available technology (BAT) of textile processing. Journal of Cleaner Production 15, 1259–1270.

Nimon, W., Beghin, J., 1999. Ecolabels and international trade in the textile and apparel market, ecolabels and international trade in the textile and apparel market. American Journal of Agricultural Economics 81 (5), 1078–1083. Proceedings Issue.

Nowack, M., Hoppe, H., Guenther, E., 2012. Review and downscaling of life cycle decision support tools for the procurement of low-value products. International Journal of Life Cycle Assessment 17, 655–665.

Phau, I., Ong, D., 2007. An investigation of the effects of environmental claims in promotional messages for clothing brands. Marketing Intelligence & Planning 25 (7), 772–788.

Pammi, S., Rohit, S., 2010. Creating a global vision for sustainable textiles. In: Textiles: A Global Vision, the Textiles Institute Centenary World Conference, 3–4 November 2010.

Ruževičius, J., Waginger, E., 2007. Eco-labelling in Austria and Lithuania: a comparative study. Engineering Economics Commerce of Engineering Decisions. ISSN: 1392-2785 4 (54), 96–102.

Saicheua, V., Cooper, T., Knox, A., 2012. Public Understanding Towards Sustainable Clothing and the Supply Chain. macromarketing.org.

Sarkis, J., 2003. A strategic decision framework for green supply chain management. Journal of Cleaner Production 11, 397–409.

Seuring, S., 2004. Integrated chain management and supply chain management comparative analysis and illustrative cases. Journal of Cleaner Production 12, 1059–1071.

Shishoo, R., 2012. "The Global Textile and Clothing Industry", Technological Advances and Future Challenges. Woodhead Publishing Limited, Cambridge.

Slater, K., 2003. "Environmental impact of textiles", Production, processes and protection. Woodhead Publishing.

Søndergard, B., Hansen, O.E., Holm, J., 2004. Ecological modernisation and institutional transformations in the Danish textile industry. Journal of Cleaner Production 12, 337–352.

Thai, V.N., Tokai, A., Yamamoto, Y., Ngugen, D.T., 2011. Eco-labeling criteria for textile products with the support of textile flows: a case study of the Vietnamese textile Industry. Journal of Sustainable Energy & Environment 2, 105–115.

Toke, L.K., Gupta, R.C., Dandekar, M., 2012. An empirical study of green supply chain management in Indian perspective. International Journal of Applied Sciences and Engineering Research 1 (2), 372–383.
Url-1: http://www.cool-organic-clothing.com/green-clothing.html (date received: 01.09.14.).
Url-2: www.oecotextiles.com/.../textile_industry, Textile Industry Poses Environmental Hazards, Ecotextiles (date received: 23.03.12.).
Url-3: http://www.humananova.org/en/columns/0/2/reusing-waste-textile-and-clothing-products-in-a-way-which-avoids-throwing-the-items-away/ (date received: 18.08.14.).
Url-4: http://www.bir.org/industry/textiles/ (date received: 20.09.14.).
Url-5: http://www.unep.org/about/ (date received: 18.08.14.).
Url-6: www.iso.org (date received: 18.08.14.).
Url-7: http://www.astm.org/Standards/environmental-standards.html (date received: 28.08.14.).
Url-8: http://www.global-standard.org/certification.html (date received: 28.08.14.).
Url-9: http://www.naturtextil.com/.
Url-10: http://www.ecolabelindex.com/ecolabels/?st=category,textiles (date received: 28.08.14.).
Url-11: http://ec.europa.eu/environment/ecolabel/the-ecolabel-scheme.html (date received: 28.08.14.).
Url-12: http://ec.europa.eu/environment/ecolabel/products-groups-and-criteria.html (date received: 28.08.14.).
Url-13: https://www.oeko-tex.com (date received: 28.08.14.).
Url-14: https://www.oeko-tex.com/en/manufacturers/test_methods/test_methods.xhtml, (date received: 28.08.14.).
Url-15: https://www.oekotex.com/en/manufacturers/concept/oeko_tex_standard_100plus/oeko_tex_standard_100plus.xhtml (date received: 29.08.14.).
Url-16: http://www.nordic-ecolabel.org/portals/paper/what-is-in-a-label/blue-angel/20.08.2014 (date received: 29.08.14.).
Url-17: http://www.blauer-engel.de/ (date received: 29.08.14.).
Url-18: http://www.ecolabelindex.com/ecolabel/blue-angel (date received: 29.08.14.).
Url-19: http://www.eco-institut.de/en/product-testing/textiles/ (date received: 29.08.14.).
Url-20: http://suchdichgruen.de/ (date received: 29.08.14.).
Url-21: http://www.tuv.com/de/deutschland/home.jsp (date received: 29.08.14.).
Url-22: http://www.vaude.com/de-NL/Verantwortung/Green-Shape/ (date received: 29.08.14.).
Url-23: http://www.marque-nf.com// (date received: 29.08.14.).
Url-24: http://www.skal.nl/ (date received: 29.08.14.).
Url-25: http://www.nordic-ecolabel.org/ (date received: 30.08.14.).
Url-26: http://www.ecolabelindex.com/ecolabel/nordic-ecolabel-or-swan (date received: 30.08.14.).
Url-27: http://www.krav.se/ (date received: 30.08.14.).
Url-28: http://www.bluesign.com/ (date received: 30.08.14.).
Url-29: http://www.coop.ch/pb/site/uebersichtsseiten2012/node/73468143/Lde/index.html (date received: 30.08.14.).
Url-30: http://www.imo.ch/ (date received: 30.08.14.).
Url-31: http://textileexchange.org/GRS (date received: 30.08.14.).
Url-32: http://www.ecomark.jp/ (date received: 30.08.14.).
Url-33: http://www.ecolabelindex.com/ecolabel/singapore-green-label-scheme-sgls (date received: 30.08.14.).
Url-34: http://www.sec.org.sg/web/ (date received: 30.08.14.).
Url-35: http://www.bmpcotton.com.au/product/ (date received: 30.08.14.).
Url-36: http://www.geca.org.au/ (date received: 30.08.14.).
Url-37: http://www.tei.or.th/greenlabel/ (date received: 30.08.14.).

Url-38: http://www.sepa.gov.cn/ (date received: 30.08.14.).
Werf, H.M.G., 2004. Life cycle analysis of field production of fibre hemp, the effect of production practices on environmental impacts. Euphytica 140, 13–23.
Werf, H.M.G., Turunen, L., 2008. The environmental impacts of the production of hemp and flax textile yarn. Industrial Corps and Products 27, 1–10.
Woolridge, A.C., Ward, G.D., Phillips, P.S., Collins, M., Gandy, S., 2006. Life cycle assessment for reuse/recycling of donated waste textiles compared to use of virgin material: an UK energy saving perspective. Resources, Conservation and Recycling 46, 94–103.
You, S., Cheng, S., Yan, H., 2009. The impact of textile industry on China's environment. International Journal of Fashion Design, Technology and Education 2 (1), 33–43.

Environmental impact assessment methods for textiles and clothing

8

B. Resta, S. Dotti
University of Bergamo, Dalmine, Bergamo, Italy

8.1 Introduction

The textile and clothing industry is one of the world's largest industries, while also one of the most polluting (European Commission, 2013). According to the EIPRO study (Tukker et al., 2006), the textile and clothing sector alone is responsible for 2–10% of the European life cycle environmental impacts, reaching the fourth position in the ranking of product categories that cause the greatest environmental impact, just after food and drinks, transport and housing. The precise environmental impact of textile and clothing goods varies significantly depending on the type of fibre the product is made from (Muthu et al. (2012), Sandin et al. (2013), Shen et al. (2010), and Karthik and Gopalakrishnan (2014)) and the production processes it is subjected to (Beck et al. (2000), Gabarrell et al. (2012), Karthik and Gopalakrishnan (2014), Muthu (2014), Nieminen-Kalliala (2003), and Ren (2000)). However, generally speaking the negative impacts associated with this sector can be summarised into five main categories (DEFRA, 2008; Draper et al., 2007; Fletcher, 2008; Gardetti and Torres, 2013; Gwilt and Rissanen, 2011): (1) energy consumption in the production of primary materials (especially man-made fibres), in yarn manufacturing, in yarn and fabric finishing, in washing and drying clothes, as well as in marketing and sales processes; (2) water consumption associated with raw materials growth, production stage pretreatment chemicals, dyes and finishes, as well as with laundry; (3) use of chemicals (especially in wet pretreatment, dyeing, finishing and laundry) and their release in water; (4) solid waste arising from yarn, fabric and final product manufacturing, packaging and disposal of products at the end of their life; and (5) direct CO_2 emissions, particularly related to transportation processes.

A comprehensive quantification of these aspects was completed in May 2006 by the European Commission's Joint Research Centre (JRC) and included the Environmental Impact of Products (EIPRO) study (Beton et al., 2011), where the potential environmental impacts of textile products consumed in the EU-27 have been estimated along with their life cycle stages and compared considering indicators identified in the ReCiPe method[1] (Goedkoop et al., 2008) (Figure 8.1), also broken down by product and fibre types.

[1] Midpoint indicators: Climate change, Ozone depletion, Photochemical oxidant formation, Particulate matter formation, Ionising radiation, Terrestrial acidification, Human toxicity, Terrestrial ecotoxicity, Freshwater ecotoxicity, Marine ecotoxicity, Metal depletion, Fossil depletion, Water depletion, Freshwater eutrophication, Marine eutrophication, Agricultural land occupation, Urban land occupation and Natural land transformation. Endpoint indicators: Human health, Ecosystem diversity and Resource availability.

Handbook of Life Cycle Assessment (LCA) of Textiles and Clothing. http://dx.doi.org/10.1016/B978-0-08-100169-1.00008-3

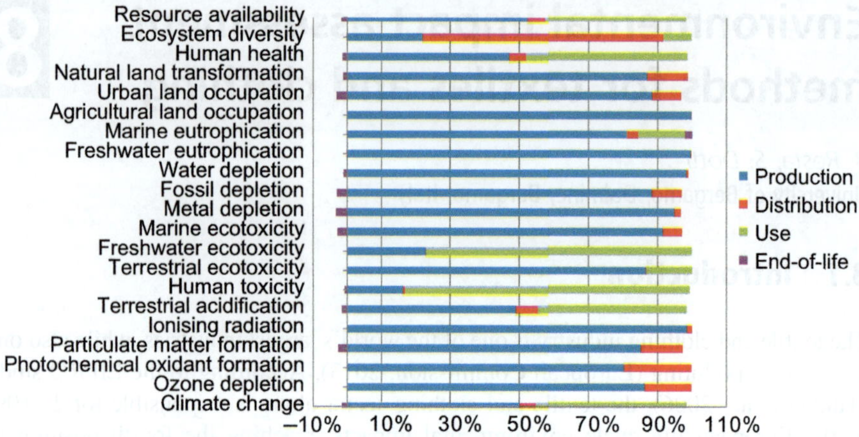

Figure 8.1 Environmental impacts of textile consumption in the EU-27 according to the ReCiPe indicators.
Data source: Beton et al. (2011).

Given the high negative environmental impacts of the textile and clothing sector, the need for assessment instruments supporting policy, plan and project making have emerged throughout the world. In particular, one of the most successful and widely practiced environmental management tools in the world, either through legislation or through regulations and guidelines (Glasson et al., 2013), is the environmental impact assessment (EIA). It is a systematic process designed to identify and predict the potential impacts of human activity on the biophysical and human environment, which originated from the United States' National Environmental Policy Act of 1969 (which became law on January 1, 1971). As discussed by Tromans (2012), the nature of EIA has been well explained by Christopher Wood as encompassing several key elements (Wood, 1995):

> *[…] EIA refers to the evaluation of the effects likely to arise from a major project (or other action) significantly affecting the natural and man-made environment. Consultation and participation are integral to this evaluation.*
>
> *The EIA process should supply decision-makers with an indication of the likely consequences of their actions. Properly used, EIA should lead to informed decisions about potentially significant actions, and to positive benefits to both proponents and population at large.*
>
> *In principle, EIA should lead to the abandonment of environmentally unacceptable actions and to the mitigation to the point of acceptability of the environmental effects of proposals which are approved. EIA is thus an anticipatory, participatory environmental management tool […].*

Over 100 countries now have a formal EIA system, also covering the textile and clothing sector,[2] as well as many nongovernmental organisations (e.g. United Nations Environment Programme, World Bank). However, only a few studies have developed methodologies and tools specifically for the textile and clothing sector (i.e. Patashar et al., 1997); existing environmental assessment methods are mostly too general, and the assumption that all the business are facing a similar environmental problem is rarely appropriate. Consequently, there is a need to develop an industry-specific assessment method to manage its green issues, suitable for its own practices and technologies (You et al., 2009).

Beyond EIA, other methodologies and tools are available for assessing environmental impacts of human activities, as illustrated by Ness et al. (2007) in their framework for sustainability assessment tools. Examples include the system of economic and environmental accounting (SEEA), environmental auditing, life cycle assessment (LCA) and material flow analysis (MFA). As analysed by Finnveden and Moberg (2005), different tools (both analytical and procedural[3]) can be described in relation to a number of different characteristics, among which the most relevant have been demonstrated as the object of the study and the impacts of interest (Figure 8.2). Since different tools focus on different objects, in general, they cannot easily replace each other; for example, LCA is traditionally used for products and EIA is traditionally used for projects.

In this chapter, the use of EIA as a procedural tool to assess the natural resources and environmental impacts at project level for the textile and clothing sector will be investigated.

8.2 Review of EIA process and methods

The system on how the different phases of an EIA process are fulfilled varies in different countries, but the steps are roughly the same no matter where the EIA is conducted (Glasson et al., 2013). In particular, EIA is intended to be an iterative process to follow up to projects post-implementation to determine actual environmental outcomes, interpret and communicate information about those outcomes and investigate opportunities for improved project environmental performance. There is a large component of iteration in the process. The first iterative block consists of a screening to decide if an EIA is needed, a scoping to identify impacts and issues to consider, project and site/environment description and an identification of key impacts. Then, impact prediction, evaluation and reviewing follow.

[2] For example, in Europe textile and leather industries are included in Article 4(2) of Directive 97/11/EC (projects for which the national authorities have to decide whether an EIA is needed).

[3] Procedural tools focus on the procedures and the connections to their societal and decision context, whereas analytical tools focus on technical aspects of the analysis (Wrisberg et al., 2002). Analytical tools can be used within the framework of procedural tools.

Objects \ Impacts	Natural resources	Environmental impacts	Natural resources and environmental impacts	Economic aspects, natural resources and environmental impacts
Policy, plan, programme and project	En, EF, MFA	RA-accidents	SEA and **EIA**	CBA
Region and nation	En, EF, MFA, TMR, DMI, DMC	-	SEEA incl. IOA	SEEA incl. IOA
Organisation	En, EF, MFA	-	EMS with environmental auditing	
Product / function	En, EF, MFA, MIPS	-	LCA	LCC
Substance	En, EF, MFA, SFA	SFA, RA-chemical		

CBA, cost–benefit analysis; DMC, direct material consumption; DMI, direct material input; EF, ecological footprint; EIA, environmental impact assessment; EMS, environmental management system; En, energy analysis; IOA, input–output analysis; LCA, life cycle assessment; LCC, life cycle costing; MFA, material flow accounting; MIPS; material intensity per unit service; RA, risk assessment; SEA, strategic environmental assessment; SEEA; system of economic and environmental accounts; SFA, substance flow analysis; TMR, total material requirement.

Figure 8.2 Environmental assessment tools.
Modified from Finnveden and Moberg (2005).

More specifically, the fundamental components of an EIA would necessarily involve the following stages (Glasson et al., 2013):

- **Screening** is the process of determining whether or not EIA is required for a particular project. It may be partly determined by the EIA regulations operating in a country. For example, in Europe, the EIA Directive (85/337/EEC) has been in force since 1985 and applies to a wide range of defined public and private projects, which are defined in Annexes I (Mandatory EI) and II (Discretion of Member States).
- **Scoping** is a process for identifying environmental impacts of the project. This process, therefore, determines the limits and the scope of the environmental degradation involved with the project, and consequently the cost and time of the assessment study.
- The **consideration of alternatives** aims at ensuring that the proponent has considered other feasible approaches to avoid, mitigate or compensate adverse impacts on biodiversity, including the 'no action' option.
- A **baseline study** is the study of the original status of the environment in the area before the development work of the project is started. This study serves the purpose of a base reference against which the changes due to implementation of the project are measured. Baseline studies are based on the experience with respect to environmental aspects and cover everything important about the environmental impacts of the project.
- During the **identification of the main impacts**, the previous steps are brought together and all potentially significant environmental impacts are identified and taken into account.
- The **prediction of impacts** aims to identify the magnitude and other dimensions (nature, extent/location, timing, duration, reversibility, likelihood and significance) of identified change on the environment caused by the project.
- The **evaluation and assessment of significance** evaluates the relative importance of the predicted impacts to allow a focus on the main adverse impacts.

- **Mitigation** refers to the introduction of measures to avoid, reduce and/or compensate for any significant adverse impact.
- Reporting the **Environmental Impact Statement (EIS)**, or EIA report, includes an environmental management plan (EMP) and a nontechnical summary for the general audience.
- The **review** of the EIS is based on the terms of reference (scoping) and public (including authority) participation.
- **Decision-making** involves the consideration of whether to approve the project or not, and under what conditions.
- **Monitoring** and **environmental auditing** is related to the control of whether the predicted impacts and proposed mitigation measures occur as defined in the EMP. It ensures that unpredicted impacts or failed mitigation measures are identified and addressed in a timely fashion.

All through the process **public consultation and communication** between different parties involved occurs. The key objectives of public involvement are to

- acquire knowledge that may be useful for decision-making;
- assist the evaluation of alternatives and mitigation measures;
- minimise conflicts through the early identification of contentious issues;
- provide a chance for the public to influence project design and increase its confidence in the EIA process;
- improve transparency of decision-making.

Different types of public involvement exist and could be organised as steps of increasing intensity and interaction: informing, consulting, participating and negotiating (Table 8.1). Information and notification are preconditions of meaningful public involvement and, on their own, are not a sufficient provision in public involvement for an EIA of an important proposal. Consultation represents a two-way exchange of information designed to include the views of stakeholders on a proposal and its impacts. Participation is a more interactive process of involving the public in addressing the issues and trying to reach shared positions. Negotiation among stakeholders is an alternative dispute resolution mechanism, based on consensus building and mutual accommodation of different interests.

A wide array of methods have been utilised over the last 40 years to accomplish the various activities required in the conduction of an EIA study (Gough et al., 1998; Ness

Table 8.1 Levels and forms of public involvement

Level	Form of involvement
Informing	One-way flow of information from the proponent to the public.
Consulting	Two-way flow of information between the proponent and the public with opportunities for the public to express feedback on the proposal.
Participating	Interactive exchange between the proponent and the public.
Negotiating	Face-to-face discussion between the proponent and key stakeholders to build consensus and reach a mutually acceptable resolution of issues.

Adapted from Bass et al. (1995).

et al., 2007). Overall, EIA methods should be based on equality, openness, cost-effectiveness and efficiency approach and should also be

- Comprehensive: recognise intricate systems and bound complex interrelationships;
- Selective: pinpoint critical (significant) impacts and eliminate as early as possible unimportant impacts;
- Comparative: determine environmental changes resulting from the project compared by that occur under existing conditions;
- Objective: provide unbiased measurements free from political and external influences.

8.2.1 Methods for impact identification

A number of EIA methods and tools are available for identifying impacts. Some are also useful for scoping and/or presenting the results of the EIA or assigning significance. Methods for impact identification can be positioned on a continuum, from simple to advanced methods, often involving the application of mathematical models.

The simplest approach is a **checklist** of potential impacts that should be considered. Checklists are comprehensive lists of environmental effects and impact indicators designed to stimulate the analyst to think broadly about possible consequences of contemplated actions. These can be of different types:

- *Simple checklists* list the components or aspects, usually of the environment, that might be considered by the assessor, but no other assistance is provided to guide the impact identification process.
- *Descriptive checklists* provide additional assistance by indicating, for example, the specific variables to be measured to characterise each component.
- *Scaling checklists* go a step further and include simple devices for assessing importance or significance of suspected impacts. This might be done through the use of letter or numeric scales, assigned after comparison with criteria supplied in the checklist, to indicate the importance of an impact. Another approach is to use threshold values based on statutory criteria (e.g. for water quality standards) or on derived measures (e.g. visitor carrying capacity for a given locality). The suspected impact can be estimated in broad terms and given a value to represent its significance. On that basis, a start can be made on comparing and ranking alternative project options.
- *Questionnaire checklist* is a form of scaling checklist but uses a series of carefully directed questions to elicit information about possible impacts and their likely importance.

Checklists help to organise the work and identify important issues in the initial stages of assessment. The risk of using checklists is that important issues not included in the checklist may exist. Moreover, checklists cannot represent the interdependence, connectivity or synergism between interacting environmental components.

Leopold et al. (1971), in response to the National Environmental Policy Act of 1969, were the first to suggest using a **matrix** method for EIA. Basically, matrices combine a checklist of environmental conditions likely to be affected with a list of project activities, the two lists arranged in the form of a matrix. The possible cause-and-effect relationships between activities and environmental features are then identified and evaluated cell by cell. Several types of matrices are used in EIA studies:

- *Simple matrix* refers to a display of project actions or activities along one axis, with appropriate environmental factors listed along the other axis of the matrix. When a given action or activity is anticipated to cause a change in an environmental factor, this is noted at the intersection point in the matrix.
- As the simple matrix, the *assessment matrix* has environmental factors in the columns and activities of the project proposal in the rows. In addition, each cell of the matrix is given a rating criteria based on the level of impact of each element of the proposal on the corresponding environmental indicator. The rating criteria are determined by the person undertaking the assessment and are visually represented by a symbol/colour/number in the cell. Further comments can be added to the assessment matrix providing suggestions for alternatives or mitigation proposals. The Leopold Matrix (Leopold et al., 1971) is a particular assessment matrix, where the intersections between rows and columns are filled in to indicate the magnitude (from −10 to +10) and the importance (from 1 to 10) of the impact of each activity on each environmental factor. The number of actions listed on the horizontal axis is 100, while the number of environmental factors (characteristics and conditions) listed on the vertical axis is 88, for a total of 8800 interactions.
- In the *weighted matrices*, relative weights are assigned to each development activity. If the relative priority of development activity is determined, the total value of a particular activity is the sum of the vertical column represented by that in the matrix multiplied by the priority value. Finally, the total value of all the interactions is the sum of all horizontal values in the matrix. This method is particularly helpful in identifying major activities and in defining areas where attention is mostly needed in the process of analysis.

Impact matrices can clearly identify the potential impact areas, predict the impact severity, specify the corresponding mitigation measures and help in identification of agencies responsible for implementing mitigation measures. However, matrices are limited to identifying cause-and-effect linkages (direct impact). On the contrary, a **network diagram** visually describes these linkages, providing some indication of how an ecosystem operates and a guide to identification of second- and third-order effects (indirect impact). The different types of network diagrams are reviewed in Perdicoúlis and Glasson (2006):

- *Digraphs* – or directed graphs – are perhaps the simplest form of causal networks. Their elements are nodes and directional links (unidirectional arrows), with optional additional information marked directly on these elements.
- *Cause-and-effect diagrams* are directed graphs, but their elements are stated textually in various shapes – mostly rectangles. Causal relationships are marked by unidirectional arrows, usually carrying no quantitative information. In general, therefore, cause-and-effect diagrams are graphically more elaborate but less rich in information than the simpler digraphs.
- Unlike the cause-and-effect diagrams, instead of tracing actions and their consequences, *flow diagrams* trace flows of materials and/or energy.
- *Tree diagrams* present several development pathway options simultaneously. Decision trees are used for tracing actions and consequences and outlining the corresponding decision options, while event trees are used for exploring alternative scenarios.

One of the two limitations of causal networks is their difficulty to deal with time and space (Canter, 1999). Their second drawback appears to be their potential risk for increased complexity (UNEP, 2002) beyond the optimum level of simplification (Holling4, 2001). It is suggested that when causal relationships appear too complex, people tend to either simplify in their own way or ignore the causal model altogether (Goldvarg and Johnson-Laird, 2001).

Overlay techniques have a long history of use in environmental planning being ideally suited for consideration of spatial aspects. They rely on a set of maps of environmental characteristics (physical, social, ecological and aesthetic) that are overlaid to produce a composite characterisation of the local environment. A modern version of the overlay method is the computer-based geographical information system (GIS). In simple terms, a GIS stores, retrieves, manipulates and displays environmental data in a spatial format. A set of maps or overlays of a given area provide different types of information and scales of resolution. The main drawbacks are the lack of appropriate data and the expense of creating a usable system.

Quantitative methods attempt to produce a composite index by weighting, standardising and aggregating the importance of all environmental impacts. The environmental evaluation system (EES), devised by the Battelle Columbus Laboratories for the US Bureau of Land Reclamation to assess water resource developments, highways, nuclear power plants and other projects (Dee et al., 1973), is the best-known quantitative method. It is a checklist of 78 environmental, social and economic parameters that may be affected by a project or an activity. It assumes that these parameters can be expressed numerically and that they represent an aspect of environmental quality. For each component, the authors have developed an index of environmental quality, normalised to a scale ranging from 0 to 1 (degraded-high quality), using a value function method. Each impact indicator is then given as the difference in environmental quality between the states with and without action. To enable impacts to be compared directly, each parameter is given an importance weighting, which is then multiplied by the appropriate environmental quality score. A composite score for the beneficial and adverse effects of a single project, or for the net impact of alternative projects, can be obtained by adding up the weighted impact scores. The system also incorporates a warning system based on a series of red flags used to indicate that the value of an impact indicator cannot be estimated because of inadequate data, or that the value of a particular impact indicator is unacceptable, even though the weighted index suggests that the project may be given approval to proceed on environmental grounds.

The most advanced and sophisticated methods for impact identification rely on **expert systems** (Rodriguez-Bachiller and Glasson, 2003), defined as computer systems that emulate the decision-making ability of a human expert (Jackson, 1998). The basic idea behind expert systems is that expertise, which is the vast body of task-specific knowledge, is transferred from a human to a computer and then stored in the computer and users call upon the computer for specific advice as needed (Liao, 2005). Several expert systems have been proposed in the literature (Liu and Lai, 2009). Among them, two categories are noteworthy. The use of analytic hierarchy process (AHP) and its variants is the first group due to their capability for supporting multi-criteria decision-making. The second category exploits the fuzzy logic method to inference the environmental impacts or significances.

In general, the simpler methods for impact identification are easier to use, more reliable and more effective in presenting information in the EIS, but their coverage of impact significance, indirect impacts or alternatives is very limited. More complex models incorporate these aspects, but at the cost of immediacy.

8.2.2 Methods for impact prediction

As for impact identification, several methods are used for predicting the characteristics of impacts.

Mathematical models are used to represent the behaviour of the environment by means of mathematical functions, based upon scientific laws and/or statistical analysis, and are often computer based. The underlying functions can range from simple input—output relationships to more complex dynamic mathematical models, characterised by a large number of interrelationships. Mathematical models can be divided into deterministic and stochastic models. Deterministic models are built on fixed relationships. On the contrary, stochastic models specify 'the degree of probability of the occurrence of a certain event by specifying the statistical probability that a certain number of events will take place in a given area and/or time interval' (Loewenstein, 1966). Statistical models use statistical techniques such as regression or principal components analysis to describe the relationship between data, to test hypotheses or to extrapolate data. Mathematical models can also be divided into empirical or 'black-box' models, where the relationships between inputs and outputs are established from analysis of observations in the environment, and internally descriptive models, where the mathematical relationships within the model are based on some understanding of the mechanisms of the processes occurring in the environment.

Field and laboratory experimental methods utilise existing data inventories to predict impacts. Field tests are carried out in unconfined conditions, typically at the same scale as the predicted impact. Laboratory tests are usually cheaper to run but may not represent all the conditions of a natural system.

Analogue models rely on the judgement of experts, making predictions based on analogous situations. Such models can be developed from site visits, literature searches or the monitoring of similar projects and include

- comparing the impacts of a proposed solution with a similar existing solution
- comparing the environmental conditions at one site with those at similar sites elsewhere
- comparing an unknown environmental impact with a known environmental impact.

Additionally, some of the impact identification methods previously discussed may also be of value in impact prediction (i.e. **checklists**, **assessment matrix** and **networks**).

Generally, the sophistication of prediction methods used should be in proportion to the scope of the EIA and relevant to the importance of the particular impact.

8.2.3 Methods for impact significance evaluation

Once impacts have been predicted, their relative significance needs to be assessed to evaluate whether the impacts may be considered acceptable. Evaluation methods can be of various types, which can be grouped into two main sets of methods: those that assume a common utilitarian ethic with a single evaluation criterion and those that are based on the measurement of personal utilities, including multiple criteria.

The **cost–benefit analysis (CBA)** approach (Hundloe et al., 1990), which seeks to express impacts in monetary units, falls into the first category. It is a method of assessing the consequences of projects and actions, in which the estimated benefits are weighed against the costs. For this purpose, all consequences must be measured in the same unit, and the traditional choice of unit is money. To be explicitly included in a CBA, then, environmental changes must be valued in monetary terms.

Multi-criteria and multi-attribute methods seek to overcome some of the limits of CBA. In particular, they allow for a pluralist view of society, composed of diverse 'stakeholders' with diverse objectives and with different values concerning environmental issues. Most of the methods use some kind of simple scoring (where the expected consequences of each option are assigned a numerical score on a strength of preference scale for each option for each criterion) and weighting (numerical weights are assigned to define, for each criterion, the relative valuations of a shift between the top and bottom of the chosen scale) scheme. An overview of different multi-criteria and multi-attribute methods is included in the manual published by the UK's Department for Communities and Local Government (Dodgson et al., 2009).

8.2.4 Selection of methods

As described in this section, numerous types of methods have been developed and used in the EIA process for projects, plans, programmes and policies. However, no single type of method can be used to satisfy the variety of activities in an EIA study and no single system of methods is currently available.

Table 8.2 reports the 11 types of methods previously presented arrayed against the three main activities that are typically associated with an EIA study. An 'x' in the table indicates that the listed method is or may be directly useful for accomplishing an activity. However, the absence of an 'x' does not mean that the method is not useful for the activity; it merely suggests that it may be indirectly related to the activity.

As previously discussed, such methods present some advantages, but, at the same time, they also suffer from major shortcomings. Attempts to overcome these limits have led to the search for newer techniques and development of newer methods. Thus far, most of the emerging methodologies have not been able to stand the test of extensive field trials and are largely confined to paper. The reasons for this are that most of the newer methods either require information on the precise nature of inter-parameter relationships (which are not available in most cases), or they are computationally or experimentally too demanding and/or costly to be of much practical utility. When carrying out an EIA study, the most appropriate

Table 8.2 Synopsis of EIA methods and study activities

Types of methods in EIA	Impact identification	Impact prediction	Impact significance evaluation
Analogue models		x	x
Checklists (simple, descriptive, questionnaire)	x	x	
Expert systems	x	x	x
Field and laboratory experimental methods		x	
Matrices	x	x	
Mathematical models		x	x
Networks	x	x	
Overlay mapping (GIS)	x	x	
Quantitative methods	x	x	
Cost–benefit analysis (CBA)			x
Multi-criteria and multi-attribute methods			x

perspective is to consider methods as 'tools' that can be selected and modified as appropriate to aid the EIA process. However, there are no specific guidelines supporting the selection and modification of methods mostly appropriate for a given sector (i.e. textile and clothing). As a general rule, simpler methods that are less demanding in terms of data, timing and technical requirements (i.e. assessment matrices) are probably more useful in the EIA process (Canter and Sadler, 1997; Toro et al., 2010).

Therefore, as already highlighted in the chapter's introduction, the key issue involves building a complete system of tools, developed from simple methods, coherently with the specific needs of the textile and clothing sector.

8.3 An EIA tool kit for the textile and clothing industry: the SustainTex system

In order to support textile and clothing companies in effectively undertaking an EIA process, a kit of EIA tools, called SustainTex, has been developed by the authors and is presented herein. Such tools derive from standard EIA methods that have been adapted to the environmental, technological, technical and process characteristics of the textile and clothing sector. From a methodological perspective, two methods have been employed for building the SustainTex system tool kit: field surveys (conducted in several textile and clothing companies) and the consultation of an assessment panel. Field surveys comprised walkthrough surveys, consultations and interviews with company informants (including the managing director, plant manager, shop floor supervisors and workers, and representatives from Corporate Social Responsibility (CSR) Committee, manufacturing, quality and accounting), and site inspections at eight Italian textile and clothing companies. For the identification of the assessment panel, composed by both academic and industrial experts, the guidelines reported in Noble (2004) were followed. The resulting panel was composed of 37 experts covering the entire textile and clothing production chain.

The rest of this section describes the tools in the SustainTex system tool kit, while their use is illustrated in the next section by means of two real cases. In particular, three tools have been developed to support the different EIA process phases:

- the Textile & Clothing Assessment Matrix (TCAM), based on the combination of the Textile & Clothing Inventory Matrix (TCIM) with the Sustainable Textile & Clothing Scorecard (STCS), can be used for the identification of key impacts, and their prediction and evaluation and the assessment of their significance. The STCS can also be used in the post-decision phase for the audit of predictions and mitigation procedures;
- the Textile & Clothing Mitigation Solution Database useful for the identification of mitigation measures; and
- the Mitigation Solution Evaluation Method for the decision-making stage.

The comprehensive SustainTex system is illustrated in Figure 8.3.

8.3.1 The Textile & Clothing Assessment Matrix

In line with a general assessment matrix, the TCAM has activities in the columns and environmental factors in the rows, as well as a rating score in each cell based on the level of impact of each element of the proposal on the corresponding environmental indicator. For the identification of activities, the TCIM was built, while the environmental factors are described in the STCS tool. The rating score is determined by collecting actual inventory data and using kilograms of carbon dioxide (CO_2) equivalent as common measurement unit. The resulting TCAM supports companies for impact identification and prediction phases, as well as for significance evaluation.

8.3.1.1 The Textile & Clothing Inventory Matrix

The TCIM aims at producing an inventory for all the primary and support processes along the textile and clothing production chain. Two approaches were merged for securing this comprehensive inventory: the collection of primary data through field surveys and the use of models from the technical literature. The first task in establishing a process inventory is to list all the macro-processes that should be incorporated in the tool. Twelve macro-processes were considered: spinning, weaving, knitting, desizing, scouring, bleaching, dyeing, printing, finishing, cutting, making and trimming. Then, each macro-process was decomposed into processes and all the inputs (raw materials and auxiliaries, water and energy), outputs (environmental releases in the form of air emissions, water emissions or solid waste) and resources (including tools and equipment that support the processes execution) that relate to each process

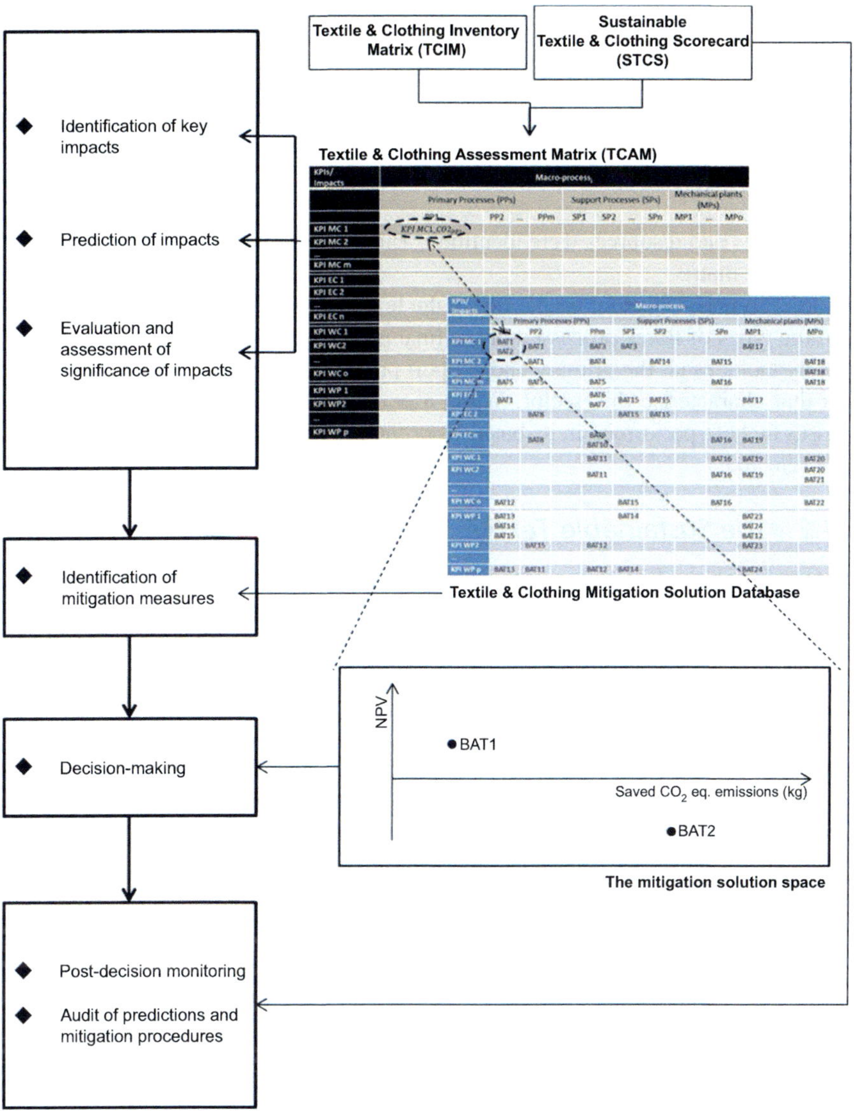

Figure 8.3 The SustainTex system.

were represented through an input/output graph (Figure 8.4), and then coded and listed in a matrix.

Additionally, principal mechanical plants at the service of the manufacturing plant, such as electrical system, water system, heating and cooling system, pneumatic system, lighting system, etc., were considered. As for primary and support processes,

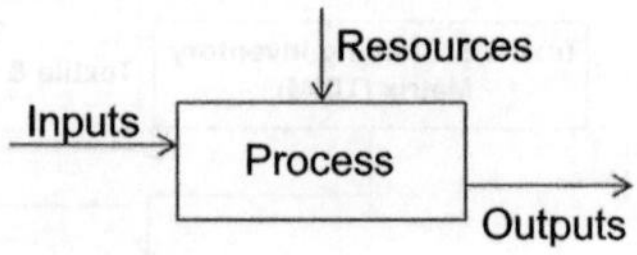

Figure 8.4 Input/output/resource diagram of a process.

inputs, outputs and resources were identified. Table 8.3 shows the general structure of the inventory matrix.

In order to make this tool usable for all the textile and clothing companies, regardless of the types of fibres processed or the final product produced, the TCIM comprehensively includes all the primary and support processes as well as all the mechanical plants that characterise this sector. When a company applies such a tool, only the processes that are performed in its business or the processes that need to be controlled can be selected.

8.3.1.2 The Sustainable Textile & Clothing Scorecard

The second tool, the STCS, includes the significant impacts related to the textile and clothing sector. A first list was compiled by analysing both general (i.e. GRI, 2013; Figge et al., 2002; Epstein and Wisner, 2001; Schaltegger and Wagner, 2006) and textile/clothing-focussed (Nieminen et al., 2007; Nowack et al., 2012; Nieminen-Kalliala, 2003) guidelines and tools. Additionally, the 199 sustainability reports of the textile and apparel sector retrieved from the Sustainability Disclosure Database (as of September 2014) were examined to build the STCS. Finally, impacts included in the Leopold matrix and in the Battelle EES were considered. As a result, a scaling checklist of 30 key performance indicators (KPIs), each one representing a specific environmental impact, was compiled and then included in a structures questionnaire (an extract is shown in Table 8.4) and judged by a panel of experts via the conduction of a formalised Delphi study (Linstone and Turoff, 1978). A Delphi survey is a consensus-based group process for systematically soliciting, collating and refining a set of informed judgements on issues determined by a small number of variables. The technique usually consists of a set of sequential questionnaires. With each subsequent questionnaire, information and feedback from results of earlier questionnaires are provided, allowing a structured dialogue among experts. During the Delphi survey, the panel members were asked to rate each KPI (impact) on a 5-point scale (1 = essential, 2 = important, 3 = do not know/depends, 4 = unimportant, 5 = should not be included) as to whether they believed the impact should be included in the scorecard (Table 8.4). Any comment that was added by panel members, and that was judged by the research group to be an original KPI, was included as a new item to be rated in the second round of the survey.

The survey responses were analysed by obtaining percentages for each item. The following cutoff points were used:

- If at least 80% of the panel members rated a KPI as essential or important, it was included in the scorecard;

Table 8.3 The Textile & Clothing Inventory Matrix (TCIM) – general macro-process

		Macro-process$_j$										
		Primary processes (PPs)				**Support processes (SPs)**				**Mechanical plants (MPs)**		
		PP1	**PP2**	…	**PPm**	**SP1**	**SP2**	…	**SPn**	**MP1**	…	**MPo**
Inputs	I1	x			x				x			
	I2				x		x					
	…											
	Ii		x		x	x				x		x
Outputs	O1		x		x			x		x		x
	O2	x			x	x	x			x		x
	…											
	Oo	x			x				x	x		x
Resources	R1	x	x			x						x
	R2				x							
	…											
	Rr	x			x		x		x	x		

Table 8.4 Scaling checklist of environmental KPIs (impacts) for area 1 and importance scale (Linstone and Turoff, 1978)

Scale Reference	Definitions	Area 1: Material consumption					
1. Very important	A most relevant point First-order priority Has direct bearing on major issues Must be resolved, dealt with or treated	Total material consumption	1	2	3	4	5
2. Important	Is relevant to the issue Second-order priority Significant impact, but not until other items are treated Does not have to be fully resolved	Green material consumption	1	2	3	4	5
3. Moderately important	May be relevant to the issue Third-order priority May have impact May be a determining factor to major issue	Recycled material consumption	1	2	3	4	5
4. Unimportant	Insignificantly relevant Low priority Has little impact Not a determining factor to major issue	Recyclable material consumption	1	2	3	4	5
5. Most unimportant	No priority No relevance No measurable effect Should be dropped as an item to consider	Toxic material consumption	1	2	3	4	5
		Others (specify)					

- If 60–79% of panel members rated a KPI as essential or important, we asked all panel members to re-rate that item.
- Any KPIs that did not meet the above two conditions were excluded.

The resulting STCS (Table 8.5) includes four areas (material consumption, energy consumption, water consumption and waste production) and 18 KPIs (or environmental impacts).

8.3.1.3 The Textile & Clothing Assessment Matrix

By combining the TCIM and the STCS tool, the TCAM was developed in a two-step process. During the first step, each KPI of the STCS was linked to the TCIM indicating the inputs, the outputs and the resources involved in the KPI calculation. In Table 8.6 a general example is reported.

In the second step, the method for rating the significance of the environmental impacts previously defined was defined. The assessment of the significance of environmental aspects can be very subjective when a given expert performs it for a specific organisation. Therefore, an objective approach is required. Specifically, in

Table 8.5 **The sustainable Textile & Clothing Scorecard (STCS)**

Area	#	KPIs/impacts
Material consumption (MC)	KPI MC 1	Total material consumption
		of which:
	KPI MC 2	*Recycled materials*
	KPI MC 3	*Refurbished materials*
	KPI MC 4	*Biodegradable materials*
Energy consumption (EC)	KPI EC 1	Natural gas consumption
	KPI EC 2	Electric energy consumption
		of which:
	KPI EC 3	*Energy consumption from nonrenewable resources*
	KPI EC 4	*Energy consumption from renewable resources*
Water consumption (WC)	KPI WC 1	Total water consumption
		of which:
	KPI WC 2	*Recovered water consumption*
Waste production (WP)	KPI WP 1	Wastewater production
	KPI WP 2	Unsorted waste
	KPI WP 3	Special waste
	KPI WP 4	Waste to urban recycling
		of which:
	KPI WP 5	*Iron*
	KPI WP 6	*Paper*
	KPI WP 7	*Plastic*
	KPI WP 8	Reusable waste

the SustainTex system the scoring system for the prediction and evaluation of impact significance is based on the collection of primary inventory data (material/energy/water consumption and waste production) that are then placed in the TCIM (replacing the ticks in Table 8.3 with inventory data) and used for calculating the KPIs included in the STCS.

Directly comparing the KPI values to define the most critical impact(s) is difficult since the inventory data have widely disparate units (i.e. kilowatt hour, litre and kilogram). Quantifying all the KPIs in amount of CO_2 equivalent, as prescribed by

Table 8.6 Linking TCIM and STCS

Area	KPIs/ Impacts	Inputs	Outputs	Resources			
				R1	R2	…	Rr
Material consumption (MC)	KPI MC 1	I1, I2, I3	–	I1	–		I2, I3
	KPI MC 2	I2, I6, I10	–	–	I2, I6, I10		–
	KPI MC 3	I11	–	I11			–
	KPI MC 4	I1, I4, I5, I7, I8, I9	–	–	–		I1, I4, I5, I7, I8, I9
Energy consumption (EC)	KPI EC 1	I11	–	I11	–		
	KPI EC 2	I13	–	I13	–		–
	KPI EC 3	I13	–	I13	–		–
	KPI EC 4	I12	–	–	I12		–
Water consumption (WC)	KPI WC 1	I15	–	–	–		I15
	KPI WC 2	I14, I15	–	–	–		I14, I15
Waste production (WP)	KPI WP 1	–	O1, O2, O3, O5		O1, O2, O3, O5		
	KPI WP 2	–	O4	O4			O4
	KPI WP 3	–	O1	O1			
	KPI WP 4	–	O3, O4		O3		O4
	KPI WP 5	–	O4, O5	O5			O4
	KPI WP 6	–	O5	O5			
	KPI WP 7	–	O1		O1		
	KPI WP 8	–	O1, O5, O6, O7		O1, O5, O6		O6, O7

a carbon foot-printing approach (Finkbeiner, 2009), is an effective way to put all measures on an equal dimension. This choice was based on the EIPRO study (Beton et al., 2011), where impacts of the textile sector along the different life cycle stages (production, transport, use and end of life) are assessed. Focussing on the production stage, as object of the EIA study, the predominant impact category is 'Climate Change', where all the contributions are expressed in kilogram CO_2 equivalents (Hauschild et al., 2013). The adopted formula is

$$\text{KPI } i\_\text{CO}_{2_p} = \sum_s \text{GWP}_s * Q_s$$

where

- KPI $i\_\text{CO}_{2_p}$ is the KPI indicator i calculated for the process p, indicated in amount of CO_2 equivalent;
- GWP_S is the global warming potential (characterisation factor) for substance s (input or output) calculated in terms of CO_2 equivalent;
- Q_s the quantity of substance s.

$\text{GWP}_s * Q_s$ is then summed up for all the substances (input/output) linked to KPI_i (as defined in Table 8.5) and involved in process p (as defined in Table 8.3). Several databases retrieving characteristic factors for a number of substances are available: CML-IA developed by the Universiteit Leiden (http://cml.leiden.edu/software/data-cmlia.html); ReCiPe created by RIVM, CML, PRé Consultants and Radboud Universiteit Nijmegen (http://www.lcia-recipe.net/home); ELCD developed by the Institute for Environment and Sustainability in the European Commission Joint Research Centre (JRC), in co-operation with the Environment DG (http://eplca.jrc.ec.europa.eu/?page_id=86); ecoinvent Database by ecoinvent, an association founded by ETHZ, EPFL, PSI, Empa and Agroscope (http://www.ecoinvent.org/); GaBi Databases provided by PE International (http://www.gabi-software.com/databases/gabi-databases/); SimaPro Databases developed by PRé Consultants (http://www.pre-sustainability.com/databases).

For each KPI, the impact significance (IS) is calculated as

$$\text{KPI\_IS}_i = \frac{\text{KPI\_CO}_{2_i}}{\sum_i \text{KPI\_CO}_{2_i}} * 100$$

The general logic of the rating process is reported in Table 8.7.

Once impact significances are determined for each process, for each macro-process and for the whole company, considerable information becomes available to provide guidance on how environmental performance may be improved. The information may indicate that one of the processes in the assessment matrix is dominating the environmental impact or that the total impact is largely due to one KPI, suggesting where attention should be focussed for improvement actions. The scores included in the assessment matrix then serve as a baseline against which alternatives may be judged.

Table 8.7 The rating process and the assessment matrix structure

KPIs/ **Impacts**	**Macro-process$_j$**											**KPI_CO_2**	**KPI_IS**
	Primary Processes (pps)				Suppрt Processes (SPs)				Mechanical Plants (MPs)				
	PP1	PP2	…	PPm	SP1	SP2	…	SPn	MP1	…	MPo		
KPI MC1	$KPI\ MC1\_CO_{2_{PP1}}$											$KPI\_CO_{2_{MC1}}$	$KPI\_IS_{MC1} = \frac{KPI\_CO_{2_{MC1}}}{Total\ KPI\_CO_2}\%$
KPI MC 2												$KPI\_CO_{2_{MC2}}$	$KPI\_IS_{MC2} = \frac{KPI\_CO_{2_{MC2}}}{Total\ KPI\_CO_2}\%$
…													
KPI MC m												$KPI\_CO_{2_{MCm}}$	$KPI\_IS_{MCm} = \frac{KPI\_CO_{2_{EC1}}}{Total\ KPI\_CO_2}\%$
KPI EC 1												$KPI\_CO_{2_{EC1}}$	$KPI\_IS_{EC1} = \frac{KPI\_CO_{2_{EC1}}}{Total\ KPI\_CO_2}\%$
KPI EC 2												$KPI\_CO_{2_{EC2}}$	$KPI\_IS_{EC2} = \frac{KPI\_CO_{2_{EC2}}}{Total\ KPI\_CO_2}\%$
…													
KPI EC n												$KPI\_CO_{2_{ECn}}$	$KPI\_IS_{ECn} = \frac{KPI\_CO_{2_{ECn}}}{Total\ KPI\_CO_2}\%$
KPI WC1												$KPI\_CO_{2_{WC1}}$	$KPI\_IS_{WC1} = \frac{KPI\_CO_{2_{WC1}}}{Total\ KPI\_CO_2}\%$
KPI WC2												$KPI\_CO_{2_{WC2}}$	$KPI\_IS_{WC2} = \frac{KPI\_CO_{2_{WC2}}}{Total\ KPI\_CO_2}\%$
…													
KPI WC o												$KPI\_CO_{2_{WCo}}$	$KPI\_IS_{WCo} = \frac{KPI\_CO_{2_{WCo}}}{Total\ KPI\_CO_2}\%$
KPI WP1												$KPI\_CO_{2_{WP1}}$	$KPI\_IS_{WP1} = \frac{KPI\_CO_{2_{WP1}}}{Total\ KPI\_CO_2}\%$
KPI WP2												$KPI\_CO_{2_{WP2}}$	$KPI\_IS_{MC1} = \frac{KPI\_CO_{2_{WP2}}}{Total\ KPI\_CO_2}\%$
…													
KPI WP p												$KPI\_CO_{2_{WPp}}$	$KPI\_IS_{MC1} = \frac{KPI\_CO_{2_{WPp}}}{Total\ KPI\_CO_2}\%$
Total	$Total\ KPI\ MC1\_CO_2$											$Total\ KPICO_2$	100%

In addition, the general structure of the resulting assessment matrix is reported in Table 8.6: the columns represent the textile and clothing processes (divided into primary processes, support processes and mechanical plants); the rows represent the environmental impacts (KPIs) and each cell reports the significance of the specific environmental impact for the specific process. Aggregate ratings are also possible.

8.3.2 The Textile & Clothing Mitigation Solution Database

Once the most critical elements have been identified, mitigation solutions can be proposed. Mitigation is defined in EC Directive 97/11 as 'measures envisaged in order to avoid, reduce and, if possible, remedy significant adverse effects' (Council Directive 97/11/EC, 1997).

Basically, potential solutions could refer to

- technical solutions, such as the installation of a solar plant that allows green production of electricity and/or of hot water for the heating system;
- organisational solutions to drive processes towards sustainability, such as the application of lean manufacturing principles that can be adopted to optimise the workshop efficiency in using inputs and resources;
- material supply and waste disposal, including reuse, recycling and recovery.

With particular reference to the textile and clothing sector, a list of best available techniques (BATs) for supporting the identification of potential technical, organisational and supply/waste disposal solutions is available (European Commission, 2003). BATs are generically defined under the scope of the European IPPC (Integrated Pollution Prevention and Control) Directive (European Commission, 2008, p. 24) as 'the most effective and advanced stage in the development of activities and their methods of operation which indicate the practical suitability of particular techniques for providing in principle the basis for emission limit values (ELV) designed to prevent and, where that is not practicable, generally to reduce emissions and the impact on the environment as a whole'.

The BATs identified for the textile and clothing sector by the European Commission have been included in a database organised in accordance with the TCAM, as represented in Table 8.8. In particular, each BAT has been attributed to one or more cells (KPI-Process) of the matrix, if possible.

Thus, a direct link between criticalities and potential mitigation solutions was established.

8.3.3 The Mitigation Solution Evaluation Method

All the identified potential mitigation solutions are then evaluated to support the decision-making phase. The evaluation phase includes both environmental and economic aspects. First, the saved material/energy/water consumption and avoided waste production are assessed and quantified in amount of CO_2 equivalent. It is important to

Table 8.8 **The Textile & Clothing Mitigation Solution Database**

KPIs/ impacts	Macro-process$_j$										
	Primary processes (PPs)				Support processes (SPs)				Mechanical plants (MPs)		
	PP1	**PP2**	**…**	**PPm**	**SP1**	**SP2**	**…**	**SPn**	**MP1**	**…**	**MPo**
KPI MC 1	BAT1 BAT2	BAT1		BAT3	BAT3				BAT17		
KPI MC 2		BAT1		BAT4		BAT14		BAT15			BAT18
KPI MC 3											BAT18
KPI MC 4	BAT5	BAT5		BAT5				BAT16			BAT18
KPI EC 1	BAT1			BAT6 BAT7	BAT15	BAT15			BAT17		
KPI EC 2		BAT8			BAT15	BAT15					
KPI EC 3											
KPI EC 4		BAT8		BAT9 BAT10				BAT16	BAT19		
KPI WC 1				BAT11				BAT16	BAT19		BAT20
KPI WC 2				BAT11				BAT16	BAT19		

										BAT20 BAT21
KPI WP 1	BAT13 BAT14 BAT15				BAT14				BAT23 BAT24 BAT12		
KPI WP 2		BAT15		BAT12					BAT23		
KPI WP 3	BAT12									BAT12	
KPI WP 4				BAT25	BAT25	BAT26					
KPI WP 5									BAT27, BAT28, BAT29		
KPI WP 6		BAT1, BAT5									
KPI WP 7					BAT3						
KPI WP 8	BAT13	BAT11		BAT12	BAT14				BAT24		

consider a systematic perspective, considering also the negative environmental burden that a solution could bring. For example, a technical solution able to decrease natural gas consumption could require an additional energy absorption. Consequently, for each solution, all the saved CO_2-equivalent emissions (exclusive of additional CO_2 emissions) should be estimated ceteris paribus, considering a constant production volume.

Afterwards, the identified solutions, potentially advantageous to mitigate the environmental impact, are assessed with financial tools to evaluate their investment returns and cash flow impact. In particular, the financial analysis must also capture all relevant and significant environmental costs related to the alternatives, as prescribed by the total cost assessment (TCA) method (Epstein, 1996). According to such method, the environmental costs on which the solution has an impact need to be identified and estimated. TCA is similar to traditional capital budgeting techniques except that it attempts to include all costs and benefits associated with each alternative, including environmental expenditures and savings. In accordance with Curkovic and Sroufe (2007), four tiers of costs are considered (Table 8.9): direct costs, hidden costs, contingent costs and less tangible costs.

Once all the costs (and savings) associated with each solution are identified, financial tools for rating investments, familiar to many businesses, are then used to evaluate the economic added value of each option: net present value (NPV), internal rate of return (IRR) and payback period. Additionally, CO_2-equivalent savings are assessed to identify environmental benefits of each solution.

Economic and environmental added value can then be represented in a mitigation solution space, structured as a Cartesian coordinate plane, for example (saved

Table 8.9 Tiers of costs

Direct costs	**Hidden costs**
Buildings	Regulatory compliance
Equipment installation	Environmental monitoring
Project engineering	Legal support
Material	Sampling and testing
Labour	Education and training
Waste management	Utilities
Contingent liability costs	***Less tangible costs***
Accidental releases	Corporate image
Legal damages	Community goodwill
Settlement for remedial actions	Customer acceptance

From Curkovic and Sroufe (2007).

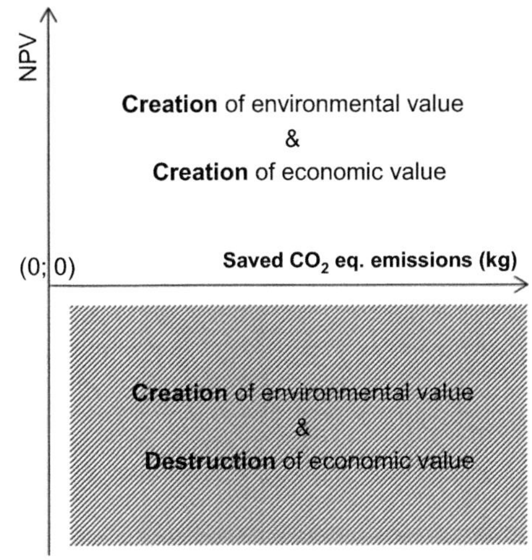

Figure 8.5 The mitigation solution space (saved CO_2-eq. emissions; NPV).

CO_2-eq. emissions; NPV), to identify the solution(s) with a meaningful combination of economic and environmental benefits (Figure 8.5). Since the solutions are selected to decrease the environmental impact, the negative x-axis (expressing additional CO_2-eq. emissions) is not considered in the Cartesian representation.

8.4 Illustrative example

Following are two illustrative case examples.

8.4.1 Case 1: spinning company

DoreTex is a yarn-spinning company that focuses on high-end yarns. While 60% of the company's output is synthetic, the company's core focus is cotton. This is both because the DoreTex production technology is based on the system used for cotton and because the natural fibre, along with linen, hemp, silk, wool and cashmere, is what ensures the company's privileged relationship with the fashion industry. Consequently, the application of the decision-making process was focussed on the cotton division.

Firstly, the scope of the EIA process was limited to the spinning macro-process. Then, the processes and mechanical plants associated with the company's business were selected from the TCIM (Table 8.10): in total 16 processes (12 primary and 4 support), 4 mechanical plants, 25 inputs, 20 outputs and 29 resources have been considered.

For each process/mechanical plant, inputs, outputs and resources were measured. Afterwards the KPIs included in the STCS were calculated and then assessed in terms of CO_2-equivalent emissions, as presented in Table 8.11.

Table 8.10 TCIM for the Doretex spinning macro-process

		Macro-process: spinning																			
		Primary processes (PPs)												Support processes (SPs)				Mechanical plants (MPs)			
		PP1	PP2	PP3	PP4	PP5	PP6	PP7	PP8	PP9	PP10	PP11	PP12	SP1	SP2	SP3	SP4	MP1	MP2	MP3	MP4
Inputs	Ball bearing[a]																X				
	Battery[a]																X				
	Belt[a]																X				
	Bush[a]																X				
	Circular comb[a]																X				
	Clothing fixed flat[a]																X				
	Compressed air		X		X	X			X	X	X	X	X		X						
	Corrugated board														X						
	Cylinder[a]																X				
	Cylinder wiring[a]																X				
	Distilled water																X				
	Electric energy		X	X	X	X	X	X	X	X	X	X	X		X			X	X	X	X
	Filter[a]																X				
	Grease																X				
	Light bulb																				X
	Lubricating oil																X				
	Packaging film														X						
	Pallet														X						
	Paper cone												X								
	Paraffin wax												X								
	Plastic cone												X								
	Rag															X					
	Ring[a]																X				
	Solvent															X					
	Water																	X	X		

Outputs	Bale packaging	X																			
	Ball bearing[b]																X				
	Battery[b]																X				
	Belt[b]																X				
	Bush[b]																X				
	Circular comb[b]																X				
	Cylinder[b]																X				
	Cylinder wiring[b]																X				
	Clothing fixed flat[b]																X				
	Dust			X								X					X	X	X		
	Filter[b]																X				
	Iron		X																		
	Light bulb[b]																				X
	Plastic				X																
	Rag[b]															X	X				
	Ring[b]																X				
	Strapping band	X																			
	Trash		X																		
	Waste fibres				X																
	Wastewater																	X	X		
Resources	Air handling unit																	X			
	Aspiration cage				X	X															
	Bale plucker		X																		
	Blending machine			X																	
	Broom															X					
	Can					X	X		X	X	X										
	Carding machine					X															
	Centrifugal fan																	X			

Continued

Table 8.10 Continued

	Macro-process: spinning																			
	Primary processes (PPs)												Support processes (SPs)				Mechanical plants (MPs)			
	PP1	PP2	PP3	PP4	PP5	PP6	PP7	PP8	PP9	PP10	PP11	PP12	SP1	SP2	SP3	SP4	MP1	MP2	MP3	MP4
Colour scanning cameras				X																
Combing machine								X												
Compressor																			X	
Drawing frame						X			X											
Dust separator		X		X	X															
Forklift truck	X													X						
Lapping frame							X													
Lighting system																				X
Opening and cleaning machine				X																
Packaging machine														X						
Palletising machine														X						
Pressure switch					X															
Pump																	X	X		
Repair tools																X				
Roving frame										X										
Scissors	X																			
Spark detector		X																		
Spinning machine											X									
Static filter system																				
Strapping band cutting machine	X																			
Winding machine												X								

Primary processes: Bale laydown (PP1), Bale plucking (PP2), Blending (PP3), Opening and cleaning (PP4), Carding (PP5), Drawing (PP6), Lapping (PP7), Combing (PP8), Post-combing drawing (PP9), Roving (PP10), Spinning (PP11), Winding (PP12).
Support processes: Picking and handling (SP1), Palletising (SP2), Facility cleaning (SP3), Maintenance (SP4).
Mechanical plants: Conditioning system for roving (MP1), Conditioning system for spinning (MP2), Pneumatic system (MP3), Lighting (MP4).
[a]Spare parts.
[b]Used.

Table 8.11 STCS for the Doretex spinning macro-process

Area	#	KPIs/impacts	CO_2 eq. emissions (kgs)
Material consumption (MC)	KPI MC 1	Total material consumption	320.459.22
		of which:	
	KPI MC 2	*Recycled materials*	–
	KPI MC 3	*Refurbished materials*	–
	KPI MC 4	*Biodegradable materials*	–
Energy consumption (EC)	KPI EC 1	Natural gas consumption	–
	KPI EC 2	Electric energy consumption	2.846.058.18
		of which:	
	KPI EC 3	*Energy consumption from nonrenewable resources*	*2.846.058.18*
	KPI EC 4	*Energy consumption from renewable resources*	–
Water consumption (WC)	KPI WC 1	Total water consumption	508.474.00
		of which:	
	KPI WC 2	*Recovered water consumption*	–
Waste production (WP)	KPI WP 1	Wastewater production	n.a.
	KPI WP 2	Unsorted waste	7.415.20
	KPI WP 3	Special waste	990.05
	KPI WP 4	Waste to urban recycling	257.045.20
		of which:	
	KPI WP 5	*Iron*	*10.192.00*
	KPI WP 6	*Paper*	*21.853.20*
	KPI WP 7	*Plastic*	*22.000.00*
	KPI WP 8	Reusable waste	48.994.29

The 'Energy Consumption' KPI was recognised as the most critical, since it was responsible for the major contribution to climate change. 'Energy' is the only input linked to energy consumption.

As represented in the TCIM (Table 8.10), energy is an input in 11 primary processes (PP2-Bale plucking, PP3-Blending, PP4-Opening and Cleaning, PP5-Carding, PP6-Drawing, PP7-Lapping, PP8-Combing, PP9-Post-combing drawing, PP10-Roving, PP11-Spinning, PP12-Winding), 1 support process (palletising) and 4 mechanical plants (pneumatic system, lighting system, conditioning system for roving and for spinning). Further details of the contribution of each process to 'Energy Consumption' are shown in Table 8.12.

The most critical process is 'Spinning' since it mostly contributes to the CO_2-equivalent emissions of the company related to energy consumption. For the energy consumption of this specific process, the Mitigation Solution Database suggests the installation of a high-efficiency motor on each spinning machine. Technical parameters and costs related to such solution are reported in Table 8.13, while Table 8.17 includes the financial analyses of the investment (with 10% discount rate). The latter refers to the nine spinning machines owned by the company.

The second critical process is the 'Conditioning system (for roving)'. In this case, the installation of an inverter on the centrifugal fan for the air handling unit (AHU) is the proposed solution. Table 8.14 shows technical parameters and costs characterising this technical solution.

In Table 8.17, economic and financial evaluation of the investment are reported (with 10% discount rate).

The third criticality is related to the 'Winding' process. As for the 'Spinning' process, the best solution found in the database is the installation of high-efficiency motors on the winding machines (Table 8.15 for technical parameters and costs, and Table 8.17 for financial analyses of the investment). The latter refers to the two winding machines installed in the factory.

Additionally, a fourth option related to the lighting system was proposed by the experts: the substitution of neon lighting with LED technology (Table 8.16 and Table 8.17).

The four proposed solutions are compared in Table 8.17. The estimated saved energy is also calculated and converted into CO_2 equivalent.

It could be noted that the best solution both from an environmental and economic perspective is the installation of high-efficiency electric motors for spinning machines, also related to the most critical process. The second best solution, economically and environmentally, is the substitution of the neon lighting system with the LED technology (Figure 8.6). With the implementation of these two options it is possible to save 35,403 kg of CO_2 equivalent (3% of the emission deriving from the consumption of electric energy).

8.4.2 Case 2: weaving and finishing company

Itex is a weaving and finishing company that produces high-quality fabrics for clothing applications. Firstly, the scope of the EIA process was limited to three

Table 8.12 **Contribution to energy consumption**

Process/mechanical plant	CO_2 eq. emission (kg)	Contribution to CO_2 eq. emission
Spinning	656.588.83	23%
Conditioning system (for pre-spinning processes)	593.645.79	21%
Winding	348.764.34	12%
Pneumatic system	301.009.01	11%
Conditioning system (for spinning)	296.145.24	10%
Carding	173.724.13	6%
Bale plucking	118.711.49	4%
Lighting system	99.646.96	4%
Roving	84.906.73	3%
Combing	40.328.81	1%
Post-combing drawing	34.970.44	1%
Lapping	33.691.95	1%
Palletising	24.441.71	1%
Opening and cleaning, blending	22.561.57	1%
Drawing	16.921.18	1%
Total	**2.846.058.18**	**100%**

Table 8.13 **Technical parameters and costs of high-efficiency electric motors for spinning machines**

Working hours/year	5000
Energy cost	0.15 €/kWh
Initial investment for motor purchasing	1600 €/motor
Initial investment for motor installation	200 €/motor
'Traditional' motor power	40 kW
High-efficiency motor power	45 kW
Energy saving	2.2%

Table 8.14 Technical parameters and costs of installing an inverter on the centrifugal fan of the conditioning system

Working hours/year	5000
Energy cost	0.15 €/kWh
Energy consumption/hour	73 kWh
Inverter purchasing	3925 €
Electric system upgrade	5000 €
Dampers for centrifugal fan	5000 €
Energy saving	4%

Table 8.15 Technical parameters and costs of high-efficiency electric motors for winding machines

Working hours/year	5280
Energy cost	0.15 €/kWh
Motor power	45 kW
Mean energy consumption/hour	41 kWh
Motor purchasing	21 500 €
Energy saving	1.3%

Table 8.16 Technical parameters and costs of LED lighting technology

Energy cost	0.15 €/kWh
Working hours/year	5000
Neon lamp lifetime	15 840 h
LED lamp lifetime	50 000 h
Neon lamp power	60 W
LED lamp power	30 W
Number of lamps	150
Neon lamp purchasing cost	2.6 €/Neon lamp
Neon lamp power supply and starter purchasing cost	5 €/Neon lamp
LED lighting fixture installation cost	10 €/LED lighting fixture
LED lighting fixture purchasing cost	55 €/LED lighting fixture

Table 8.17 Environmental and economic comparison of the proposed solutions

	Solution#1. Installation of high-efficiency electric motors on spinning machines	Solution#2. Installation of inverter on the centrifugal fan of the conditioning system	Solution#3. Installation of high-efficiency electric motors for winding machines	Solution#4. LED technology for the lighting system
NPV (10 years)	11 357 €	−1809 €	−29.956.80 €	5879 €
IRR	25%	10%	−11%	25%
Payback period	5 years	15 years	>100 years	5 years
Saved CO_2-eq. emissions	22 167 kg	8169 kg	2982 kg	13 236 kg

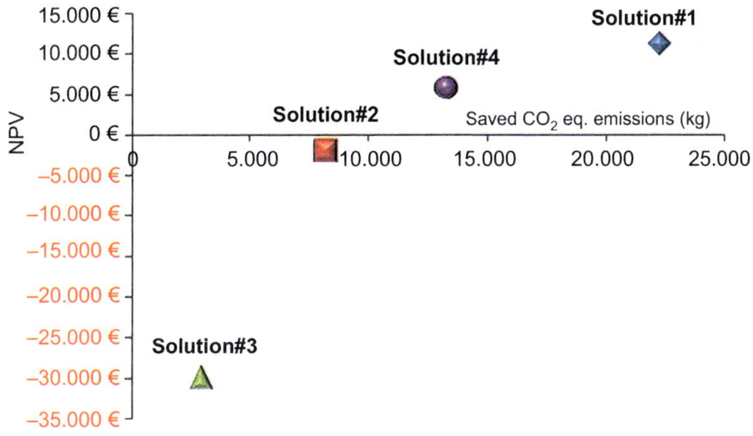

Figure 8.6 Doretex mitigation solution space.

macro-processes: weaving, dyeing and finishing. Afterwards, the processes and mechanical plants associated with the company's business were selected from the TCIM: in total 41 processes (30 primary and 11 support), 8 mechanical plants, 48 inputs, 33 outputs and 63 resources have been considered. The TCIM of the finishing macro-process is reported in Table 8.18.

For each process/mechanical plant, inputs, outputs and resources were measured. Afterwards the KPIs included in the STCS were calculated and then assessed in terms of CO_2-equivalent emissions, as presented in Table 8.19.

Table 8.18 TCIM for the Itex finishing macro-process

Macro-process: finishing															
		Primary processes (PPs)					Support processes (SPs)			Mechanical plants (MPs)					
		PP1	PP2	PP3	PP4	PP5	SP1	SP2	SP3	MP1	MP2	MP3	MP4	MP5	MP6
Inputs	Caustic soda			X											
	Chemicals and additives	X	X	X	X	X			X	X					
	Cleaner						X								
	Compressed air						X								
	Electric energy	X	X	X	X	X	X	X	X	X				X	
	Filter[a]						X								
	Light bulb													X	
	Metal spare part[a]						X								
	Natural gas	X									X	X	X		
	Peroxide		X	X	X	X				X					
	Plastic spare part[a]						X								
	Steam		X	X		X									
	Wastewater														X
	Water	X	X	X	X	X	X			X	X	X	X		
Outputs	Filter[b]						X								
	Metal spare part[b]						X								
	Packaging film[b]	X													
	Paper package[b]								X						
	Plastic package[b]								X						
	Plastic spare part[b]						X								

	Treated wastewater														X
	Waste fabrics	X				X									
	Waste sludge														X
	Wastewater	X	X	X	X		X								
Resources	Bleaching machine		X												
	Container									X					
	Forklift truck					X		X							
	Lighting system													X	
	Mercerising machine			X											
	Rameuse				X										
	Repair and cleaning tools						X								
	Sanfor					X									
	Sewing machine	X	X	X	X	X									
	Singeing tool	X													
	Wastewater treatment plant														X

Primary Processes: Singeing (PP1), Bleaching (PP2), Mercerising (PP3), Drying (PP4), Dimensional stabilisation (PP5).
Support Processes: Cleaning and maintenance (SP1), Picking and handling (SP2), Chemical kitchen (SP3).
Mechanical Plants: Chemical supplying system (MP1), Steam generating system (MP2), Water supply system (MP3), Heating system (MP4), Lighting (MP5), Wastewater treatment (MP6).
[a]Spare parts.
[b]Used.

Table 8.19 STCS for the Itex macro-processes (weaving, dyeing and finishing)

Area	#	KPIs/impacts	CO_2 eq. emissions (kgs)
Material consumption (MC)	KPI MC 1	Total material consumption	786.922.34
		of which:	
	KPI MC 2	*Recycled materials*	*46.500*
	KPI MC 3	*Refurbished materials*	–
	KPI MC 4	*Biodegradable materials*	–
Energy consumption (EC)	KPI EC 1	Natural gas consumption	11.288.710
	KPI EC 2	Electric energy consumption	7.480.446
		of which:	
	KPI EC 3	*Energy consumption from nonrenewable resources*	*2.925.714*
	KPI EC 4	*Energy consumption from renewable resources*	*4.554.732*
Water consumption (WC)	KPI WC 1	Total water consumption	1.686.200
		of which:	
	KPI WC 2	*Recovered water consumption*	–
Waste production (WP)	KPI WP 1	Wastewater production	449.442
	KPI WP 2	Unsorted waste	n.a.
	KPI WP 3	Special waste	n.a.
	KPI WP 4	Waste to urban recycling	2.393.602
		of which:	
	KPI WP 5	*Iron*	*1.714.708*
	KPI WP 6	*Paper*	*204.340*
	KPI WP 7	*Plastic*	*474.554*
	KPI WP 8	Reusable waste	–

The 'Gas Consumption' KPI was recognised as the most critical, since it was responsible for the major contribution to climate change. 'Natural gas' is the only input linked to this KPI.

As described by the TCIM (in Table 8.18 only the section related to the finishing macro-process is reported), natural gas is an input in one primary processes (singeing) and three mechanical plants (steam generating system, water supply system and heating system). Most of the gas is consumed by the steam generating system that provides energy (in the form of thermal energy of the steam) to drive seven primary processes: sizing, dyeing preparation, dyeing, post-dyeing activities, bleaching, mercerising and dimensional stabilisation (Table 8.20).

As indicated in the Mitigation Solution Database, the technical solution potentially able to decrease the steam consumption refers to the installation of rotating heat exchangers to successfully recover heat from heavily polluted water effluents, from both bleaching and mercerising processes. A rotating shaft with stainless steel discs (the heat-exchanging surfaces) creates the necessary turbulence to keep the pollutants of hot discharged effluent water from sticking to the exchanger surface. The resulting unit is a self-cleaning and maintenance-free accessory, making a great tool to recover energy with a guaranteed constant high efficiency, contrary to what happens with all other types of heat exchangers, such as shell-and-tube or plate exchangers, which are day after day strongly affected by the fouling effect. More specifically, the hot discharge water (primary fluid) with both chemical and physical pollutants, coming from a tank or directly from overflows of continuous machines, is introduced with a constant flow rate in the external area through flanged connections. The flow of discharge waters passes through the outside container using the pressure difference generated by gravity and it is flown around the rotor by means of specially shaped deflectors. The fresh clean water (secondary fluid), coming from the main network at a maximum pressure of 3 bar, enters the rotor, which is composed of a series of shell-shaped elements. Within the rotor a canalisation is created so that water circulates in a counterflow pattern against the discharged water. The rotor is activated by a motor reducer; the rotation of the elements inside the water causes a movement that increases the thermal exchange properties and avoids the physical pollutants to deposit on the

Table 8.20 Contribution to natural gas consumption

Process/mechanical plant	CO_2 eq. emission (kg)	Contribution to CO_2 eq. emission
Singeing	564.435.50	11%
Steam generating system	7.111.887.30	68%
Water supply system	790.209.70	7%
Heating system	2.822.177.50	14%
Total	**11.288.710**	**100%**

exchanger walls. The clean water, after having passed through the rotor, exits the rotating joint as heated water. The discharged water, on the opposite side, cools down as it has given its thermal content to the fresh water. A rotating heat exchanger brings advantages compared to a static heat exchanger when it is used with polluted effluents. Its rotation creates turbulence in both the primary and secondary flow, which enhances heat transfer. Rotation also prevents deposits from the dirty effluent since the turbulence due to the rotation creates a centrifugal separation effect that keeps the contaminants away from the exchanger surfaces. Thus maintenance for cleaning is practically nonexistent. Fitting costs and maintenance costs play an important role too: in a traditional plate-and-frame exchanger installation the need to counteract fowling determines the need to add supporting equipment, such as buffer tanks, pumping stations, filtering units and service valves. On the contrary, a rotating heat exchanger can be directly fitted on the process with minimal expenditure and does not need any ancillary equipment.

The installation of one rotating heat exchanger on the bleaching system and of two rotating heat exchangers on the mercerizing system was evaluated from both an economic and environmental perspective.

Technical parameters and costs related to such solutions are confidential data and cannot be reported.

Table 8.21 includes the financial evaluation of the investment (with 10% discount rate).

With the implementation of the three heat exchangers it is possible to achieve significant economic returns, while at the same time saving 736,951 kg of CO_2-equivalent (20% of the emissions deriving from the consumption of natural gas), as depicted in Figure 8.7, and providing empirical evidence against the claim that protection of the environment and economic growth are competing aims.

8.5 Conclusions and future trends

Sustainability is gaining more and more relevance on managers' agendas since it can positively contribute to the firm's value creation process. The benefits are numerous, ranging from cost reduction, through risk management and business innovation, to revenue and brand value growth. In the textile sector, several

Table 8.21 Environmental and economic evaluation of the proposed solution

	NPV (6 years)	Internal return rate	Payback period	Saved CO_2 eq. emissions
Installation of three rotating heat exchangers	342.754€	108%	2 years	736.951 kg

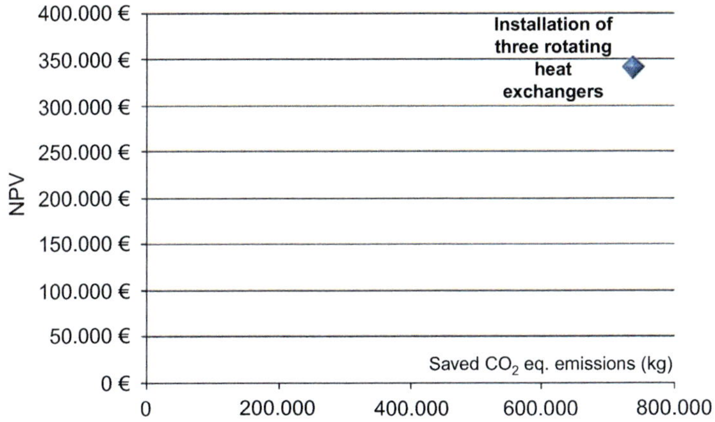

Figure 8.7 Itex mitigation solution space.

companies are starting to pave the way towards sustainability through a number of different approaches. In such a context, EIA represents a structured process for the systematic inclusion of sustainability principles and approaches into strategies, business models and operations, supporting companies in building their own unique roadmap to sustainability.

The application of the SustainTex system in an EIA process offers a set of advantages for textile and clothing companies, including

- identification of the significant environmental aspects;
- enhancement of environmental awareness for companies and their stakeholders;
- assessment of the environmental management concerning each significant environmental aspect;
- provision of a base for the implementation of an environmental management system with potential for certification;
- prioritisation of actions in environmental management.

However, there is still room for research in the development of EIA systems, methods and tools for the effective and efficient implementation of EIA studies in the textile and clothing sectors. In particular, the SustainTex system could be encoded in software to structure the decision-making effort as an expert system and assist with calculations. Such software could be linked both to the existing ERP system and to the metering equipment for directly and instantaneously calculating the KPIs and detecting criticalities. Moreover, the software could also be able to access the mitigation solution database to identify and propose potential solutions based on the criticalities that emerge from the expert system. Additionally, the system boundaries could be enlarged and a supply chain perspective should be introduced, considering also the indirect impacts of an activity. Finally, the third pillar of sustainability (social sustainability) should be addressed and considered in the sustainable scorecard.

References

Bass, S., Dalal-Clayton, B., Pretty, J., 1995. Participation Strategies for Sustainable Development. IIED, London.

Beck, A., Scheringer, M., Hungerbühler, K., 2000. Fate modelling within LCA. The International Journal of Life Cycle Assessment 5 (6), 335–344.

Beton, A., Dias, D., Farrant, L., Gibon, T., le Guern, Y., Desaxce, M., Perwueltz, A., Boufateh, I., 2011. Environmental Improvement Potential of Textiles (IMPRO-Textiles). JRC Scientific and Technical Reports. Retrieved from: http://susproc.jrc.ec.europa.eu/textiles/docs/120423%20IMPRO%20Textiles_Publication%20draft%20v1.pdf.

Canter, L., 1999. Cumulative effects assessment. In: Petts, J. (Ed.), Handbook of Environmental Impact Assessment, vol. 1. Blackwell Science, Oxford.

Canter, L.W., Sadler, B., 1997. A Tool Kit for Effective EIA Practice: Review of Methods and Perspectives on Their Application: A Supplementary Report of the International Study of the Effectiveness of Environmental Assessment. Environmental and Ground Water Institute, University of Oklahoma.

Council Directive 97/11/EC, 1997. Council Directive 97/11/EC of 3 March 1997 amending Directive 85/337/EEC on the assessment of the effects of certain public and private projects on the environment. Official Journal L 073, 5–15.

Curkovic, S., Sroufe, R., 2007. Total quality environmental management and total cost assessment: an exploratory study. International Journal of Production Economics 105 (2), 560–579.

Dee, N., Baker, J., Drobny, N., Duke, K., Whitman, I., Fahringer, D., 1973. An environmental evaluation system for water resource planning. Water Resources Research 9 (3), 523–535.

Department for Environment, Food and Rural Affairs (DEFRA), 2008. Sustainable Clothing Roadmap Briefing Note December 2007: Sustainability Impacts of Clothing and Current Interventions. DEFRA, London.

Dodgson, J.S., Spackman, M., Pearman, A., Phillips, L.D., 2009. Multi-criteria Analysis: A Manual. Department for Communities and Local Government, London.

Draper, S., Murray, V., Weissbrod, I., 2007. Fashioning Sustainability: A Review of the Sustainability Impacts on the Clothing Industry. Forum for the Future, London.

Epstein, M.J., 1996. Improving environmental management with full environmental cost accounting. Environmental quality management 6 (1), 11–22.

Epstein, M.J., Wisner, P.S., 2001. Using a balanced scorecard to implement sustainability. Environmental Quality Management 11 (2), 1–10.

European Commission, 2003. Integrated Pollution Prevention and Control (IPPC). Reference Document on Best Available Techniques for the Textiles Industry. Retrieved from: http://eippcb.jrc.ec.europa.eu/reference/BREF/txt_bref_0703.pdf.

European Commission, 2008. Directive 2008/1/EC of the European parliament and of the council of 15 January 2008 concerning integrated pollution prevention and control. Official Journal of the European Union 8–29. L 24, 29/01/2008.

European Commission, 2013. Sustainability of Textiles. Issue Paper n° 11. Retail Forum for Sustainability. August 2013. Retrieved from: http://ec.europa.eu/environment/industry/retail/pdf/issue_paper_textiles.pdf.

Figge, F., Hahn, T., Schaltegger, S., Wagner, M., 2002. The sustainability balanced scorecard–linking sustainability management to business strategy. Business strategy and the Environment 11 (5), 269–284.

Finkbeiner, M., 2009. Carbon footprinting - opportunities and threats. International Journal of Life Cycle Assessment 14 (2), 91–94.
Finnveden, G., Moberg, Å., 2005. Environmental systems analysis tools–an overview. Journal of Cleaner Production 13 (12), 1165–1173.
Fletcher, K., 2008. Sustainable Fashion and Textiles: Design Journeys. Earthscan, London.
Gabarrell, X., Font, M., Vicent, T., Caminal, G., Sarrà, M., Blánquez, P., 2012. A comparative life cycle assessment of two treatment technologies for the Grey Lanaset G textile dye: biodegradation by Trametes versicolor and granular activated carbon adsorption. The International Journal of Life Cycle Assessment 17 (5), 613–624.
Gardetti, M.A., Torres, A.L., 2013. Introduction. In: Gardetti, M.A., Torres, A.L. (Eds.), Sustainability in Fashion and Textiles: Values, Design, Production and Consumption. Greenleaf Publishing, Sheffield, UK, pp. 1–20.
Glasson, J., Therivel, R., Chadwick, A., 2013. Introduction to Environmental Impact Assessment (The Natural and Built Environment Series), fourth ed. Routledge.
Global Reporting Initiative (GRI), 2013. G4 Sustainability Reporting Guidelines. Retrieved from: https://www.globalreporting.org/resourcelibrary/GRIG4-Part1-Reporting-Principles-and-Standard-Disclosures.pdf.
Goedkoop, M.J., Heijungs, R., Huijbregts, M., De Schryver, A., Struijs, J., Van Zelm, R., 2008. ReCiPe 2008, A Life Cycle Impact Assessment Method Which Comprises Harmonised Category Indicators at the Midpoint and the Endpoint Level, first ed. Report I: Characterisation; 6 January 2009. Retrieved from: http://www.lcia-recipe.net.
Goldvarg, E., Johnson-Laird, P.N., 2001. Naive causality: a mental model theory of causal meaning and reasoning. Cognitive Science 25, 565–610.
Gough, C., Castells, N., Funtowicz, S., 1998. Integrated assessment: an emerging methodology for complex issues. Environmental Modeling and Assessment 3, 19–29.
Gwilt, A., Rissanen, T., 2011. Shaping Sustainable Fashion: Changing the Way We Make and Use Clothes. Earthscan, London.
Hauschild, M.Z., Goedkoop, M., Guinée, J., Heijungs, R., Huijbregts, M., Jolliet, O., Margni, M., De Schryver, A., Humbert, S., Laurent, A., Sala, S., Pant, R., 2013. Identifying best existing practice for characterization modeling in life cycle impact assessment. The International Journal of Life Cycle Assessment 18 (3), 683–697.
Holling, C.S., 2001. Understanding the complexity of economic, ecological, and social systems. Ecosystems 4, 390–405.
Hundloe, T., McDonald, G.T., Ware, J., Wilks, L., 1990. Cost-benefit analysis and environmental impact assessment. Environmental Impact Assessment Review 10 (1), 55–68.
Jackson, P., 1998. Introduction to Expert Systems, third ed. Addison Wesley, Harlow, England.
Karthik, T., Gopalakrishnan, D., 2014. Environmental analysis of textile value chain: an overview. In: Muthu, S.S. (Ed.), Roadmap to Sustainable Textiles and Clothing. Springer, Singapore, pp. 153–188.
Leopold, L.B., Clarke, F.E., Hanshaw, B.B., Balsley, J.R., 1971. A Procedure for Evaluating Environmental Impact. U.S. Geological Survey Circe, Washington.
Liao, S.H., 2005. Expert system methodologies and applications—a decade review from 1995 to 2004. Expert Systems with Applications 28 (1), 93–103.
Linstone, H.A., Turoff, M., 1978. The Delphi Method—Techniques and Applications. Addison-Wesley Publishing Company, Massachusetts Reading, MA.

Liu, K.F., Lai, J.H., 2009. Decision-support for environmental impact assessment: a hybrid approach using fuzzy logic and fuzzy analytic network process. Expert Systems with Applications 36 (3), 5119–5136.

Loewenstein, L.K., 1966. On the nature of analytical models. Urban Studies 3 (2), 112–119.

Muthu, S.S., 2014. Assessing the Environmental Impact of Textiles and the Clothing Supply Chain. Elsevier.

Muthu, S.S., Li, Y., Hu, J.Y., Mok, P.Y., 2012. Quantification of environmental impact and ecological sustainability for textile fibres. Ecological Indicators 13 (1), 66–74.

Ness, B., Urbel-Piirsalu, E., Anderberg, S., Olsson, L., 2007. Categorising tools for sustainability assessment. Ecological Economics 60 (3), 498–508.

Nieminen, E., Linke, M., Tobler, M., Vander Beke, B., 2007. EU COST Action 628: life cycle assessment (LCA) of textile products, eco-efficiency and definition of best available technology (BAT) of textile processing. Journal of Cleaner Production 15 (13–14), 1259–1270.

Nieminen-Kalliala, E., 2003. Environmental indicators of textile products for ISO (type III) environmental product declaration. AUTEX Research Journal 3 (4), 206–218.

Noble, B.F., 2004. Strategic environmental assessment quality assurance: evaluating and improving the consistency of judgments in assessment panels. Environmental Impact Assessment Review 24 (1), 3–25.

Nowack, M., Hoppe, H., Guenther, E., 2012. Review and downscaling of life cycle decision support tools for the procurement of low-value products. The International Journal of Life Cycle Assessment 17 (6), 655–665.

Parashar, A., Paliwal, R., Rambabu, P., 1997. Utility of fuzzy cross-impact simulation in environmental assessment. Environmental Impact Assessment Review 17 (6), 427–447.

Perdicoúlis, A., Glasson, J., 2006. Causal networks in EIA. Environmental Impact Assessment Review 26 (6), 553–569.

Ren, X., 2000. Development of environmental performance indicators for textile process and product. Journal of Cleaner Production 8 (6), 473–481.

Rodriguez-Bachiller, A., Glasson, J., 2003. Expert Systems and Geographic Information Systems for Impact Assessment. CRC Press.

Sandin, G., Peters, G.M., Svanström, M., 2013. Moving down the cause-effect chain of water and land use impacts: an LCA case study of textile fibres. Resources, Conservation and Recycling 73, 104–113.

Schaltegger, S., Wagner, M., 2006. Integrative management of sustainability performance, measurement and reporting. International Journal of Accounting, Auditing and Performance Evaluation 3, 1–19.

Shen, L., Worrell, E., Patel, M.K., 2010. Environmental impact assessment of man-made cellulose fibres. Resources, Conservation and Recycling 55 (2), 260–274.

Toro, J., Requena, I., Zamorano, M., 2010. Environmental impact assessment in Colombia: critical analysis and proposals for improvement. Environmental Impact Assessment Review 30 (4), 247–261.

Tromans, S., 2012. Environmental Impact Assessment, second ed. Bloomsbury Professional, London.

Tukker, A., Huppes, G., Guinée, J., Heijungs, R., De Koning, A., Van Oers, L., Suh, S., Geerken, T., Van Holderbeke, M., Jansen, B., Nielsen, P., 2006. Environmental Impact of Products (EIPRO) Analysis of the Life Cycle Environmental Impacts Related to the Final Consumption of the EU25. Technical Report EUR 22284 EN. European Commission, Joint Research Centre (DG JRC), Institute for Prospective Technological Studies.

United Nations Environment Programme (UNEP), 2002. Environmental Impact Assessment Training Resource Manual. UNEP, Geneva.

Wood, C.M., 1995. Environmental Impact Assessment: A Comparative Review. Longman Higher Education, Harlow, UK.

Wrisberg, N., Udo de Haes, H.A., Triebswetter, U., Eder, P., Clift, R., 2002. Analytical tools for environmental design and management in a systems perspective. In: Tukker, A. (Ed.), Eco-efficiency in Industry and Science. Kluwer Academic Publishers, Dordrecht, Netherlands.

You, S., Cheng, S., Yan, H., 2009. The impact of textile industry on China's environment. International Journal of Fashion Design, Technology and Education 2 (1), 33–43.

Part Two

Applications of LCA in the textile industry

Life cycle assessment of cotton textiles and clothing

9

S. Rana[1], S. Karunamoorthy[2], S. Parveen[1], R. Fangueiro[1]
[1]Fibrous Materials Research Group, School of Engineering, University of Minho, Guimarães, Portugal; [2]PSG Institute of Advanced Studies, Coimbatore, India

9.1 Introduction

Cotton is member of the order Malvales, family Malvaceae and genus *Gossypium*. It consists of more than 50 wild and cultivated species all over the world; but, only four of them are grown in large scale, namely *Gossypium hirsutum*, *Gossypium barbadense*, *Gossypium arboreum* and *Gossypium herbaceum*, accounting for around 95% of the world's cotton production. Extra-long staple cotton fibres like Egyptian cotton, Pima (American and Egyptian) and Sea-island cotton belong to *G. barbadense* and account for 3% of the world's cotton production. The remaining two varieties, namely *G. arboreum* and *G. herbaceum* account for around 2% of the world's cotton production (WWF Report, 2012). Cotton fibre is traded in more than 150 countries as a vital source of foreign exchange, gross domestic product (GDP) and tax income. Among 40 billion farmers worldwide, around 270 million farmers are living by growing cotton. From the above facts, the importance and enormous use of cotton fibre can be understood. Recently, the world has been moving towards sustainability and protecting naturally available resources for successive generations. In this chapter, life cycle assessment (LCA) studies on the cultivation of different types of cotton, manufacturing of apparel and technical textiles, use phase and end-of-life phase are discussed.

9.1.1 Current production scenario

Over the last 60 years, global cotton production has quadrupled from 7 million tonnes in 1950/51 to 27 million tonnes in 2006/07 (WWF Report, 2012). Global cotton production in 2013/14 is estimated as 116.7 million bales and reported as 5% lower than the previous year's production. China, the largest producer of cotton, has been estimated to produce 32 million bales, which is 9% lower than the previous year. The farmers in eastern China have started to adopt attractive and alternative cropping methods such as organic cotton, which provides a lower yield. Therefore, as compared to last year, cotton production in China has declined. Currently, the harvested area for cotton cultivation is around 5.1 million hectares, from which the expected yield is 1380 kg/hectare, which is 4% lower than the previous year (Johnson et al., 2014).

India, the second largest producer of cotton fibre, was estimated to produce around 29 million bales in 2013/14, which is 2% higher than the previous year. India harvested

Handbook of Life Cycle Assessment (LCA) of Textiles and Clothing. http://dx.doi.org/10.1016/B978-0-08-100169-1.00009-5

cotton in an 11.7 million hectare area, which is 2.5% less than the previous year. Since some parts of the land, particularly in Madhya Pradesh, have been shifted to soybean production, the land for cotton vegetation has decreased. Production of cotton fibre per hectare has been estimated to be 540 kg/hectare, which is 4% higher than the last year's production (Johnson et al., 2014).

The United States, the third largest producer of cotton fibre, produced around 13.2 million bales, which is 24% lower compared to last year's production. The reduction is due to the decrease in cotton planted acreage by 15% in 2013 as compared to last year (Johnson et al., 2014).

Pakistan is expected to produce 9.5 million bales from 3 million hectares of harvested area, with a yield of 689 kg/hectare. Brazil, the largest producer in the Southern hemisphere, is expected to produce 7.4 million bales from a 1.1 million hectare area (Johnson et al., 2014).

The world's cotton stock doubled in 2013/14 from 2009/10. The cotton fibre stock has the potential to exceed around 100 million bales. There is a probability for the stock to fall in 2014/15, due to the Chinese government's reduced support towards the increase in stock (Johnson et al., 2014). For the past three years, China has absorbed excess supplies from other countries, which is estimated as approximately 75 million bales. China has built a reserve stock of approximately 58.3 million bales for the 2013 marketing year. In 2014, China announced that they would no longer build their reserve stock. This may lead to reduction in exports to the Chinese cotton market (Adams et al., 2014).

Figures 9.1–9.4 show the details of production and consumption of cotton fibre all over the world for the past 6 years. Figure 9.1 shows the production trend of cotton fibre. Note that the production of fibre increased worldwide from 2008/09 to 2011/12, after which it started to decrease slightly. The reduction in cotton fibre production may be due to the practice of some advanced agricultural methods such as organic cotton, reduction in fertilizer use, etc., which can be seen in Figure 9.4.

Figure 9.2 shows the mill use of cotton fibre for the past few years. It is clear that the consumption of cotton fibre decreased from 2009/10 to 2011/12, and then from 2011 to 2012 it started to increase again (Adams et al., 2014). This may be due to the increase in consumption by the world market.

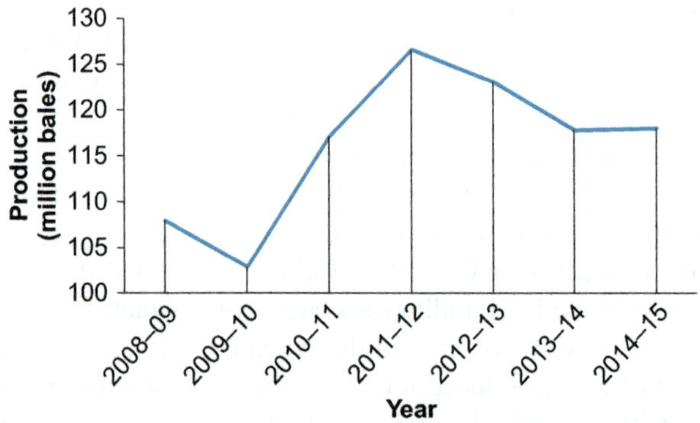

Figure 9.1 Production of cotton fibre worldwide in million bales (Adams et al., 2014).

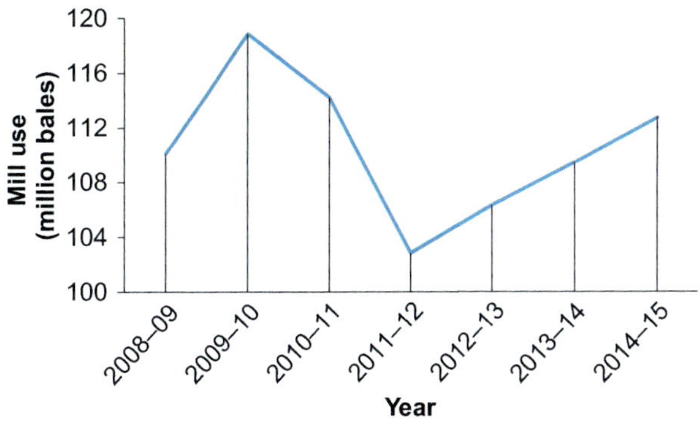

Figure 9.2 Mill use of cotton fibre worldwide in million bales (Adams et al., 2014).

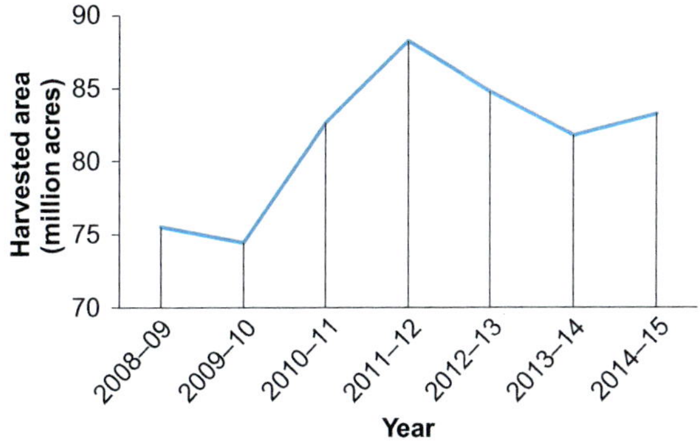

Figure 9.3 Harvested area of cotton fibre worldwide in million acres (Adams et al., 2014).

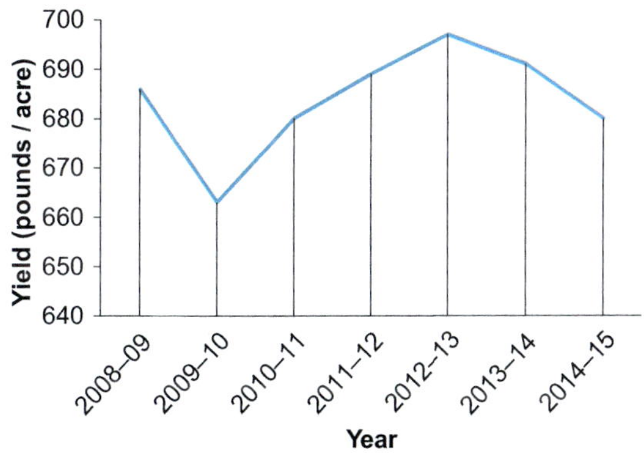

Figure 9.4 Yield of cotton fibre worldwide in pounds/acre (Adams et al., 2014).

Bangladesh, Indonesia, Turkey and Vietnam together accounted for the increase in imports to 30% in 2012/13 due to increase in their consumption (ICTSD, 2013).

The second largest exporter of cotton, India, also reduced its exports from 19% in 2007/08 to around 10% in 2012/13 due to increased consumption in the domestic market. Similarly, in Central Asia, exports are down to 10% in 2012/13 due to decreased cotton production and increased consumption (ICTSD, 2013).

Figure 9.3 shows the area used for cotton harvesting over the past few years. It can be observed that the harvesting area continuously increased until 2011/12 but after that it starts to decrease slightly (Adams et al., 2014). This is due to the fact that some parts of the land used for cotton harvesting have been used for other vegetation such as soybeans, etc. as mentioned earlier (Johnson et al., 2014).

Figure 9.4 shows the yield of cotton fibre worldwide. It can be noted from the figure that the yield of cotton fibre starts to decrease after 2012/13 (Adams et al., 2014) due to the adaptation of some different methods of cultivation such as organic cotton in many parts of the world. This decrease was also attributed to environmental factors such as rainfall, temperature, etc.

9.1.2 Application areas

Cotton is generally called the 'King of fibres' and has a wide variety of applications in general apparel and technical textiles. In apparel, cotton fibre is used to produce shirts, T-shirts, inner wear, trousers, jeans, etc. In most of the apparel products, cotton fibres are used alone or in blends with other fibres such as polyester, viscose, etc. In apparel purposes, cotton fibre is preferred because of its higher thermal and tactile comfort, which is suitable for a range of weather conditions.

With the advances in technological developments, cotton fibre also plays a vital role in technical textiles in some of the applications. Cotton fibres with a length of greater than 12 mm can be used for producing nonwovens. Such fibres can be obtained from card waste, comber noil, etc. Nonwovens made up of cotton fibre are widely used in personal hygiene, wipes and medical textile applications. In personal hygiene applications, baby diapers, feminine hygiene products, adult incontinence products, nursing pads, disposable underwear, etc. are some of the important products. Two different categories of wipes are mainly produced from cotton fibres, namely industrial wipes and surgical wipes. In medical textiles, many products are made up of cotton fibres. Disposable caps, gowns, masks, shoe covers, drapes, sponges, dressings, bed linens, contamination control gowns, examination gowns, etc. are some of the important cotton products used in medical applications (Rupp, 2009).

9.1.3 Sustainability aspects

The United Nations defines *sustainable development* as development that meets the needs of the present, without compromising the ability of future generations to meet their own needs. Basically, this definition is supported by three main pillars of sustainable development. They are:

- Economic sustainability
- Environmental protection
- Social security

The quadrupled increase in production of cotton from the 1950s to 2013 was achieved with the help of technological developments in the field of agriculture, mechanical and chemical engineering, etc. Use of such advancements also led to many environmental impacts. Some of these nonsustainable practices include (WWF Report, 2012):

- Indiscriminate use of pesticides and fertilizers
- Extensive use of irrigated water, with no regard to water quantity and quality
- Use of applications that contribute to soil erosion
- An unbalanced (quantity, time of use) use of resources in some areas

In the present scenario, sustainability is one of the world's key issues. For all products, either newly developed or existing, sustainability has to be evaluated. LCA has been used as a tool to evaluate the environmental impacts and sustainability of various products including cotton textiles.

9.1.4 Importance of LCA

Cotton has been reported as one of the most water- and pesticide-intensive crops (Chapagain et al., 2005). Use of pesticides affects both humans and wildlife. In the United States, the damage to society due to the use of pesticides was found to be 9.6 billion USD per year (Pimental, 2005). Around 53% of the cotton fields are irrigated worldwide and from that 73% of cotton fibre is grown. Irrigated cotton is mostly grown in warm regions of the world where freshwater supply is scarce. Irrigated cotton is mainly located in Egypt, Uzbekistan and Pakistan. In Pakistan, around 31% of irrigation water is taken from ground water (Chapagain et al., 2005). Moreover, the manufacturing and use of cotton products involves significant use of nonrenewable energy and also has different environmental impacts. Therefore, LCA studies on cotton fibre and textiles are extremely important to understand the environmental impacts of different practices followed in cultivation, harvesting and other processes used for manufacturing and use of cotton products, so that new strategies can be developed to reduce environmental impacts and enhance sustainability of cotton textiles.

9.1.5 Methodology of LCA

The International Organization for Standardization (ISO) defines LCA as 'the compilation and evaluation of the inputs, outputs and potential environmental impacts of a product system throughout its life cycle'. The United States Environmental Protection Agency (EPA) defined LCA to be composed of four separate but interrelated components (Smith and Barker, 1995):

1. Goal and scope
2. Life cycle inventory
3. Life cycle impact analysis
4. Interpretation

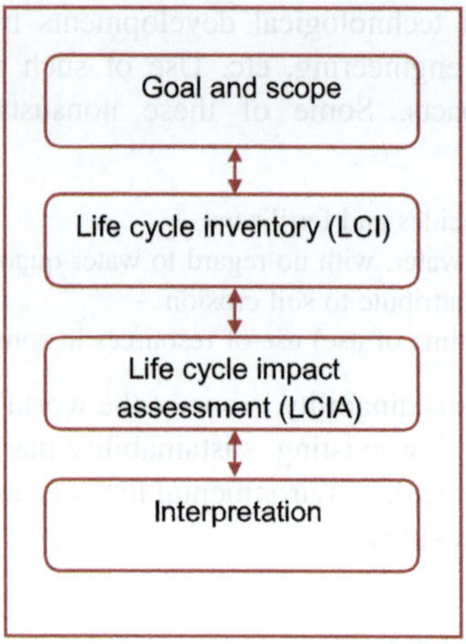

Figure 9.5 Diagrammatic view of LCA model (Life cycle assessment, 2006).

The diagrammatic representation of the different phases of LCA and their interaction with each other are shown in Figure 9.5.

The researchers generally follow one of two different methods available for LCA (Tukker, 2004). The first method is the cradle-to-grave method. The second method is the cradle-to-factory gate or cradle-to-gate method. In the former method, the entire life of the product is considered and this includes the following processes:

- Cultivation of cotton (preparation of the land, seeding, fertilizer and pesticides, watering, harvesting, etc.)
- Ginning, baling, transportation
- Spinning of yarns (from godown to post-spinning operations)
- Conversion of material from yarn to fabric (weaving or knitting or braiding)
- Nonwoven processes (if the fibres are not spun as nonwovens)
- Chemical processing (scouring, bleaching, mercerization, dyeing, printing and finishing)
- Conversion of fabric to garment (pattern making, cutting, sewing, pressing, etc.)
- Inspection, packing, transportation, distribution
- Use of the product by consumer (washing, drying, ironing and use)
- Disposal of the product

Figure 9.6 shows the simplified life cycle of cotton yarn.

In the second method, the steps until the product reaches a certain stage of the whole process or part of the process are considered, for example, perhaps until the cultivation or ginning or yarn production or until the process of inspection and packing.

In both methods, impact on the environment is studied in many aspects. The term *environment* can be defined as the circumstances surrounding an organism or group of organisms. Connell (1995) categorized environment into four different stages:

1. Immediate environment
2. Local environment

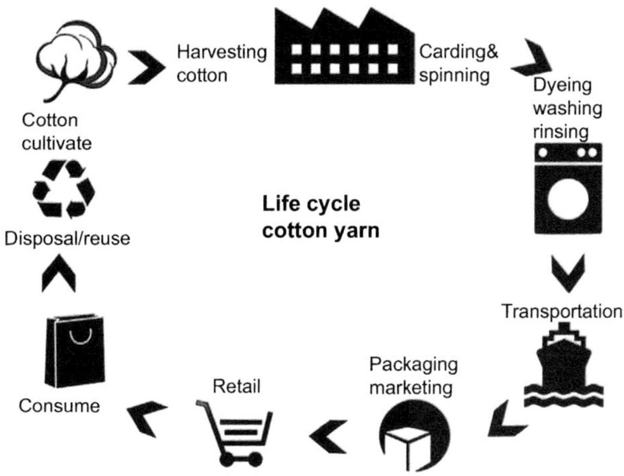

Figure 9.6 Different phases of LCA model of cotton yarn based on cradle-to-grave approach. http://buddhajeans.com/dictionary/life-cycle-assessment-cotton-yarn-lca/.

3. Regional environment
4. Global environment

In most of the studies, many of the following environmental parameters are considered for evaluation:

- Climate change
- Ozone layer depletion
- Terrestrial acidification
- Human toxicity
- Terrestrial ecotoxicity
- Freshwater ecotoxicity
- Freshwater eutrophication
- Agricultural land occupation

9.2 LCA of cotton textiles

Textile products made from cotton fibre are used extensively all over the world. Numerous cotton products are being manufactured and used every day. Therefore, it is extremely important to understand the impacts of these products on environment and sustainability. In the following sections, various LCA studies conducted on cotton fibre, yarn and fabrics are discussed.

9.2.1 *Fibre and yarn*

Murugesh Babu and Selvadass (2013) conducted a study to analyse the impacts of conventional and organic cotton fibres. In this study, the cradle-to-gate approach was used and the functional unit was defined as the production of 1 kg of cotton fibres from both conventional and organic cotton seeds. Various stages of cotton fibre growth, that is, from cultivation to harvesting, fertilizers and pesticide use and their production and transportation, irrigation and electricity consumption were considered. It was found that organic seed cotton fibre has a lower environmental impact compared to conventional seed cotton fibre. This is due to the practices followed in organic farming such as promotion of natural products for pest control, spraying of neem and mass trapping, etc. It was also reported that the majority of the pesticides that are used for conventional cotton are classified as hazardous (Murugesh Babu and Selvadass, 2013) and the use of these pesticides results in higher environmental impacts.

Further, Beton et al. (2014) performed LCA and compared three varieties of cotton, namely conventional cotton, organic cotton and genetically modified cotton. Figure 9.7 shows relative ratio of yield, use of pesticides and fertilizers for organic and genetically modified cotton with respect to the conventional cotton. Even though the yield of organic cotton was slightly lower than that of conventional cotton, the eutrophication and terrestrial ecotoxicity impacts were significantly lower. Organic cotton reduced the freshwater and marine eutrophication by 12 and 15%, respectively, as compared to conventional cotton. In addition, it reduced 75% of terrestrial ecotoxicity owing to the absence of pesticides in organic farming. From Figure 9.14, it can be seen that the yield of organic cotton was slightly lower and the yield of genetically modified cotton was slightly higher as compared to conventional cotton. The difference in the yield was due to the difference in farming practices such as absence of pesticides in the case of organic cotton.

Even though the genetically modified cotton such as Bt cotton has attracted farmers, it has not been much preferred by environmentalists. Although it reduces the use of pesticides and other pollutants, the release of genes may be harmful to the ecosystem and this is objectionable due to the uncertainty of the risk (Jan, 2009).

Cotton made in Africa, called CmiA cotton, has been reported to have much lower environmental effects as compared to other varieties of cotton. LCA studies reported

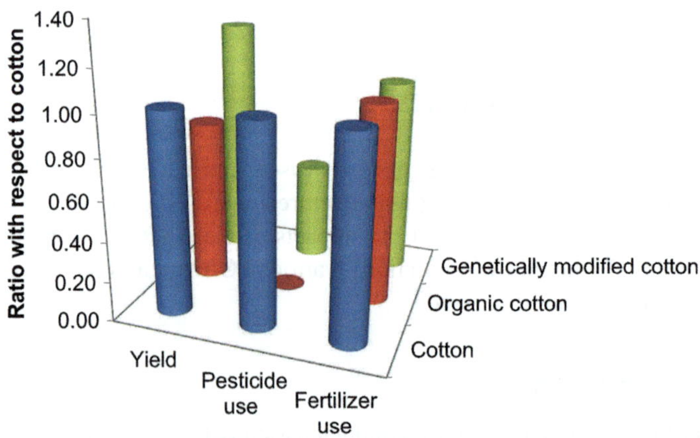

Figure 9.7 Comparison of three different types of cotton (Beton et al., 2014).

much lower emission of greenhouse gases from CmiA cotton than other types of cultivated cotton. It emits only 1.04 kg CO_2 equivalent as compared to 1.8 kg CO_2 equivalent in the case of conventional cotton (CmiA, 2014). This has been possible by effectively utilizing the resources and rain-fed agriculture.

When comparing the environmental impacts of conventional cotton fibre with other commercially available textile fibres such as polyester, viscose, etc., Chapman concluded from the LCA study that cotton fibre consumes less than half of the energy (49 MJ/kg) required for the production of polyester (109 MG/kg) and much lower energy than the energy required for the production of viscose (71 MJ/kg) (Chapman, 2010). On the other hand, water consumption in case of cotton was found to be very high due to the cultivation process. Transport of cotton fibre also has had a large environmental impact, next to energy consumption for washing and drying of cloths. The highest toxicity impact was found due to cotton cultivation owing to use of chemicals.

Similarly, global warming potential due to cotton fibre production was compared with other textile fibres by Shen and Patel (2010), as presented in Figure 9.8. It is clear that cotton has the lowest global warming potential (GWP) among the four commercially available fibres shown in the graph.

The impacts of cotton yarn production by spinning mills were studied by Emmenegger et al. (2013). For this purpose, questionnaires were provided and data were collected from the spinning mills and the entire supply chain of the product. The received data were verified and processed into the database for modelling. The results were normalized and converted into functional units. Then the effects were analysed based on various environmental parameters such as GWP, water depletion, etc. Among all these factors, electricity and transports of spinning mill were identified as the most important factors affecting almost all environmental impact categories significantly. Packaging and auxiliaries were reported to affect not more than 17% of the impact in all indicators. GWP of the studied spinning mills was 2.46 kg CO_2 eq/kg of cotton and 1.25 CO_2 eq/kg of cotton for Asia and Europe, respectively. This difference

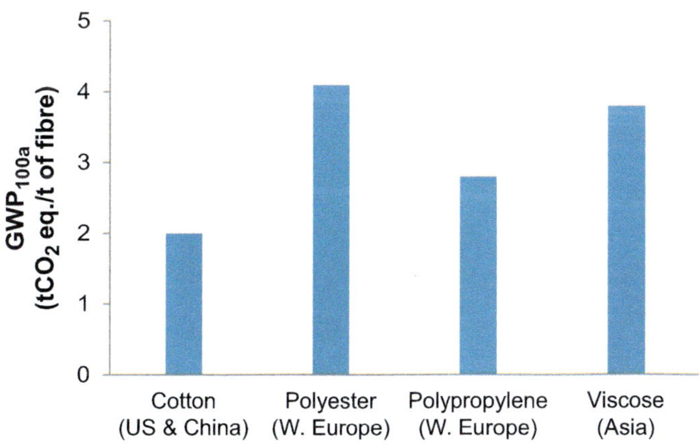

Figure 9.8 Global warming potential of different types of fibres (Shen and Patel, 2010).

between two places was attributed to the difference in power generation system, cultivation method, process sequence, chemicals used, etc. (Emmenegger et al., 2013).

9.2.2 Cotton fabrics for apparel

Several LCA studies have been conducted on cotton fabrics used in the apparel sectors. An LCA study was conducted by Cotton Incorporated and PE International to investigate the impacts of cotton knitwear and woven garments. In this study, a cradle-to-grave approach was followed. Three different phases of a cotton garment were modelled: agricultural phase, textile manufacturing phase and consumer phase including disposal. This study was conducted by selecting four regions of the world, namely Turkey, China, India and Latin America, which together account for 66% and 51% of world's total knitwear and woven fabric production, respectively. The following figures (Figures 9.9–9.13) show the different environmental effects of 1000 tonnes of cotton knitwear and woven pants.

Among almost all environmental impacts, consumer use has a higher effect than the agriculture and manufacturing phases. Between the knitwear and woven fabrics, on average, the latter has a higher environmental impact, particularly in the consumer phase (Executive summary, 2012).

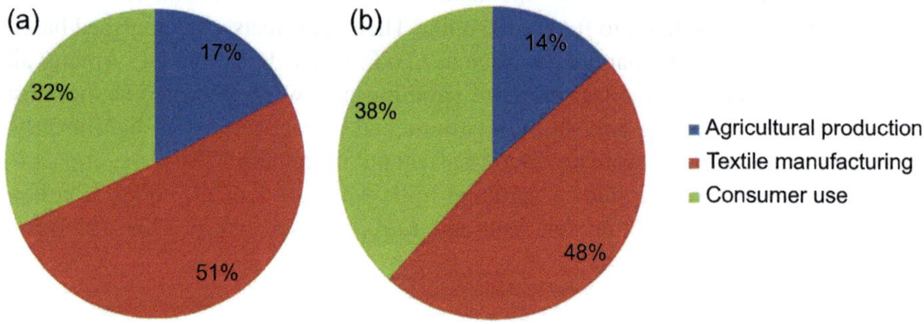

Figure 9.9 Acidification potential for 1000 kg of cotton fabric: (a) knitwear (b) woven pants.

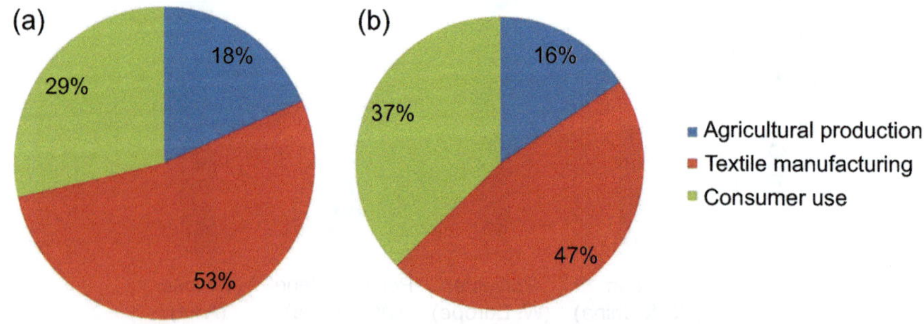

Figure 9.10 Eutrophication potential for 1000 kg of cotton fabric: (a) knitwear (b) woven pants.

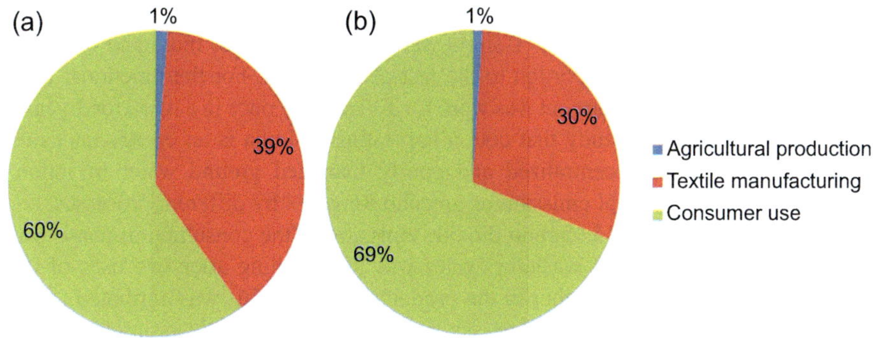

Figure 9.11 Global warming potential for 1000 kg of cotton fabric: (a) knitwear (b) woven pants.

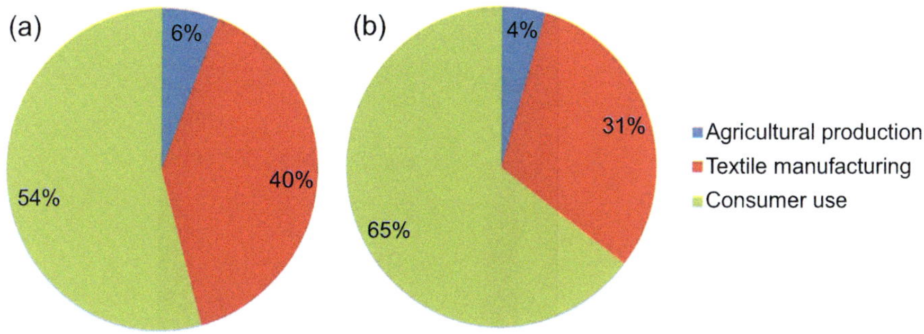

Figure 9.12 Energy demand by consumer for 1000 kg of cotton fabric: (a) knitwear (b) woven pants.

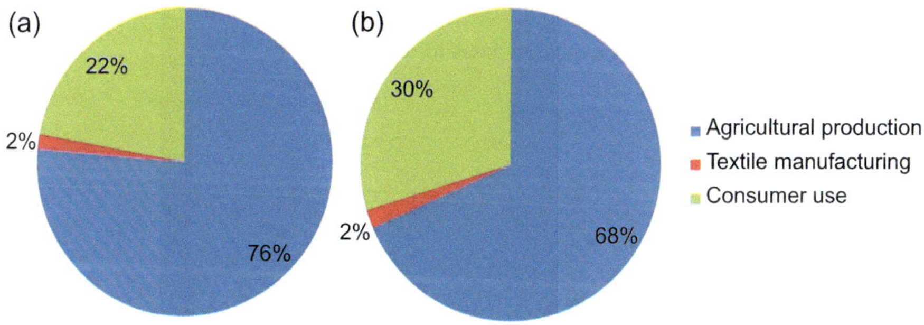

Figure 9.13 Water consumption per 1000 kg of cotton fabric: (a) knitwear (b) woven pants.

The consumer phase shares more than 50% in global warming potential and energy demand (Figure 9.12). This is due to higher energy consumption and frequent laundering of the fabrics by the consumers. Among all three phases, the agricultural phase has shown lower environmental impact except consumption of water (Executive summary, 2012).

Steinberger et al. conducted a case study on cotton T-shirts and polyester jackets (Steinberger et al., 2009). Cotton T-shirts were manufactured in India and exported to Germany and a small proportion to the rest of the world. For the functional unit, it was considered that the garment was worn for 100 days or once in a week for 2 years. It was concluded in this study that cotton harvesting in India is an energy-intensive process owing to the decentralized and poorly managed ground water irrigation. Figure 9.14 shows the total emission of greenhouse gases by different processes.

As shown in Figure 9.15, carbon dioxide emission in the consumption phase was higher since the number of washing cycles was 50, washing after two uses of the T-shirt. NO_x emission was higher in the production stage. This was attributed to the emissions from the agricultural processes and to the use of fertilizers in cultivation. There were also chances of emission in the dyeing process due to burning of wood for heat on a large scale.

Nelson et al. have studied LCA of T-shirts produced from cotton fibre (Nielson and Nielson, 2009). In this study, they have used two different approaches, namely

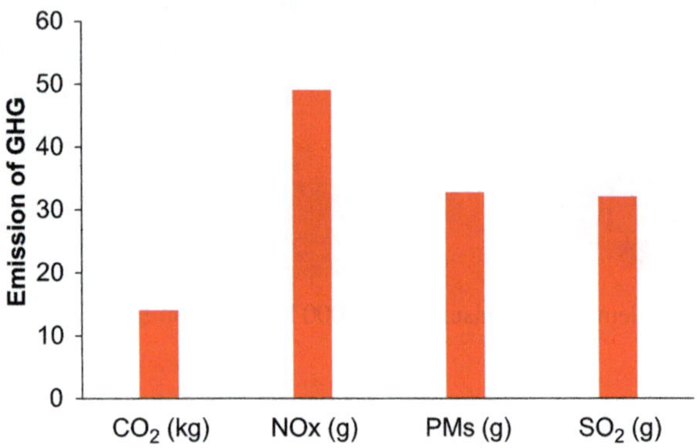

Figure 9.14 Total emission of greenhouse gases for cotton T-shirt.

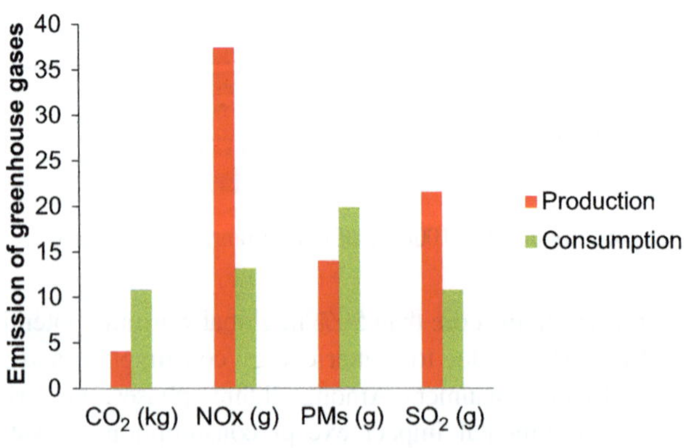

Figure 9.15 Emission of greenhouse gases during production and consumption stages.

conventional dyeing and treatment using biotechnology for two different cotton products: navy T-shirt and blue T-shirt. In conventional treatment of the navy T-shirt, the processing sequence was scouring, dyeing, soaping and bio-polishing. In the bioprocess of navy T-shirt, the process sequence was bio-scouring, combined dyeing and bio-polishing and enzymatic rinse process. In conventional treatment of blue T-shirt, the process sequence was bleaching, peroxide removal, dyeing, soaping and bio-polishing. In the bio-process of blue T-shirt, bio-bleaching, combined dyeing and bio-polishing and enzymatic rinse process were performed. The study was conducted by treating 40 kg of textile material in the factory. They followed the cradle-to-grave approach and evaluated the environmental impact using the following formula,

$$EP(j)_i = Q_i \times EF(j)_i$$

where,

Q_i is the emitted quantity of substance i
j is the environmental impact (e.g. global warming)
i is CO_2, CO, CH_4, N_2O, NO_3, PO_4, SO_2, etc.
EF is an equivalency factor

The outcome of their research shows a net reduction in environmental impact in the case of bio-processing for both navy and blue T-shirts. The net reduction in the global warming was around 1400 and 1100 kg CO_2 eq per tonne for navy and blue T-shirts, respectively. The reduction in global warming was attributed to a much lower contribution of chemical production in the case of bio-processing. Similarly, the reduction in freshwater use for navy and blue T-shirts was 99 and 72 m^3, respectively. Reduction in several other parameters such as acidification, eutrophication, photochemical smog formation, fossil energy, etc. was also reported because of enzyme use in the bio-processing of cotton T-shirts.

In another study, two identical shirts made of cotton and linen fibre were studied by LCA. It was ensured in this study that both the fibres consumed around 80% water and primary energy in use phase. Among these, the cotton shirt (83 kWh) was found to consume much lower primary energy as compared to the linen shirt (100 kWh) due to less ironing time. On the other hand, the linen shirt was reported to consume only one-quarter of the water (6.4 L) that was used for the cotton shirt (24 L) (Chapman, 2010).

9.3 LCA of technical cotton textiles

Kalliala and Nousiainen (1999) have studied LCA of hotel textiles, which include bed sheets, towels, etc. They have compared the life cycle of bed sheets made of 100% cotton and 50/50 polyester/cotton blend. They have also compared different products like double bed sheets, organic bleached material, etc. The following figures (Figures 9.16–9.18) show the energy consumption and environmental impacts of different types of products used in hotel textiles. It was observed that production of polyester

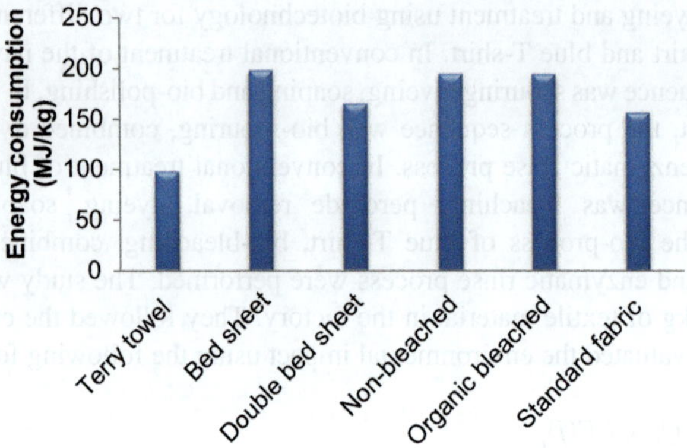

Figure 9.16 Energy consumption of different types of fabrics.

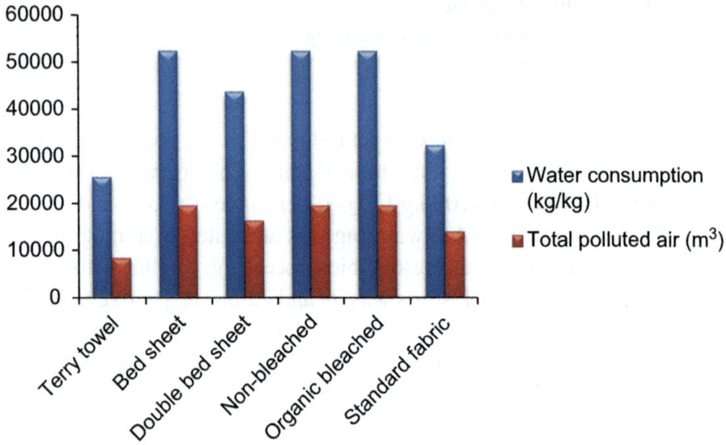

Figure 9.17 Water consumption and air pollution of different types of fabrics.

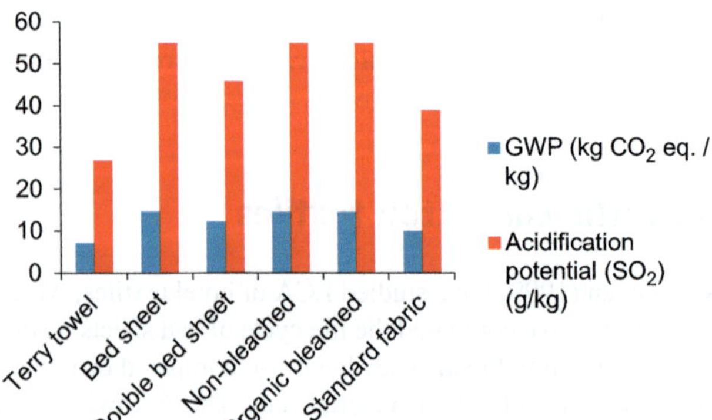

Figure 9.18 Environmental impact of different types of fabrics.

consumed 63% higher primary energy than the required energy for cotton production due to the manufacturing of polyester from nonrenewable resources such as fossil raw materials. Cotton growing consumed 1.65 kg and emitted around 4.7 kg CO_2 per kg of cotton fibre, and, therefore, net 3 kg of CO_2 was emitted per kg of cotton fibre. Moreover, this study revealed that the 100% cotton fabrics consumed 20% higher energy in the laundering process than 50/50 cotton/polyester fabrics.

In their study, it was considered that the lifetime of the textiles except the terry towel was high. The life of the terry towel was considered, on an average, to be 44% of the lifetime of bed sheets since the former showed a high tendency to heavy soiling and higher stealth rate and as a consequence, their impact on the environment was lower.

The use of cotton in diapers started before 1936 with cotton ribbed towelling or muslin cloth. Recently, disposable diapers and also absorbent core materials are partially made of cotton or viscose or a mixture of these fibres (Sustainability Report, 2007–2008). Meseldzija et al. (2013) analysed the environmental effects of disposable and reusable or cloth diapers. They have studied the effect of these two diapers on natural resource consumption, air and water pollution and waste disposal. It was reported in this study that 95% of parents in Canada prefer the disposable diapers than cloth diapers. This leads to the manufacturing of 1.5 billion disposable diapers. It was estimated that the production of diapers for a baby per year has to consume 300 pounds of wood, 50 pounds of petroleum and 20 pounds of chlorine. To produce only one disposable diaper, 360 mL or 1 cup of crude oil is consumed. Also, 200–400 kg of fluff is needed for 1 baby per year consuming around 4.5 trees per baby and requiring 1 billion trees per year globally. However, when a cotton diaper is used, less than 10 kg of cotton is used for 2 years of diaper use. However, the reusable diapers consume more water and energy than the disposable diapers due to washing and drying of clothes (Lehrburger, 1991). It is reported that 22,455 gallons of water is required for a human being in his lifetime, if reusable diapers are used. But, disposable diapers consume six times higher energy in the manufacturing phase than the reusable diapers (Meseldzija et al., 2013). A similar study has also been conducted by Aumônier and Collins to compare the performance of disposable and reusable cotton cloth diapers (Aumônier and Collins, 2005). It has been reported that the total number of nappies (diapers) sold in UK alone is 2.47 billion, which would be approximately 110,000 per annum. Different surveys reported around a 5% market for reusable nappies. Figures 9.19 and 9.20 show the environmental impact of disposable and reusable nappies in the UK. Both disposable and reusable nappies (i.e. nappies produced from cotton cloth) have performed similarly as far as the environmental effect is concerned, except the freshwater aquatic ecotoxicity, which was higher in the case of reusable nappies. This was due to the high number of washing cycles for reusable nappies as compared to one-time use and disposal of the disposable nappies. It was also expected that the freshwater aquatic ecotoxicity would go still higher up to 400 kg 1.4 DB eq depending on the detergents used for washing. The energy consumption for the reusable nappies was higher than the disposable nappies; on the other hand, the municipal solid waste was higher for disposable nappies than the conventional nappies (Aumônier and Collins, 2005).

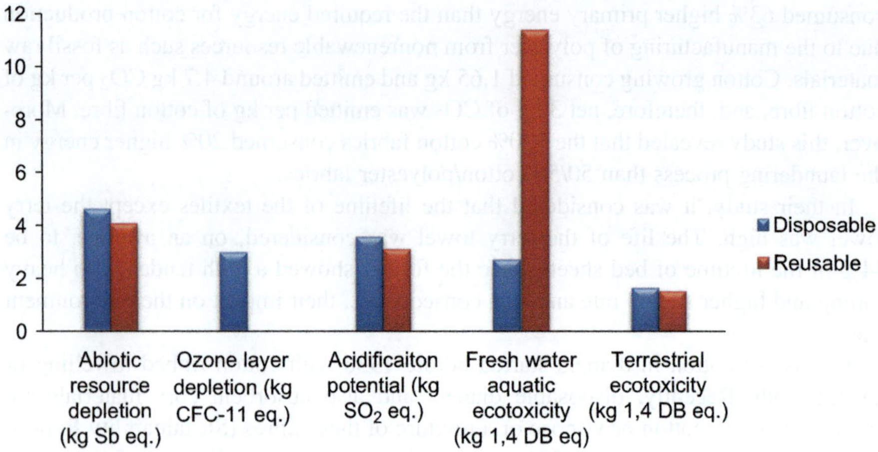

Figure 9.19 Environmental impact of disposable and reusable nappies in the UK (Aumônier and Collins, 2005).

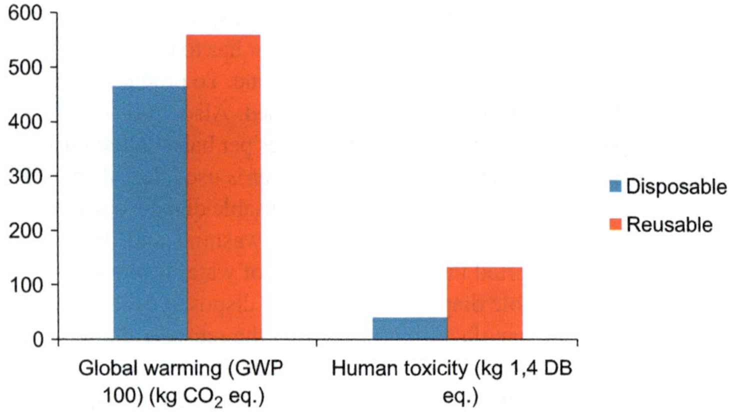

Figure 9.20 Environmental impact of disposable and reusable nappies in the UK (Aumônier and Collins, 2005).

Similar studies have also been conducted by Weisbrod and Hoof (2012) environmental impact of a registered product of Pampers diapers.

LCA studies conducted by Manufacturers Committee of EDANA (Sustainability Report (2007−2008)) compared the impacts of modern-day baby wipes and conventional cloths made of cotton fibre. These products are manufactured for entirely different applications and duration of usage. The modern-day wipes are meant for short-term use, whereas conventional cotton fibre washable cloths are used for longer duration. For the normalization of two different entities, 1000 ordinary nappy changes were considered as the functional unit. According to the results of LCA analysis, in terms of freshwater consumption, energy-related issues and global warming potential, modern baby wipes performed better. On the other hand, in terms of land use, acidification, aquatic and terrestrial eutrophication, conventional cotton cloths were found to have much lower impacts.

9.4 Discussion of LCA results

From the LCA studies presented in the previous section, it is quite clear that cotton fibres and textiles have significant environmental impacts. It is, therefore, necessary to identify different parameters during the whole life cycle of cotton products influencing different environmental parameters and to follow the suggestions from different LCA studies to reduce the environmental impacts of cotton textiles.

In the cultivation phase of conventional cotton fibre, use of pesticides and fertilizers and use of water are the primary factors affecting the environment. One effective approach to reduce the environmental effects is the cultivation of organic seed cotton fibres, although the yield is lower than the conventional cotton. The genetically modified cotton uses lesser amounts of pesticides and provides higher yield as compared to conventional cotton fibre, but its cultivation is not preferred by the environmentalists due to the risks associated with the release of genes. On the other hand, cotton produced in Africa utilizing resources and rain-fed agriculture (known as CmiA cotton) produces much lower GHG emissions as compared to conventional cotton fibres. LCA also reveals that cotton fibre produces much lower GHG emissions as compared to other commercial fibres such as polyester, polypropylene, viscose, etc. and also has much lower global warming potential.

During manufacturing of cotton yarn in the spinning mill, electricity use and transport are the main factors affecting different impact categories. Among different types of fabrics, knitted fabrics have lower environmental impacts as compared to woven fabrics. However, LCA studies suggest that for cotton textiles, the consumer phase involving frequent laundering and drying has higher impact as compared to the agriculture and manufacturing phase. However, the consumption of water is very high in the agriculture phase. Also, NO_x emission is usually higher in the cultivation stage due to agricultural practices and use of fertilizers. Bio-processing of cotton textiles using enzymes is an effective approach to reduce the environmental impacts of cotton textiles.

Washing, laundering and ironing are important processes in the consumer phase of cotton textiles, which lead to significant environmental impacts. Water and electricity consumption are the main factors. Cotton fabrics consume much higher water as compared to linen cloths in the consumer phase and higher energy as compared to cotton/polyester blends.

However, a recently performed LCA benchmarking study by van der Velden et al. (2014) on textiles made from cotton, polyester, nylon, acryl and elastane revealed that the environmental effects of different products depend not only on the base materials but also on the thickness (linear density) of the yarn. The environmental burden of different textile processes such as spinning, weaving and knitting decreases with the increase in the linear density of yarn. According to the cradle-to-grave analysis, cotton fibre textiles have the highest environmental impacts, followed by nylon, elastane, polyester and acryl. According to this study, the use phase has less relative impact, which is not in agreement with most of the previous studies. This contradiction is attributed to the change in the laundering, drying and ironing practices in recent times and use of more energy-saving appliances than the past. Among the various textile processes, knitting has a lower environmental impact as compared to spinning and weaving for the linear density lower than 100 dtex. However, it was found that the results of LCA study are highly case sensitive depending on the followed practices, if the dyeing and finishing, use phase and end-of-life phase are considered, and it is, therefore, really difficult to make general conclusions on the impacts of various products and processes.

9.5 Factors influencing sustainability of cotton textiles

As mentioned earlier in this chapter, sustainability represents the type of developments that meet the needs of present generations without comprising the ability of future generations to fulfil their needs. The use of natural resources and the generated outputs and environmental effects are two important aspects of sustainability. Products or processes that use higher nonrenewable resources as well as produce detrimental effects on the environment are considered to be less sustainable. As discussed in the previous sections, LCA has been used to identify, quantify and evaluate the environmental impacts (inputs and outputs) of products made from cotton textiles and this information can be utilized to determine the sustainability of cotton textiles and to compare that with other textile fibres (Rana et al., 2014a,b).

According to the World Health Organization, around 20,000 deaths occur in the developing countries due to pesticide poisoning. Many of these deaths are related to the cotton sector (Pal, 2009). With the use of organic cotton, the use of fertilizers can be reduced and pesticide use can be avoided. There are a number of factors that are governing the sustainability of cotton textiles. The factors related to the cultivation of cotton fibre follow in the next sections.

9.5.1 Soil fertility

To improve soil fertility and improve yield, nitrogen fertilizers are used in conventional cultivation methods. As synthetic fertilizers have environmental impacts, in the case of organic cultivation, livestock manure and legumes can be used to improve the soil condition, replacing the synthetic nitrogen fertilizers. It can be animal manure, green manure and compost. The type of organic fertilizer may be recommended based on the type and properties of soil (WWF Report, 2012).

9.5.2 Pests and disease

In cotton cultivation, 15% of the yield loss is due to disease and insect damage to plants. Therefore, different types of hazardous pesticides are used, leading to significant environmental impacts. To overcome this, in organic farming of cotton, the pests are managed by crop rotation, mixed cropping system and maintaining density of plants at optimum level (Kooistra et al., 2006).

9.5.3 Water

High use of water is a primary environmental concern for cotton cultivation. Worldwide, around 53% of cotton-cultivated area is under irrigation. In India, most of the cotton-cultivated zones are rainfall zone (69%), whereas a small proportion of land is under irrigation (31%) (WWF Report, 2012). The ratio of water obtained from rainfall for cultivation purposes is 405:133 (WWF Report, 2012). There are different types of irrigation systems for the cultivation of cotton. The flood-or-furrow irrigation system is one of the well-known systems followed for the irrigation process in cotton cultivation. Around 95% of the irrigated cotton field is subjected to flood-or-furrow irrigation. Mobile irrigation systems are used in a smaller proportion of cotton fields, which is approximately 2% of the entire irrigated cotton fields worldwide. As the flood-or-furrow irrigation system leads to wastage of a great deal of water due to evaporation and transpiration into the fields, more efficient irrigation systems should be followed to improve the sustainability of cotton fibre.

In spite of the above facts, cotton fibre production is found to be more sustainable as compared to other commercial textile fibres. In order to compare the sustainability of cotton fibre production with other commercial fibres, Muthu et al. (2012) performed life cycle inventory (LCI) and life cycle impact assessment (LCIA) considering the amount of energy and water consumed and greenhouse gases emitted. An environmental impact index (EI) was determined from a scoring system based on these factors influencing ecological sustainability. Subsequently, EI was used to calculate the ecological sustainability index (ESI) of different fibres. The estimated EI and ESI values for different fibres are presented in Figure 9.21. It is clear that cotton fibre has lower EI and higher ESI than most of the fibres, except flax fibres. The organic cotton has the lowest EI and highest ecological sustainability. Due to use of the lowest amount of energy among all studied fibres and less water as compared to cotton, flax

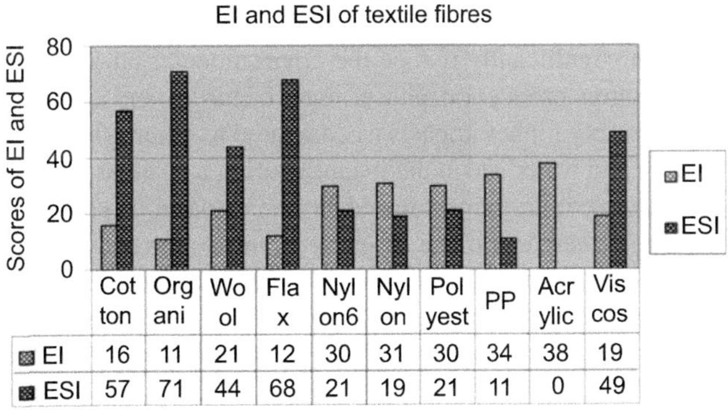

	Cotton	Organi	Wool	Flax	Nylon6	Nylon	Polyest	PP	Acrylic	Viscos
EI	16	11	21	12	30	31	30	34	38	19
ESI	57	71	44	68	21	19	21	11	0	49

Figure 9.21 EI and ESI values of textile fibres.
Muthu, SS., Li, Y., Hu, JY., Mok, PY., 2012. Ecological Indicators. 13, 66–74.

fibres also have very low EI and high ESI values. Highest ecological sustainability of organic cotton fibre was due to no use of synthetic pesticides and fertilizers. In spite of using less water than conventional cotton, the reasons for higher EI of viscose are more damage to human health due to use of many chemicals in the viscose production process, ecosystem quality, use of more resources and lower capacity to absorb CO_2 than that of plants.

Further, the sustainability of cotton fibre production can be improved through suitable utilization of the by-products. Linter, cotton seed and stalks are the main by-products of cotton fibre. From cotton seeds, oil can be produced. Cotton stalks can be used for the production of various materials. Nowadays, research on production of composites using cotton stalks is under consideration (Jan, 2009).

Nevertheless, as discussed earlier, cotton fibre products have a high environmental burden and, therefore, lower sustainability (as compared to other commercial fibres such as nylon, polyester, elastane, acryl, etc.) due to the contribution of the manufacturing, use and end-of-life phases. One of the important approaches to reduce the environmental impacts of the manufacturing phase is to use thicker cotton yarns (van der Velden et al., 2014). On the other hand, the impacts of the use phase can be reduced through use of more energy and water-saving laundering and drying machines. However, there are many uncertainties regarding the use and end-of-life phases of the products due to insufficient data available, and further studies are extremely essential to obtain better conclusions.

9.6 Conclusions

In this chapter, the first few sections discuss the production and consumption of cotton fibre in a global perspective. The importance of LCA for the cotton products, LCA methodology used by most of the researchers and environmental impact assessments are discussed briefly. This chapter also discusses the studies conducted on LCA of different cotton products for apparel and technical textile purposes. It can be concluded from these LCA studies that cotton cultivation has significant environmental impacts due to use of pesticides, fertilizers and water. Nevertheless, cotton production has lower impacts as compared to synthetic and man-made cellulosic fibres. Further, cultivation of organic cotton can significantly reduce the environmental burden of cotton fibres. In contrast to the cotton production phase, manufacturing and use phase of cotton textiles show considerably higher impacts as compared to other synthetic fibres mainly due to higher energy and water use in the manufacturing as well as use phase of cotton textiles. Use of thicker cotton yarns can reduce the impacts of cotton textiles in the manufacturing phase, whereas the use of energy- and water-efficient appliances can significantly reduce the environmental burden of cotton products. However, the results of these various LCA studies are highly case-sensitive depending on the type of product, application, duration of use, user practices, etc. and the data available in the literature, therefore, are not sufficient to draw any general conclusions. It is therefore highly mandatory to carry out further studies to evaluate the environmental impact and sustainability of various cotton textiles and to compare them with other types of products.

References

Adams, G., Boyd, S., Huffman, M., 2014. The Economic Outlook for U.S. Cotton. National Cotton Council of America.

Aumônier, S., Collins, M., 2005. Life Cycle Assessment of Disposable and Reusable Nappies in the UK. Environment Agency, Bristol, ISBN: 1 84432 427 3.

Beton, A., Dias, D., Farrant, L., Gibon, T., Le Guern, Y., Desaxce, M., Perwueltz, A., Boufateh, I., 2014. Environmental improvement potential of textiles (IMPRO-textiles). In: Wolf, O., Cordella, M. (Eds.), JRC Scientific and Technical Reports. European Commission.

Chapagain, A.K., Hoekstra, A.Y., Savenije, H.H.G., Gautam, R., (September 2005). The water footprint of cotton consumption, value of water, Research Report Series No. 18.

Chapman, A., 2010. Mistra Future Fashion – Review of Life Cycle Assessments of Clothing (A Report of MISTRA, Sweden).

Connell, D., 1995. The environmental impact of the textiles industry. In: Carr, C.M. (Ed.), Chemistry of the Textiles Industry. Springer, Netherland, pp. 333–354.

Cotton Made in Africa (News report: 06/11/2014): http://www.cotton-made-in-africa.com.

Emmenegger, M.F., Meyer, K., Gioacchini, M., Massetti, D., Zeller, H., Zah, R., 2013. Managing environmental sustainability in the textile supply chain. In: The 6th International Conference on Life Cycle Management, Gothenburg.

Executive Summary, 2012. The Life Cycle Inventory & Life Cycle Assessment of Cotton Fibre & Fabric – LCA Executive Summary. Cotton Incorporated and PE International. (Executive Summary, LCA for Baby Wet Wipes, EDANA).

ICTSD, 2013. Cotton: Trends in Global Production, Trade and Policy. Information Note. International Centre for Trade and Sustainable Development, Geneva, Switzerland.

Jan, E.G., 2009. Environmental benefits of natural fibre production and use. In: Proceedings of the Symposium on Natural Fibres, pp. 3–17.

Johnson, J., MacDonald, S., Meyer, L., Norrington, B., Skelly, C., 2014. The World and United States Cotton Outlook. Agricultural Outlook Forum 2014.

Kalliala, E.M., Nousiainen, P., 1999. Life cycle assessment, environmental profile of cotton and polyester-cotton fabrics. AUTEX Research Journal 1 (1), 8–20.

Kooistra, K.J., Pyburn, R., Termorshuizen, A.J., 2006. The Sustainability of Cotton, Consequences for Man and Environment. Science Shop Wageningen University & Research Centre. Report 223. ISBN: 90-6754-90-8585-000-2.

Lehrburger, C., Mullen, J., Jones, C.V., 1991. Diapers: Environmental Impacts and Lifecycle Analysis. The National Association of Diaper Services. Life Cycle Assessment: Principles and Practice. EPA/600/R-06/060. May 2006.

Meseldzija, J., Poznanovic, D., Frank, R., 2013. Assessment of the differing environmental impacts between reusable and disposable diapers. Dufferin Research.

Murugesh Babu, K., Selvadass, M., 2013. Life cycle assessment for cultivation of conventional and organic seed cotton fibres. International Journal of Research in Environmental Science and Technology 3 (1), 39–45.

Muthu, S.S., Li, Y., Hu, J.Y., Mok, P.Y., 2012. Quantification of environmental impact and ecological sustainability for textile fibres. Ecological Indicators 13, 66–74.

Nielson, A.M., Nielson, P.H., 2009. Comparative life cycle assessment of the elemental T-shirt produced with biotechnology and a conventional T-shirt produced with conventional technology. Novozymes.

Pal, S., December 2009. Organic cotton: an overview. The Indian Textile Journal.

Pimentel, D., 2005. Environmental and economic costs of the application of pesticides primarily in the United States. Environment, Development and Sustainability 7 (2), 229–252.

Rana, S., Pichandi, S., Parveen, S., Fangueiro, R., 2014a. Natural plant fibers: production, processing, properties and their sustainability parameters. In: Muthu, S.S. (Ed.), Roadmap to Sustainable Textiles and Clothing: Eco-friendly Raw Materials, Technologies, and Processing Methods. Springer, Singapore, pp. 1–35.

Rana, S., Pichandi, S., Parveen, S., Fangueiro, R., 2014b. Regenerated cellulosic fibers and their implications on sustainability. In: Muthu, S.S. (Ed.), Roadmap to Sustainable Textiles and Clothing: Eco-friendly Raw Materials, Technologies, and Processing Methods. Springer, Singapore, pp. 239–276.

Rupp, J., January/February 2009. Nonwovens Made of Cotton. Textile World.

Shen, L., Patel, M.K., 2010. Life cycle assessment of man-made cellulose fibres. Lenzinger Berichte 88, 1–59.

Smith, G.G., Barker, R.H., 1995. Life cycle analysis of a polyester garment, resources. Conservation and Recycling 14, 233–249.

Steinberger, J.K., Friot, D., Jolliet, O., Erkman, S., 2009. A spatially explicit life cycle inventory of the global textile chain. International Journal of Life Cycle Assessment 14, 443–455.

Sustainability Report, 2007–2008. Absorbent Hygiene Products. Manufacturers Committee of EDANA.

Tukker, A. (Ed.), 2004. Handbook on Life Cycle Assessment, vol. (7). Kluwer Academic Publishers, Dordrecht.

van der Velden, N.M., Patel, M.K., Vogtländer, J.G., 2014. LCA benchmarking study on textiles made of cotton, polyester, nylon, acryl, or elastane. The International Journal of Life Cycle Assessment 19 (2), 331–356.

Weisbrod, Anne, V., Gert, Van Hoof, 2012. LCA-measured environmental improvements in Pampers® diapers. The International Journal of Life Cycle Assessment 17 (2), 145–153.

WWF Report, 2012. Cotton Market Sustainability in India. Published by WWF-India.

LCA of wool textiles and clothing

10

B.K. Henry[1], S.J. Russell[2], S.F. Ledgard[3], S. Gollnow[4], S.G. Wiedemann[5], B. Nebel[4], D. Maslen[6], P. Swan[7]
[1]Queensland University of Technology, Brisbane, Queensland, Australia; [2]University of Leeds, UK; [3]AgResearch Ruakura Research Centre, Hamilton, New Zealand; [4]PE International Australasia, Wellington, New Zealand; [5]FSA Consulting, Toowoomba, Queensland, Australia; [6]The New Zealand Merino Company Limited, Christchurch, New Zealand; [7]Australian Wool Innovation Limited, Sydney, Australia

10.1 Introduction

This chapter examines life cycle assessment (LCA) of wool textiles and clothing. We focus on products made from the natural fibre (wool) formed from the fleece of sheep (*Ovis aries*) and particularly on apparel, but the principles and methodology described can readily be extended to a wide range of wool textiles including not only clothing but also household fabrics and industrial products (BSI, 2014) and to products from other small ruminants producing fibre such as members of the goat and camelid families (LEAP, 2014a). We examine in more detail those aspects of the life cycle that affect the intensity of environmental impacts in a way that sets wool products apart from those made from alternative fibres. These specific life cycle features include the cradle-to-farm gate production stage, which is characterised by wide diversity and multiple products from sheep, and on the end-of-life stage where the range of options following first use of clothing by the consumer (reuse, recycling or disposal) can alter the wool supply chain impacts substantially. A case study is provided to illustrate application of LCA to quantification of environmental impacts of two items of wool apparel. This study quantifies, for the first time, the relative contributions of different stages of the complex supply chain across production, processing, manufacture and retail over multiple countries. Supply chain impact 'hotspots' are identified and the importance of considering multiple impact and resource use categories is highlighted. Finally, we discuss emerging research to resolve data and methodology challenges and future prospects for LCA of wool products.

10.1.1 The wool textile value chain

Wool has a complex protein structure that gives it unique properties that make wool fibre suitable for a wide range of products including warm outer garments, base layer wear, fashion apparel, luxury interiors and items required to have fire-retardant properties (IWTO, 2014; CSIRO, 2008). Wool is at the high value end of fibres, and although it accounted for only 1.3% of world textile consumption in 2012 (IWTO, 2013), it holds a much higher level of importance in terms of unit price and status.

Handbook of Life Cycle Assessment (LCA) of Textiles and Clothing. http://dx.doi.org/10.1016/B978-0-08-100169-1.00010-1

The value of a wool fleece and its use is largely determined by the diameter of the fibre. Merino is the finest wool with grades ranging from ultrafine (<16.5 μm) to broad wool of >23 μm (Nolan et al., 2013). Other sheep species and crossbreeds have higher diameter fibres with coarser wools (up to 35–45 μm) commonly used for floor coverings. Approximately two-thirds of the global wool harvest is used in the manufacture of apparel, and about one-third in interior products such as carpets, upholstery and rugs, with a small proportion (about 5% of the total) destined for industrial uses such as insulation (IWTO, 2014).

The wool supply chain starts with sheep farming systems. Globally, production of wool involves over a billion sheep across geographically and climatically diverse regions in around 100 countries. However, the majority of 2 million tonnes total annual harvest is concentrated in relatively few countries (FAOSTAT, 2014). Table 10.1 lists the top 12 countries ranked on clean wool production. Together these countries account for over 40% of the total sheep population and close to 70% of global greasy wool production. Australia is the leading producer of clean wool and in 2012 was responsible for about half of the world export volume. Approximately half of apparel wool is supplied by Australia (FAOSTAT, 2014), while New Zealand is also a major producer of high quality fine wool for apparel as well as stronger wool suited to interior furnishings. China accounted for 27% of wool imports in 2012 and is a major centre of

Table 10.1 Top 12 individual countries ranked for clean wool produced

Country	Sheep ('000 head)	Greasy wool (t)	Clean wool equivalent (t)
Australia	74,700	374,157	245,073
China	139,600	400,057	177,697
New Zealand	31,263	167,900	127,830
Russia	20,767	51,502	28,750
Iran	48,750	61,897	27,854
Sudan	52,428	55,221	27,611
Turkey	25,032	49,542	24,771
Uruguay	7350	32,500	24,050
South Africa	24,391	40,621	24,040
Kazakhstan	15,200	39,600	23,176
Turkmenistan	14,000	38,333	22,825
United Kingdom	32,215	34,000	22,780
World total	1110.647	1,999,284	1,109,433

2012 data; sourced from FAO and AWI, October 2014.

wool processing having almost 30% of global spinning and weaving. Italy (10%), India (8%) and the United Kingdom (8%) are also significant importing countries, with Italy and the United Kingdom together representing 20% of spinning and weaving. The Asian subcontinent, predominantly India and Pakistan (14%), is also important in wool processing and manufacturing.

The major steps in the life cycle of wool textiles and apparel are illustrated in Figure 10.1. The relative contribution of each stage to the environmental impact of wool products differs for different impact or resource use categories and between diverse production systems. Supply chain efficiency also significantly affects environmental performance. LCA is further complicated because of the need to consider not only the range of intermediate products shown in the right-hand side of Figure 10.1 but also co-products such as meat, milk and lanolin (see Figure 10.2), which are allocated a

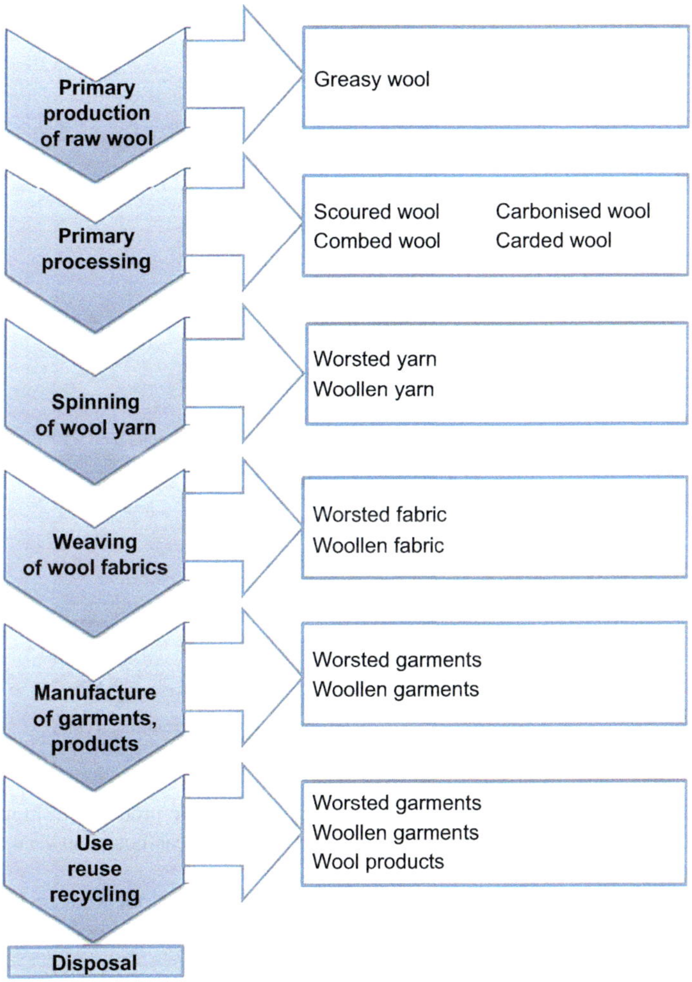

Figure 10.1 Simplified diagram of stages in the life cycle of wool apparel.

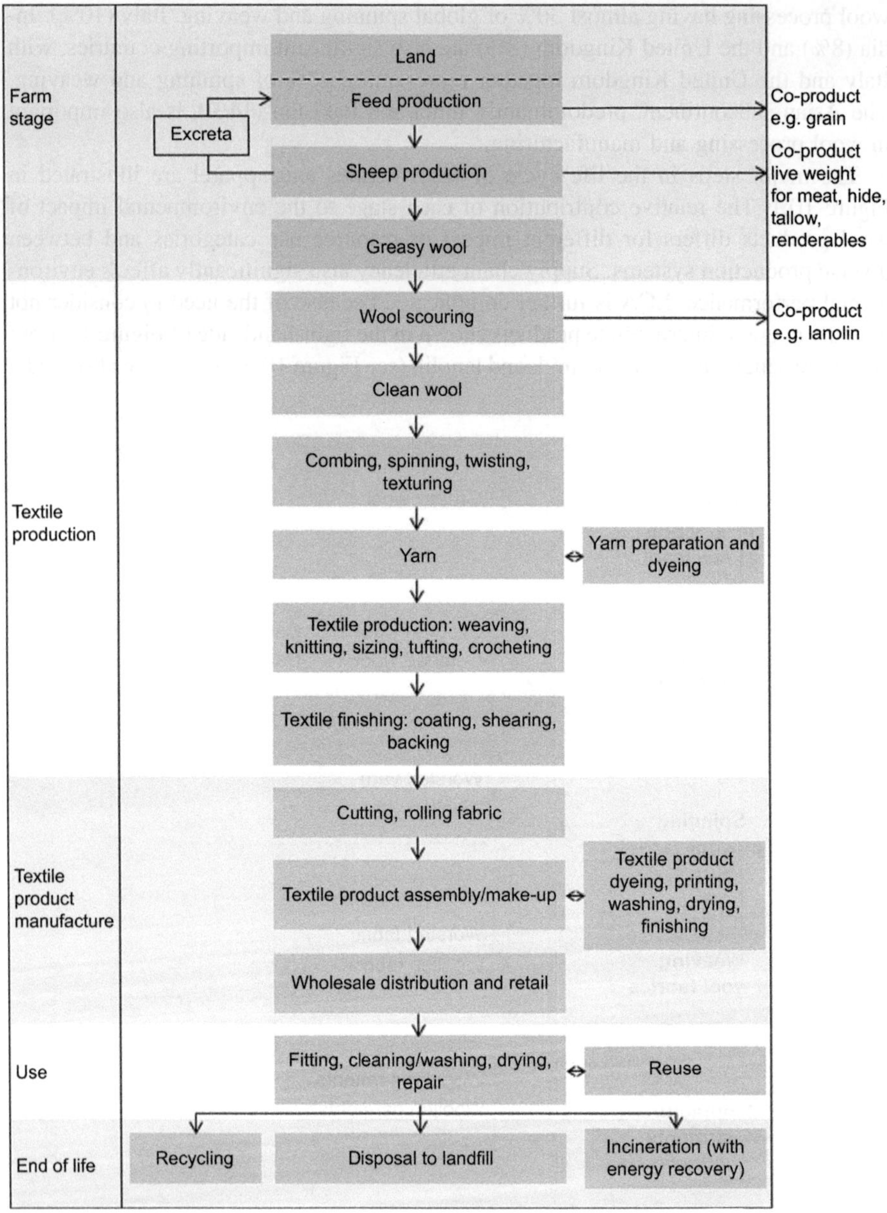

Figure 10.2 System boundary diagram for the life cycle of textile produced from wool from sheep (excluding inputs, outputs and emissions). Transportation can occur between any of the various stages outlined.
This figure is adapted with permission from BSI (2014) to account for wool production on farm.

share of the impacts in order to accurately quantify the impacts of a wool product. Wool supply chains are frequently also characterised by multiple transfers between countries and regions, and transport around the globe can involve large distances (Bevilacqua et al., 2011, Section 10.3 of this chapter). Complex social and economic interactions affect not only the volume or units of trade in wool intermediate and final products but also the quality and value of the units and how they change over time (Verikios, 2009).

The diversity and complexity of wool supply chains mean that data requirements are challenging for full LCA. For the consumer use stage, statistics on the period of garment wear and frequency and method of cleaning are known to be highly variable and are difficult to obtain. For example, average life expectancy of wool garments conservatively ranges from one or two years for items such as shirts to four years for suits and coats and 10 years for blankets (Drycleaning Institute of Australia Limited, 2014). As discussed in Section 10.2.6, information is also needed on consumer choices of 'disposal' after the period of first use – for reuse, recycling or consignment to landfill or other municipal waste streams that are recognised as significant in the full life cycle of wool products.

10.1.2 Published LCA studies of wool

There are few wool LCA studies that have attempted to quantify multiple environmental impacts across the full cradle-to-grave or cradle-to-cradle life cycle of wool textile products. Table 10.2 summarises available published LCA studies for wool. Most published studies have been restricted to primary production (cradle-to-farm gate) or to the first stage of processing to clean wool (cradle-to-primary processing). In addition, the majority of LCA studies have assessed only a single impact category, usually global warming potential (GWP), which is commonly reported as greenhouse gas (GHG) emissions in carbon dioxide equivalents (CO_2e) and often called the product carbon footprint. Other impacts and resource depletion categories of environmental significance arise from the production and use of textiles and clothing, including acidification, eutrophication, human toxicity, biodiversity impact, water use and land use. A single indicator such as GHG emissions cannot be interpreted as a measure of environmental sustainability and if used to prioritise mitigation strategies may introduce risks of perverse outcomes from unidentified trade-offs (ISO, 2013).

Inconsistencies in scope, methodology and assumptions between published wool LCA studies mean that use of the results to compare products or systems and interpretation for policy or management decisions frequently carries high uncertainty. A major source of methodological inconsistency has been the handling of co-products, notably for the on-farm production of wool from sheep where co-products, particularly meat, can be economically and socially important. Eady et al. (2012) investigated GHG emissions for a mixed farming (grain, wool and meat) system in Western Australia using a cradle-to-farm gate LCA approach. Allocation between the fine wool (for textiles) and meat from merino sheep was assessed using either biophysical or economic relationships. The study estimated 36.2 or 28.7 kg CO_2e/kg greasy wool using biophysical or economic allocation methods, respectively. Brock et al. (2013) also

Table 10.2 Summary of the scope of publicly available wool LCA studies and the impact categories that were evaluated

Published LCA study	Functional unit	System boundary	Impact categories
Eady et al. (2012)	1 kg 19.5 μm greasy wool	To farm gate	GWP
Brock et al. (2013)	1 kg 19 μm greasy wool	To farm gate	GWP
Wiedemann et al. (2015)	1 kg greasy wool	To farm gate	GWP, ED, LU
Potting and Blok (1995)	1 sq m carpet	Pre-farm to disposal	Most CML[a] indicators
Barber and Pellow (2006)	1 t dry wool top	Pre-farm to wool top at spinning mill	GWP[b], ED
Brent and Hietkamp (2003)	1 kg dyed yarn	Pre-farm to dyed yarn	Most CML indicators[a], GWP[b], LU, WU
Petersen and Solberg (2004)	Wool carpet	Post-farm gate to use	GWP
Murphy and Norton (2008)	1 sq m insulation	Farm to disposal	GWP, ED, AP, EP
Bowyer (2009)	Wool broadloom carpet	Pre-farm to disposal	GWP, ED, AP, EP, HTP, ETP, WU, ODP, smog, indoor air quality, habitat alteration
Bevilacqua et al. (2011)	1 wool sweater	Pre-farm to disposal	GWP[b]

[a]**CML impact categories (LCA Institute of Environmental Sciences NL)**

GWP_{100} = global warming potential, 100-year timeframe (kg CO_2e)
AP = acidification potential (kg SO_2 eq)
EP = eutrophication potential (kg PO_4 eq)
ODP = ozone layer depletion potential (kg CFC-11 eq)
ADP = abiotic depletion potential (kg antimony eq)
HTP = human toxicity potential (kg 1,4-dichlorobenzene eq)
ETP = ecotoxicity potential (kg 1,4-dichlorobenzene eq)
POCP = photochemical oxidant creation potential (kg ethylene eq)

Resource use

ED = Fossil energy demand (MJ)
LU = Land use (ha)
WU = Water use (L).

[b]GHG emissions exclude sheep enteric CH_4 and N_2O.

conducted a cradle-to-farm gate GHG study of merino wool production, in this case from eastern Australia, and estimated emissions to be 24.9 kg CO_2e/kg greasy wool using economic allocation between wool and meat. However, GHG emissions were found to be much lower at 14.8 kg CO_2e/kg greasy wool when the farming enterprise dominance was shifted from wool to meat by introducing meat breed rams (Dorset), resulting in higher return from sheep meat. Recently, Wiedemann et al. (2015) compared several alternative allocation and system expansion methods for dealing with co-products using case-farm systems in New Zealand and the United Kingdom that produce broad wool and two systems in Australia producing fine wool. They found only small differences (<25%) in GHG emissions per kg product (i.e. wool and meat) between the four systems but large effects of methodology for handling co-products. For example, results ranged from −26 to +37 kg CO_2e/kg greasy wool for the UK system using different methods. The study examined multiple resource use categories. It showed that for all case study farms, only a small amount of arable land was used for wool products, with most land being non-arable. On these lands grazing of perennial pastures or shrubs by ruminants was often the only viable productive use.

These published studies have recognised wool and meat as the co-products for sheep production systems of interest. However, sheep systems in some regions also provide milk with high economic value for human food. It is also recognised that sheep can provide other functions that can potentially be accounted for, thereby reducing the environmental impacts allocated to wool. For example, Ripoll-Bosch et al. (2013) studied sheep systems in Spain and noted that the benefits of sheep to ecosystem services (e.g. conservation of biodiversity and landscapes) were important in extensive mid-high altitude natural/seminatural areas and allocated up to one-half of total GHG emissions to these ecosystem services based on an economic approach.

At the other end of textile supply chains, amounts of reuse and recycling of wool products are key parameters for quantification in LCA to determine the full lifetime and extent of avoidance of the production of new raw material. However, for these activities few data are available and generalisations are difficult because practices differ between countries in terms of recycling and reuse of wool products and, ultimately, in the mode of disposal. The environmental impacts associated with recycling processes as well as transport and storage of post-use products must be included in the accounting. WRAP (2012a) provided data showing that in some cases there can be a net benefit of recycling with the burdens associated with managing the waste being outweighed by the avoidance of new garment production. Woolridge et al. (2006) reported an LCA study for textile products that found that reuse and recycling of donated clothing results in a reduction in the environmental burden compared to purchasing new clothing made from virgin materials, taking into account the energy footprint of recycling and reuse and assuming the production of new apparel was avoided. In addition, diverting clothing from landfills is economically as well as environmentally beneficial in terms of promoting material security, reducing raw material costs and creating employment and economic growth.

The dearth of full supply chain, multi-impact LCA studies for wool textiles was noted in a review of published wool LCA studies (Henry, 2011), which also identified significant data gaps and research needs for more meaningful assessment of the environmental performance of wool supply chains. A subsequent LCA study (Bevilacqua et al., 2011) sought to quantify the GHG emissions (carbon footprint) across the full value chain of a wool sweater. The estimated emissions were 1.95 kg CO_2e per sweater of median weight (estimated as 0.265 kg in the 2009 collection). This assessment of GHG emissions did not include non-CO_2 GHG emissions (sheep enteric methane and excreta nitrous oxide) and this result is consequently markedly lower than reported for cradle-to-farm gate assessments that counted all significant GHG sources (Henry, 2011). However, the strength of the study by Bevilacqua et al. (2011) was in examining a complex supply chain and covering in detail the transport structure and energy emissions across eight countries and multiple locations in Italy where the processing of the fibre occurred.

10.2 Wool LCA methodological and data challenges

The general guidelines for application of LCA were set out in ISO 14044 (ISO, 2006). Those guidelines defined the preferred process starting with the importance of defining the goal and scope of an LCA study, followed by data collection and inventory analysis and then impact assessment. More recently, LCA-based guidelines covering limited impact categories specifically for small ruminants, that is, for sheep and goats (LEAP, 2014a) and for textiles (BSI, 2014), have been published. Methodological aspects of the LCA approach are described in the following sections, with emphasis on the specific application challenges for wool fibre produced from sheep. Initially, a brief description is provided of the two distinct approaches, attributional and consequential LCA, since the fundamental choice between these has a substantial influence on methodology, results and interpretation of the study.

10.2.1 Attributional versus consequential LCA

The terms attributional LCA (ALCA) and consequential LCA (CLCA) are used (Rebitzer et al., 2004) to identify studies that aim to describe, respectively, either: (1) a static representation of an average product system and its environmental exchanges; or (2) how the environmental exchanges of the system are affected by a change in the system. The choice of which way to model product life cycle environmental impacts depends on the scope and intended goals of the research. Specific aspects of consequential modelling have been described by Ekvall and Weidema (2004), while guidelines such as those developed by LEAP (2014a) describe an attributional approach. Key differences relate to data collection and handling of multifunctionality (i.e. co-products).

Representative results in ALCA require a comprehensive dataset with respect to factors that contribute to variable efficiency and environmental impacts, while CLCA is focussed on using marginal data to assess the part of the production system most likely to change as a consequence of choices or changes, for example, in supply or demand. Hence, if studying an increase in the demand for wool, a CLCA would only study farming systems that would be expected to increase in response to market changes.

Supply chains such as those for wool textiles give a range of co-products that must be considered as part of an LCA study. In ALCA an allocation approach is commonly used to divide the environmental burden between the multiple co-products, whilst in

CLCA the system is expanded to include the consequences of a change. In contrast to allocation, system expansion includes the additional functions related to the co-products (ISO, 2006). In practice, this is frequently achieved by removing the impact of avoided processes from the system of interest, commonly termed substitution or the avoided-burden method. Within wool production systems, the substitution method has been applied to co-production of wool and meat (live weight, LW) from sheep at the farm (Wiedemann et al., 2015). In that study, applying the substitution method to two Australian farm systems gave lower estimates of emissions intensity (−3 to 12 kg CO_2e/kg wool) when increased LW from the dual purpose sheep would avoid production of LW from beef cattle or meat-only wool-shedding sheep compared to 10–37 kg CO_2e/kg wool based on allocation approaches using ALCA.

The remainder of this chapter focusses on attributional LCA, which has been most commonly applied for sheep and wool systems, often to provide benchmarking results for a given system or product and to identify ‘hotspots’ of environmental impacts in wool supply chains. ALCA avoids the need for some of the sensitive assumptions of CLCA (Finnveden et al., 2009) and is supported by recent development of specific guidelines to improve the consistency of application to wool apparel and textiles (LEAP, 2014a; BSI, 2014). However, the inherent limitations associated with ALCA that constrain the applicability of results for predicting the influence of a change in supply and demand should be noted.

10.2.2 Goal, scope and system boundary

The goal of a study is important in defining the methodology to be used and its level of detail. Considerations include the intended aims of the study, such as benchmarking a product, identifying hotspots through the life cycle and/or evaluating potential improvement in environmental performance. The scope defines the breadth of the study and identifies the production system, functions, functional unit, system boundaries and impact categories to be studied.

For textiles and clothing produced from natural fibres from animals, the production system and system boundaries must cover all aspects of the animal production system involved in fibre production, and this stage represents the main difference between wool and production of man-made or plant-derived fibres. An example of a system boundary diagram illustrating the various key contributors over the life cycle of an article of clothing from wool is given in Figure 10.2.

In practice, wool production systems are very diverse globally ranging from sheep-only systems where the sheep are housed for much of the year and fed brought-in crops through to very extensive grazing/browsing systems where sheep roam over large areas and feed on native or introduced pastures and/or shrubs. Additionally, the sheep may be in shared systems with other animal types (e.g. cattle, goats, and deer) or in mixed farming systems with crops for sale but where sheep feed on crop stubble and waste, unharvested grain. A further complication is the various co-production components that need to be accounted for (Figure 10.2).

A major methodological complexity associated with the farm stage is the need to account for all contributing factors associated with the sheep population required for wool production. This sheep population refers to all animal classes linked to production of the wool. Thus, for sheep it includes the breeding ewes and rams as well as all

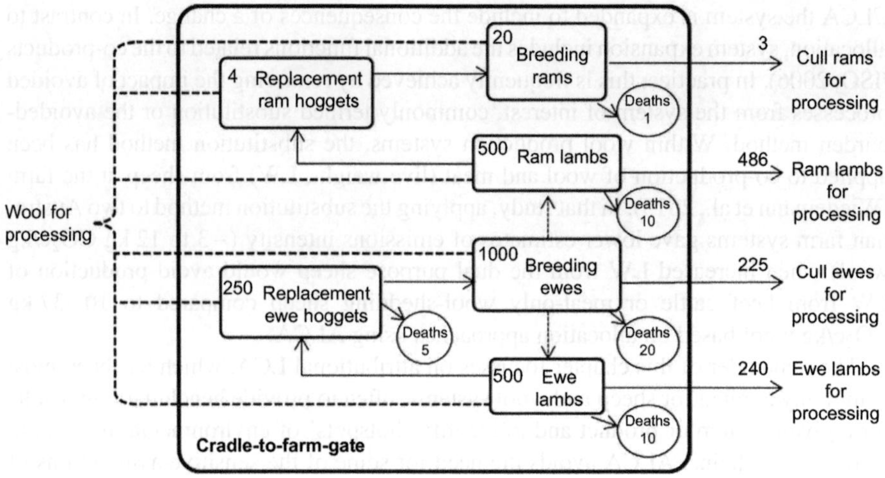

Figure 10.3 Simplified example of a sheep population illustrating relative numbers of breeding and replacement sheep on farm and surplus sheep sold for meat processing. The population is based on a breeding ewe flock of 1000, 100% lambing, 25% replacement rate, 2% death rate and first lambing at 2 years old. All sheep classes contribute to wool production, which is generally harvested by shearing sheep once or twice a year. Wool may also be harvested from hides of sheep during the meat-processing stage.
Adapted from Leap Guidelines, LEAP (2014a).

replacements required to maintain a stable population (Figure 10.3). All of these sheep produce wool, although lambs generally produce less wool of a finer fibre diameter that may have a different end use than wool from older breeding sheep.

10.2.3 Resource use and environmental impact categories

A key benefit of LCA is that it involves determination of multiple resource use and environmental impact categories so that trade-offs and interactions can be evaluated. However, most published LCA studies of sheep to date have assessed few indicators, and often only GHG emissions. General methodology for determining GHG emissions from products was provided in ISO/TS 14067 (ISO, 2013), PAS 2050 (BSI, 2011) and the GHG Protocol *Product Life Cycle Accounting and Reporting Standard* (WRI, 2011). The LEAP small ruminants, guidelines (LEAP, 2014a) give further details for assessing GHG emissions and fossil energy demand for sheep and goats covering the cradle-to-primary processing gate. Key factors covered in the LEAP guidelines include estimation of animal feed intake and handling fibre and non-fibre co-products (discussed in detail later). PAS 2395 (BSI, 2014) provides useful methodology guidelines for estimating GHG emissions for the full life cycle of textile products with details for fibre processing and later life cycle stages of textiles.

There is a wide range of other impact categories with varying levels of consensus on the most appropriate methodology for each category. For example, the ILCD Handbook (ILCD, 2011) describes various methods under the main impact categories of climate change, eutrophication, acidification, ozone depletion, human toxicity, respiratory inorganics, ionising radiation, photochemical ozone formation, ecotoxicity, land use and resource depletion (mainly fossil fuels and minerals). In practice, the relative significance of these different impact categories will vary with different types of textiles and clothing. However, for many impact categories, the cradle-to-farm gate stage has been identified as a dominant determinant of total impacts for livestock-based products (e.g. Basset-Mens et al., 2007; Ledgard et al., 2011). Production of wool has a relatively minor contribution to some resource use and environmental impact categories (e.g. those dominated by fossil fuels), while others are more important.

Land use, also referred to as land occupation, is a resource use category where results are highly dependent on the methodology. A simplistic metric of total land use/kg product results in a high value assigned to wool products made from the fibre of extensively grazed sheep. However, other important measures of land use may give relatively low values for wool. For example, use of arable land (i.e. land capable of growing crops) is low or zero in most extensive systems. In contrast, use by sheep of non-arable land can be high, but this category frequently has limited alternative uses and low intensity grazing may not affect competitive land use and may have a positive rather than negative impact on ecosystem services. A component of land use is biodiversity, and while crop production generally has a significant negative effect on biodiversity, extensive browse or grazing systems may help to enhance botanical and ecosystem diversity (e.g. Bretagnolle et al., 2011; Rippoll-Bosch et al., 2013). In addition, land use and land use change for sheep production may be either a source or sink of GHG emissions. For example, deforestation for improved grazing or cultivation for feed results in GHG emissions through loss of biogenic carbon while in some sheep farming systems reforestation and improvement of land condition may enhance carbon sequestration in vegetation and soils (e.g. LEAP, 2014b).

Freshwater consumption is another resource use of global concern. The World Health Organisation has estimated that early in the twenty-first century, 1.1 billion people do not have access to water supply sources of adequate quality (WHO, 2009), and with a growing human population, stress on water reserves will increase dramatically in the next 30–40 years (Rockström et al., 2007). Globally, agriculture is estimated to account for 85% of freshwater consumption (Shiklomanov, 2000), with the majority being for irrigation of crops. However, livestock have been identified as contributing substantially to water use (Steinfeld et al., 2006). General methodology for determining water use in LCA is provided by ISO 14046 (ISO, 2014). Water use in wool systems will require assessment of water consumption at the sheep farm for livestock drinking, irrigation and other uses, and must include assessment of water losses that occur with water supply. Water is also an important aspect in the post-farm-gate supply chain for wool processing and textiles manufacture. As specified by ISO 14046, assessment of water use must also take into account water used for the dilution of pollutants released to waterways. Best practice water assessment will apply water balance modelling throughout the supply chain to determine total use, including losses.

10.2.4 Handling multifunctional processes and co-products

Sheep can produce a range of co-products (Figure 10.2). At the farm level, this may include wool and live weight (for meat), and in some systems milk can be an important co-product. During initial processing of wool, there can also be clean wool and lanolin produced, with the latter having uses for cosmetics, skin treatment products and lubricants. Application of LCA requires that total resource use and environmental emissions are allocated between the various co-products and results can be highly dependent on the methodology selected. The ISO 14044 (ISO, 2006) guidelines provided a generic allocation procedure hierarchy and LEAP (2014a) and BSI (2014) provided further guidance for sheep production systems and products. For allocation between wool and LW, a biophysical allocation method is recommended based on the protein requirements for various animal processes including fibre production and animal growth. Fibre production is mainly determined by requirements for protein (CSIRO, 2007). Wiedemann et al. (2015) applied this approach to different sheep production systems and obtained similar results when they applied a simpler approach based on protein mass (i.e. the mass of protein in wool relative to that in LW) and suggested that it may also be a suitable alternative to biophysical allocation based on protein requirements. Studies across contrasting sheep production systems resulted in values for biophysical allocation to wool of 19–40%, whereas for economic allocation it was 1–52% (Wiedemann et al., 2015). There has been little research in this area and consequently it is recommended that sensitivity analysis is carried out using several methods. This may include the use of system expansion to understand the implications of choosing alternative products or systems, and where system change or mitigation strategies are discussed.

Where scouring of greasy wool to produce clean wool also gives lanolin co-products, LEAP (2014a) recommends the use of economic allocation since these products have greatly different end uses. In practice, the recovery of lanolin from greasy wool is relatively small and usually amounts to less than 10% by value (being higher for finer wool than coarser wool), and therefore most of the resource use and GHG emissions will be allocated to the wool.

10.2.5 Inventory data challenges

General data requirements have been described (ISO, 2006; LEAP, 2014a). However, for fibre derived from animals there is a unique requirement to fully account for resource use and environmental emissions directly from all animals associated with the production of fibre. This may require accounting for multiple farms, for example, some breeding animals reared on a different farm than the main production farm or surplus animals raised off farm and shorn before being sold for meat processing.

The key requirement is that all inputs and emissions associated with an equilibrium animal population (including all classes and ages of breeding animals) are accounted for. The most complex aspect of this is obtaining an accurate estimate of the feed consumed by the animals since this determines the production of wool (and some co-products) and environmental impacts, most directly GHG emissions. The actual amount of feed provided to animals is not often measured (except where animals are confined and feed is brought to them). Since most sheep feed outdoors on pastures or browse plants, it is necessary to estimate the feed energy and protein intake indirectly using a model. The model must be scientifically based, internationally accepted and preferably a 'tier 2' model, that is, a model that accounts for animal functions including growth, maintenance, reproduction, activity (e.g. grazing and walking) and wool production. Such models require input data on the animal population and productivity. LEAP (2014a) provided a hierarchy of acceptable models in order of preference: (1) country-specific models used in a country's national GHG inventory; (2) other peer-reviewed published models applicable to the region and country; and (3) the NRC (2007) model. The models must be supported by defensible representative data on energy and protein concentrations of the different feeds consumed.

Animals excrete urine and dung directly to land or it may be collected as manure and recycled. These different forms of excreta need to be estimated since they can be significant contributors to GHG emissions and to nutrient losses to waterways that affect eutrophication potential. For feed production, data are required on land used and external inputs, such as fertiliser and resources, for example, irrigation water. Emissions generated in feed production need to be accounted for (e.g. LEAP, 2014b) in wool LCA studies.

Most other inputs will be in common with those for other fibres used for production of textiles and clothing (BSI, 2014). An important input is energy such as fuels and electricity and their related emissions across all life cycle stages (including transportation). The scouring or cleaning process for wool can potentially use significant quantities of water and generate wastewater that must be processed to minimise impacts on eutrophication.

The functional unit (e.g. 1 m^2 of textile or an article of clothing for the lifetime of the product) includes the use, recycling and end-of-life stages of the life cycle. Methods for accounting for reuse and recycling are included in PAS 2395 (BSI, 2014). An important consideration for the use stage is the longevity or product life and its care (e.g. laundry). The relative impact of the production stage for a defined period of life of a product will vary with different lifespans, that is, there will be a higher multiplier for the production stages for shorter-lasting products. Similarly, the resource use and environmental emissions of a product are reduced where there is reuse of a product. Recycling provides a mechanism of introducing material inputs to the supply chain with lower environmental burden than virgin raw materials, for example new wool. The data challenges associated with assessing the impact of recycling and reuse of textile products have been discussed by Wolf et al. (2014). Although virgin wool accounts for about 1.5% of global fibre production, recent surveys in more economically developed countries (MEDCs) indicate that wool accounts for about 5% by weight of post-consumer clothing collected for recycling and reuse (Ward et al., 2013).

10.2.6 Consideration of the reuse and recycling phases

10.2.6.1 Overview

The majority of virgin wool-rich clothing is consumed by industrialised nations and has a relatively high selling price. Wool textiles are comparatively durable products with an intended service life of 2–10 years depending upon the type (Drycleaning Institute of Australia, 2014). Not only is wool a relatively high value material, but large quantities of post-consumer 'waste' containing the fibre, notably clothing, are collected at the end of the first-use phase, for reuse. If not suitable for reuse, collected wool clothing is recycled to manufacture other commercially valuable items. In MEDCs the recycling and reuse infrastructure is well established and is responsible for diverting large volumes of wool waste from landfill disposal or incineration such that the useful life of the fibre is many years longer than might otherwise be expected. As noted in Section 10.1.2, few studies have attempted to quantify the end-of-life treatment of wool garments, and the purpose of this section is to briefly review the flows and commercial uses of post-consumer products containing wool, particularly clothing, at the end of the first-use phase. This understanding is necessary to more accurately account for the post-first-use phase in future LCA studies.

10.2.6.2 Data

The amount of wool that is recycled and reused depends upon how much can be collected and diverted from the municipal waste stream. In practice, consumers making donations to clothing banks, take-back schemes, kerbside collections and community reuse initiatives divert used clothing from normal household waste, reducing the possibility of it being landfilled or incinerated. In Europe about 35% of textile waste is collected and diverted from landfills, which is higher than in the United States and in China. Typical pathways for used clothing composed of wool at the end of life are illustrated in Figure 10.4, which is based on a detailed study conducted in the United Kingdom (WRAP, 2012b).

For the pathways shown in Figure 10.4, GHG emissions will vary according to each particular end-of-life destination. For example, substantial GHG emissions will be associated with long-distance transportation of waste clothing to lower cost countries for sorting and recycling operations, as well as incineration and fibre decomposition in landfills. This transport also contributes to fossil fuel demand in wool LCA studies. Balanced accounting will ensure that offsets in such impacts include estimate of displacement of manufacture of the same products using entirely virgin materials. The reuse of wool-containing clothing might be expected to displace at least some purchases of new clothing, and use of recycled wool fibre as an input displaces a proportion of the virgin fibre used in certain products, albeit most likely as a blend component. This displacement is included in the accounting guidelines in PAS 2395 (BSI, 2014, p. 6). The substitution ratio between new and used garments can only be assumed because it is extremely difficult to measure.

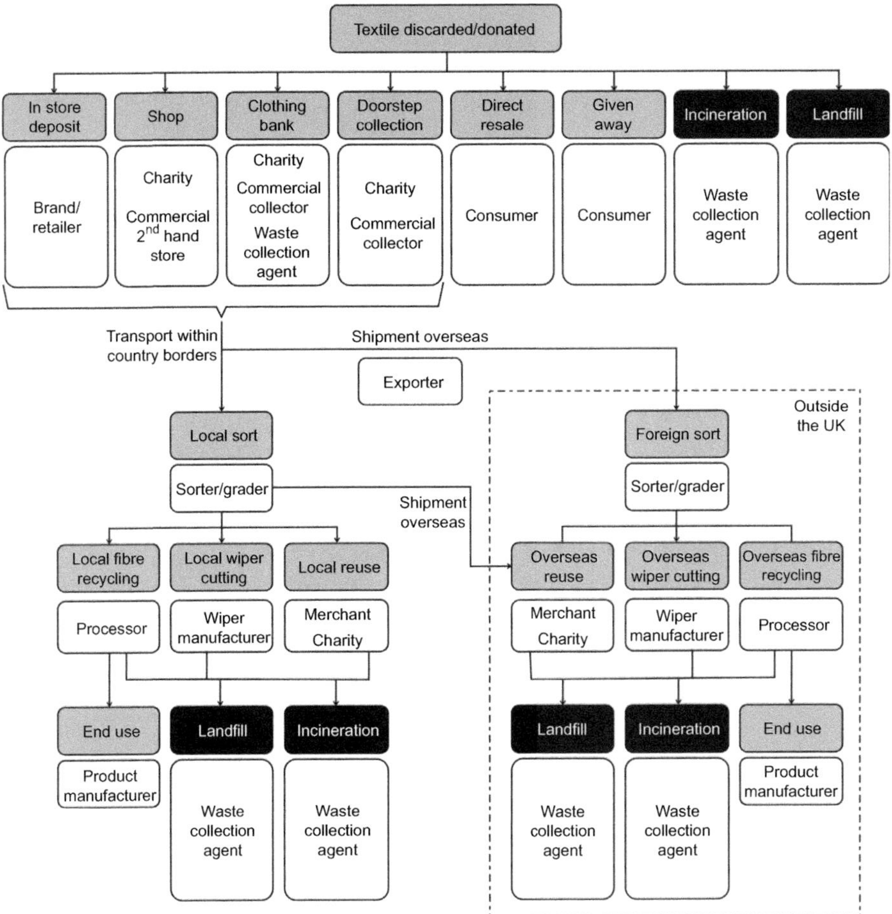

Figure 10.4 Schematic of pathways for post-consumer textiles.
This material is adapted from the Website www.wrap.org.uk of 'The Waste and Resources Action Programme', WRAP (2012b).

10.2.6.3 *Reuse and recycling processes for wool textiles*

Once collected, post-consumer wool products are sorted to identify what can be reused, recycled or remanufactured. Usually only a small amount is discarded for disposal after sorting. The different routes available from collection of waste textile products to recycling and reuse, together with the second-use products, are summarised in Figure 10.5 (WRAP, 2014). Note that this refers to materials other than just wool and pelletisation, polymerisation and extrusion relate specifically to recycling of man-made fibres. Although semi-automated systems have been developed, sorting is normally a manual process and takes place either in the same country as the items are collected or more commonly in lower labour cost countries. This means, for example, that items collected in Western Europe may be exported to Eastern Europe or other

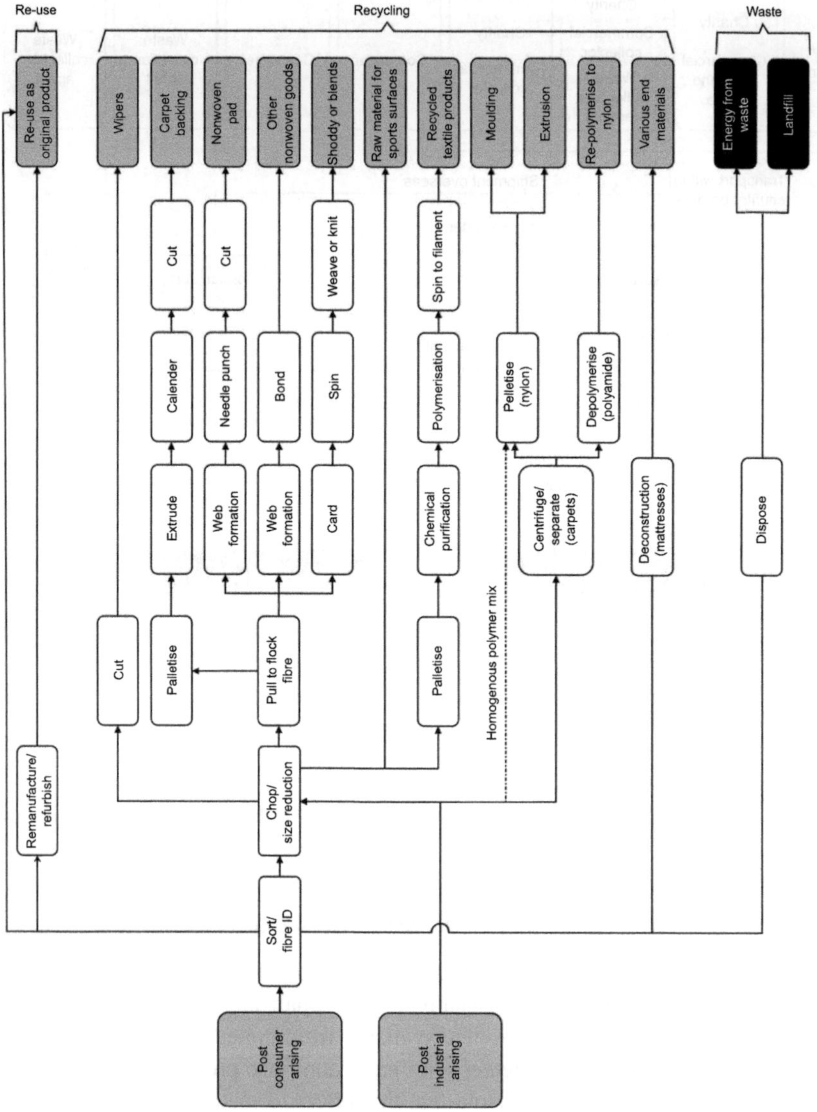

Figure 10.5 Stages involved from the collection of waste textiles to recycling and reuse.
This material is adapted from the website www.wrap.org.uk of 'The Waste and Resources Action Programme', WRAP (2014).

geographical locations for sorting, increasing the environmental impacts associated with long-distance transport. It can also have a negative economic impact on recycling companies in developed countries that may need to import sorted items that in some cases may have originally been collected in the same country.

Reuse grades containing clothing that is still wearable normally have higher resale value than the recycling grades. Wool-rich knitwear is amongst the most commercially valuable grades. Some of the clothing that is not wearable is mechanically cut up into large pieces and sold to other sectors of industry for use as wiper cloths for cleaning. The majority of the rest is mechanically recycled. Removal of wool garments for higher value reuse decreases the amount available to recyclers, and this becomes problematic when there is dependency on the supply of recycled wool to meet product performance requirements in the second-use phase product. Mechanical recycling of wool has been in commercial operation for over 200 years. Markets for recycled post-consumer clothing containing wool are therefore well established and mostly rely on open-loop recycling. Recycling involves taking wool products that are at end of life and transforming them into either the same product (closed-loop recycling) or a secondary product (open-loop recycling). Closed-loop recycling of wool is possible, but because of a reduction in the fibre length during mechanical recycling, the resulting product may require blending with at least 5–30% virgin fibre.

Post-consumer wool-rich knitwear is particularly suited to the shoddy process in which used wool clothing is mechanically pulled back into fibre using machines operating with a series of closely set rotating rollers covered in metal teeth or hooks before being converted back into yarn. The low-density yarns and loose-knitted fabric structure help to minimise the degree of fibre breakage that takes place during pulling, enabling the recovered fibre to be blended with virgin wool and spun into new yarns. Although notionally of lower grade than wool yarns made entirely from virgin fibres, shoddy yarns containing wool recyclate can still be used to make high value clothing such as wool-rich jackets and coats as well as blankets, shawls and scarves. Such consumer products frequently have a life expectancy of three to five years or more depending on frequency of use (Russell et al., 2015). One of the challenges in retrieving wool from used garment recyclate is that it may be co-mixed with other fibres that originally comprised the linings or sewing threads. In addition to manual dismantling of the garment components, carbonisation has also been used to chemically remove some cellulosic polymer components such as cotton. One of the major traditional centres for wool recycling and the production of yarns and fabrics based upon the shoddy process is the Prato district of Italy. In recent years, production of garments using the shoddy process has been to some extent rediscovered by brand owners as a possible way of reducing the overall carbon footprint and environmental impact of products containing wool (e.g. M&S, 2014; Patagonia, 2014).

Mechanical pulling of clothing containing wool can also produce fibre recyclate of short length suitable for filling cushions and soft furnishings as well as for making nonwoven insulator pads that are placed directly over the spring units in mattresses. Traditionally, when this recycled fibre material, which is known as flock or flocking, contains 40–80% wool it may be used to confer flame retardancy on mattress insulator pads. Mattresses are durable products with an expected life expectancy of five to seven years. Nonwoven fabric production is less sensitive to fibre length, and therefore wool recyclate that may be too short to convert into twisted yarns for conversion into knitted or woven structures can still be made into useful fabrics. Post-consumer wool usually forms a blend component with other fibres, a large proportion of which may also be recycled. Commercial nonwoven products containing post-consumer wool also include thermal insulation, building and construction boards, horticultural basket liners and carpet underlay. The potential use of nonwovens comprising recycled wool as sorbents for dealing with pollutants and spillages of oils or diesel fuel have also been experimentally evaluated (Radetić et al., 2003, 2008).

A relatively small proportion of post-consumer wool garments is used to make high value products for resale using the woven or knitted fabrics as component parts rather than pulling them back into fibre first. Examples include new fashion garments, accessories, bags and soft toys. These products are frequently sold at high prices because of the substantial labour and design input as well as the fact that they can be marketed as unique items.

Non-textile secondary products from wool recycling include use of wool fibre waste to enrich soil. In a closed-loop wool carpet recycling study, McNeil et al. (2007) reported that when ground-up wool carpets were used as fertiliser, elevated levels of essential elements such as nitrogen (19%), sulphur (19%) and magnesium (7%) were observed in the grass grown on it compared to the control sample. Wool-rich waste carpet is also used as mulch matting and for insulating compost heaps.

In summary, the reuse and recycling of wool products have the potential to substantially reduce environmental impacts by significantly extending material lifetime and by replacing production of new garments as well as other industrial products (Chapman, 2010). However, significant challenges have existed in modelling these impacts because of limitations in LCA methodology as well as the data gaps that exist in many countries associated with textile product waste management. Guidance for more consistent consideration of reuse and recycling in LCA GHG emissions of textiles is now available in PAS 2395 (BSI, 2014). For open-loop recycling, the GHG benefits of recycling result from the avoided emissions associated with the virgin manufacture of the secondary products that the material is recycled into (EPA, 2014). Open-loop recycling does not account for avoided emissions from manufacturing the primary material since recycling this material does not displace manufacturing of the primary material. It only displaces manufacturing of the secondary product (EPA, 2014). The potential environmental benefits associated with the shoddy process of using post-consumer recycled wool as a raw material to make new garments are particularly noteworthy. This results from sorting wool garments based on colour prior to mechanical recycling and then blending the resulting fibre recyclate appropriately prior to spinning, thus avoiding the need for a dyeing process, eliminating water usage, dyestuffs and auxiliary chemicals and wastewater effluent. LCA of building insulation products containing recycled wool has also been reported (Schmidt et al., 2004).

National campaigns aimed at improving resource efficiency, retailer-backed and incentivised take-back schemes, as well as growing consumer awareness particularly amongst the Lifestyles of Health and Sustainability (LOHAS) demographic, that encourage consumers to donate more clothing, have led to collection rates of over 40% in certain European countries. However, collection rates between individual countries vary considerably, and there is significant scope to substantially increase rates of collection in those MEDCs where most new wool garments are consumed and reduce the amount entering the municipal waste stream for landfill and incineration. Other current and emerging data challenges include storage by consumers of wool clothing without wear, which may result in overestimation of cleaning, expanded lifetimes of some high value used garments sale via the Internet and passing of wool apparel between family and friends for reuse, sometimes as valuable 'heirlooms'.

10.3 LCA case study for wool apparel

An LCA of two merino apparel products, a pair of socks and a garment, which are sold in Europe (EU) and the United States (US), is presented to illustrate the importance of key methodological issues in wool LCA and, in particular, choices for co-product allocation. This study was commissioned by the New Zealand Merino Company, Wool Research Organisation of New Zealand and an industry partner in 2014 and was conducted following ISO 14040:2006, ISO 14044:2006 and PAS 2395:2014 guidelines. The description of the case study follows the steps of defining the goal and scope of the study, important methodological choices, the life cycle inventory (LCI), life cycle impact assessment (LCIA) and, finally, interpretation of results and conclusions.

10.3.1 Goal, scope and systems boundary of the case study

The goals of the study were to:

1. Evaluate hotspots across the full life cycle of two merino outdoor apparel products (Table 10.3):
   a. A pair of socks ('Socks'), 72% merino wool, 21% nylon, 2% spandex; and
   b. A men's long-sleeved garment ('Garment'), 100% merino wool.
2. Assess the sensitivity of results to allocation methods at the farm gate where several allocation methods were applicable.

The scope of the study included all relevant processes from cradle to grave. The intended audience for the initial study was the NZ Merino Company and its industry

Table 10.3 **Declared unit for the LCA**

Declared unit	Mass excluding packaging	Mass including packaging
1 pair of socks (medium size)	0.0817 kg	0.122 kg
1 garment (men's long-sleeved base layer crew neck shirt, 250 g)	0.250 kg	0.255 kg

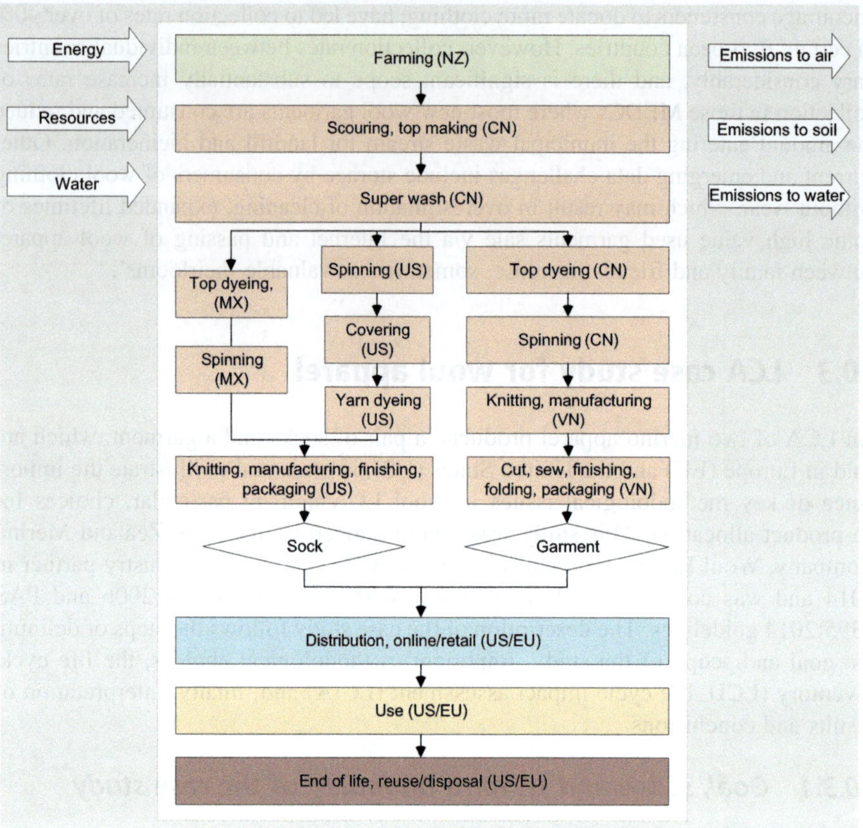

Figure 10.6 Production system and system boundaries for socks and garment (NZ = New Zealand; MX = Mexico; CN = China; VN = Vietnam; EU = Europe; US = United States).

partners. As such, no external critical review outside of the review process for this book chapter was undertaken. It was assumed that each product was washed once per week and used for one year before being discarded.

The study considered all relevant unit processes, inputs and outputs from cradle to grave, as illustrated in Figure 10.6. Inputs and outputs related to the retail location were excluded due to a lack of available data. Two scenarios were used (North American and European) to account for regional differences once the products leave the manufacturing site.

10.3.2 Data requirements and limitations

Primary data were collected for all foreground processes (e.g. farm, processing, transport, warehousing and distribution). The use phase was modelled based on literature data and background processes such as input materials, electricity, fuels, fuel combustion and end-of-life process were taken from GaBi LCA databases (PE, 2013).

The distribution information for the EU distribution centre was adapted from the primary data collected in the US. This could, in future, be improved by using primary data from a European distribution centre.

Data requirements: Production and packaging of the Garment and Socks were based on primary data. Distribution, use and end of life were modelled based on secondary data and scenarios for the US and EU.

Selected environmental indicators: This study assessed global warming potential (GWP) and two additional indicators: primary energy demand (PED) and freshwater consumption (Table 10.4). Other indicators were not included due to limitations on availability of inventory data.

Co-product allocation: As one of the goals of this case study was to understand the impacts of different product types, it was necessary to apportion environmental burdens to the relevant co-products. The stages in the wool life cycle where allocation was most relevant were:

- Farm, which produced both meat and fibre as economically significant products from sheep
- Processing, producing wool top (used in the further production stages), lanolin, noils and sweepings
- Warehousing and logistics, which were typically shared with many different product types.

For the sheep production phase, four co-product allocation methods were applied in order to test the sensitivity of the final results to the choice of method:

1. Protein required for wool and meat based on the methods of CSIRO (2007) and following the protocols of LEAP (2014a) and Wiedemann et al. (2015) for biophysical allocation from sheep (the base case) as described in Section 10.2.3
2. Mass of protein
3. Mass
4. Economic.

Modelling of biogenic carbon: Biogenic carbon stored in wool fibres was directly accounted for in the inventory as a removal of CO_2 from the atmosphere. However, the carbon sequestered in wool eventually returns to the air, and all CO_2 sequestered during growth was modelled as a release at end of life.

Table 10.4 **Environmental indicators included within this study**

Impact category	Impact assessment model	Unit	References
Climate change	Global warming potential (GWP) over a 100-year time horizon	kg CO_2 equivalent	IPCC (2007)
Consumption of primary energy	MJ (net calorific value) from nonrenewable energy sources	MJ	PE (2013)
Water consumption	Consumptive use of freshwater	kg	PE (2013)

10.3.3 Life cycle inventory

The steps in Figure 10.6 are described in further detail below.

10.3.3.1 Production

Farm production (Socks/Garment): Primary data from 27 merino farms based in New Zealand were collated, including 13 extensive farms (average stocking rate of 1.3 stock units (SU, equivalent to one 55 kg ewe rearing one lamb) per hectare), 7 medium intensive farms (average stocking rate of 3.3 SU/ha) and 7 intensive farms (average stocking rate of 9.3 SU/ha). The OVERSEER model, a tool to estimate nutrient flows and GHG emissions of farm systems (Wheeler et al., 2007), was used to calculate on-farm emissions. The three farm intensities were weighted according to their national production quantity to be representative of average New Zealand merino wool.

Scouring and top making (Socks/Garment): The input for this process was raw wool after shearing and the functional output 'wool top' was a semi-processed product of fibres in a form ready for spinning. There were five co-products considered: wool top, burrs, noils, sweepings and lanolin. Economic allocation in line with LEAP (2014a) was applied to divide the burdens between these co-products, which allocated 96.8% of burdens to wool tops.

The remaining textile manufacturing processes had only one economically valuable output and, therefore, required no allocation. These processes included:

- Super wash (Socks/Garment), a process that removed the cuticle scales from the combed fibre before applying a resin to prevent felting, imparting shrink resistance to the product to enable machine washability
- Top dyeing and spinning (Socks), in which the wool top was dyed and spun into yarn
- Spinning (Socks/Garment), in which the wool top was spun into yarn
- Covering (Socks), in which the wool was wrapped in nylon yarn
- Yarn dyeing (Socks/Garment)
- Knitting/manufacturing (Socks/Garment) in which the dyed yarn was knitted into fabric
- Cut and sew, finishing (Garment)
- Packaging (Socks/Garment) after which the finished product was packed ready for distribution
- Waste disposal and wastewater treatment was included in the production steps.

10.3.3.2 Distribution

Transport distances to warehouses and distribution centres were based on primary data collected at the production sites. Inputs and outputs were only available for the whole site and were allocated based on the total mass of apparel stored in the warehouse.

The following two distribution scenarios to the end user were included in the study:

1. Online distribution whereby the product was delivered directly to the customer; and
2. Retail distribution involving transport of the product to a retail store for pick-up by the customer. Inputs and outputs of the retail store were not available and were excluded from the study. For the shopping trip, it was assumed that the consumer bought only one item and drove 5 km each way in a medium-sized petrol car (10 km return trip).

Two location scenarios were included in the study:

1. US scenario: Based on the actual location of the warehouse/distribution centre
2. EU scenario: No data for the EU warehouse were available, and US foreground data were assumed with European background datasets.

10.3.3.3 Use

The use phase in US or EU considered a lifetime of 1 year and 52 washing cycles per year/lifetime. It was assumed that the washing and drying machines ran fully loaded. Energy, water and detergent use were allocated based on mass. As an example, given that a conventional US washing machine uses 0.21 kWh per load and one load equals 3.7 kg (equivalent to 14 garments or 45 pair of socks), 0.015 kWh would be allocated to one garment and 0.00467 kWh to one pair of socks per washing cycle.

Since the energy and water use for washing can dominate the life cycle impacts of apparel, three use-phase scenarios were applied: best case, average and worst case (Table 10.5). The average use case was created by rolling together consumer behaviour into a statistical average of reported behaviour (Cotton Inc. 2014). It is important to note that these scenarios do not directly consider different washing patterns for different fabric types (cotton vs wool). However, the best case scenario represented cold wash only and the recommended washing temperature for the analysed products was 40 °C. Energy and water use for the scenarios is summarised in Table 10.6.

Following Goodall (2012) it was assumed that a 40 °C cycle used 30% less energy than a 60 °C cycle, and from Josephy (2010) it was assumed that a 20 °C (cold wash cycle) reduced the energy use by 70% compared to a 60 °C cycle. US energy and water

Table 10.5 Description of use scenarios

	Best case		Average		Worst case	
	US	**EU**	**US**	**EU**	**US**	**EU**
Wash temperature	Cold wash	20 °C	54% Cold, 46% Heated	40 °C	Heated wash	40 °C
Washer efficiency	Energy star	A	70% Conv, 30% Energy star	B	Conventional	G
Water heater type	Nat. gas	Electric	50% Nat. gas	Electric	Electric	Electric
Drying method	Air dry	Air dry	16% Air dry, 84% Dryer	16% Air dry, 84% Dryer	Dryer	Dryer
Dryer efficiency	N/A	N/A	70% Conv, 30% Energy Star	B	Conventional	G
Air heater type	N/A	N/A	50% Nat. gas	N/A	N/A	N/A

Table 10.6 Use scenarios: energy and water use of washer/dryer per kg of product

	Best case		Average		Worst case	
	US	EU	US	EU	US	EU
Washing						
Water (l/kg)	14.71	8.33	26.66	8.33	31.79	10.00
Electricity (kWh/kg)	0.04	0.06	0.05	0.16	0.06	0.27
Nat. gas hot water (MJ/kg)	0.57	–	0.48	–	–	–
Electricity pre-heated water (kWh/kg)	–	–	0.07	–	0.17	–
Drying						
Electricity use (kWh/kg)	Air dry	Air dry	0.13	0.59	0.28	0.91
Natural gas use (MJ/kg)	–	–	0.53	–	–	–

demand for washer and dryer cycles was taken from the Department of Energy (DOE), the 'Energy Star Savings Calculator' (USEPA and DOE, 2010) and supplemented with data from the 2010 DOE Energy Conservation Program for Consumer Products (USA, 2010) and AATCC standards (AATCC, 2011).

10.3.3.4 End of life

For the end-of-life phase of the product, two scenarios were analysed within the scope of this study. The best case assumed direct release of the carbon sequestered in the product. Regardless of whether the product had one or more users, if the use phase was 10 years or less, all GHG emissions were treated as if they occurred at the beginning of the assessment period (i.e. in the first year). This approach was consistent with that recommended in ISO 14067 (ISO, 2013). A worst case scenario assumed that the products were landfilled. In this case, anaerobic decomposition of wool occurred, producing methane as well as CO_2. Methane has 25 times the GWP of CO_2 (IPCC, 2007) and landfill disposal produced a higher climate change impact as modelled based on a textile landfill dataset (PE, 2013).

10.3.3.5 Background data

Upstream processes were modelled using secondary data from the GaBi Databases 2013 (PE, 2013).

10.3.4 Case study results

For Socks, 42–55% of the total GWP was due to the farm stage and one-third due to the processing stage (Table 10.7). The use stage (average EU and average US scenario)

Table 10.7 Global warming potential, socks (CO_2e as a percentage of total life cycle emissions)

	EU scenario for socks				US scenario for socks			
Distribution	Online		Retail		Online		Retail	
End of life	Landfill	Reuse	Landfill	Reuse	Landfill	Reuse	Landfill	Reuse
Farm	52%	53%	42%	42%	54%	55%	43%	43%
Processing	25%	25%	20%	20%	25%	26%	20%	20%
Warehouse and distribution	0%	0%	0%	0%	0%	0%	0%	0%
Pick-up at retailer	0%	0%	20%	20%	0%	0%	21%	21%
Use	19%	20%	15%	16%	17%	18%	14%	14%
End of life	3%	2%	3%	1%	4%	2%	3%	1%
Total[a]	100%	100%	100%	100%	100%	100%	100%	100%

[a]The total may not be exactly 100% due to rounding errors.

contributed 14–20%. The US had slightly lower impacts due to a lower electricity consumption as most (54%) laundry was assumed to be washed cold while in the average EU scenario the washing temperature was 40 °C.

For the Garment, more than 54% of the GWP originated from the farm stage and approximately 30% from the processing stage (see Table 10.8). It should be noted that LCA impact indicators such as GWP are relative expressions only and do not predict actual impacts, the exceeding of thresholds, safety margins, or risks.

The primary energy demand was dominated for both the Socks (Table 10.9) and the Garment (Table 10.10) by the use and the processing phases. For the distribution scenario through a retail store, about one-quarter was associated with the pick-up at the store.

Water consumption quantified fresh water (from lakes, rivers or groundwater), that is, water that leaves a watershed and is lost to ecosystems and for downstream users. Most of the water was consumed during the use phase (Socks: 58–74%; Garment: 44–60%) (Tables 10.11 and 10.12). Due to the different washing machines used in the US and EU (top loader vs front loader), water consumption in the average US use scenario was substantially higher than in the average EU scenario. The estimates for the use phase have a high uncertainty relating to both data and methodology limitations relating to frequency and type of cleaning and to the water consumed during washing. The estimates of water consumption assume that all water input to washing machines is consumed, that is,100% is lost from the watershed. This depends on the water supply and wastewater system in which the washing occurs. Some cities and regional centres have water collection and infrastructure systems that facilitate return of household wastewater to watersheds, but any returns during the use phase could not be quantified due to lack of data on the proportion returned for locations of use of the two apparel products in this case study. The assumption that 100% of the water use in washing was consumed gives a high estimate, and future calculations of consumptive water use may be lower with availability of more accurate data on frequency of cleaning, lifetime of product use, and supply, management and treatment of household cleaning water.

10.3.4.1 Sensitivity of results to allocation method

As described in Section 10.3.1, four methods for co-product allocation at the cradle-to-farm gate were applied: a baseline case of biophysical allocation (protein required) and three alternative scenarios. Results for both products were found to be highly dependent on the applied allocation method. Figure 10.7 shows the GWP of the finished Socks (a) and Garment (b) for the finished product ready to be transported to a distribution centre. The cradle-to-farm gate phase contributed up to half of the total life cycle emissions, and the method of co-product allocation to the farm gate was a critical factor in the estimated environmental footprint of merino products. For example, using economic allocation for the sheep production phase resulted in estimated GHG emissions for the Socks and Garment that were, respectively, 18% and 21% higher than using the biophysical causality method based on protein required for wool and live weight (see Section 10.2.4).

Table 10.8 Global warming potential, garment (CO_2e as a percentage of total life cycle emissions)

	EU scenario for garment				US scenario for garment			
Distribution	**Online**		**Retail**		**Online**		**Retail**	
End of life	**Landfill**	**Reuse**	**Landfill**	**Reuse**	**Landfill**	**Reuse**	**Landfill**	**Reuse**
Farm	57%	57%	54%	54%	57%	58%	55%	55%
Processing	29%	30%	28%	28%	30%	30%	28%	29%
Warehouse and distribution	0%	0%	0%	0%	0%	0%	0%	0%
Pick-up at retailer	0%	0%	5%	5%	0%	0%	5%	5%
Use	11%	12%	11%	11%	10%	10%	10%	10%
End of life	2%	1%	2%	1%	2%	1%	2%	1%
Total[a]	100%	100%	100%	100%	100%	100%	100%	100%

[a]The total may not be exactly 100% due to rounding errors.

Table 10.9 Primary energy demand, socks (MJ as a percentage of total life cycle energy demand)

	EU scenario for socks				US scenario for socks			
Distribution	**Online**		**Retail**		**Online**		**Retail**	
End of life	**Landfill**	**Reuse**	**Landfill**	**Reuse**	**Landfill**	**Reuse**	**Landfill**	**Reuse**
Farm	5%	5%	3%	3%	6%	6%	4%	4%
Processing	44%	45%	31%	31%	54%	54%	35%	35%
Warehouse and distribution	1%	1%	1%	1%	1%	1%	1%	1%
Pick-up at retailer	0%	0%	31%	31%	0%	0%	36%	36%
Use	50%	50%	35%	35%	40%	40%	25%	25%
End of life	0%	0%	0%	0%	0%	0%	0%	0%
Total[a]	100%	100%	100%	100%	100%	100%	100%	100%

[a]The total may not be exactly 100% due to rounding errors.

Table 10.10 **Primary energy demand, garment (MJ as a percentage of total life cycle energy demand)**

	EU scenario for garment				US scenario for garment			
Distribution	**Online**		**Retail**		**Online**		**Retail**	
End of life	**Landfill**	**Reuse**	**Landfill**	**Reuse**	**Landfill**	**Reuse**	**Landfill**	**Reuse**
Farm	7%	7%	6%	6%	8%	8%	7%	7%
Processing	54%	54%	48%	48%	62%	62%	55%	55%
Warehouse and distribution	1%	1%	1%	1%	1%	1%	1%	1%
Pick-up at retailer	0%	0%	10%	10%	0%	0%	12%	12%
Use	39%	39%	35%	35%	29%	29%	26%	26%
End of life	0%	0%	0%	0%	0%	0%	0%	0%
Total[a]	100%	100%	100%	100%	100%	100%	100%	100%

[a]The total may not be exactly 100% due to rounding errors.

Table 10.11 Water consumption, socks (L as a percentage of total life cycle consumption)

Distribution	EU scenario for socks				US scenario for socks			
	Online		Retail		Online		Retail	
End of life	Landfill	Reuse	Landfill	Reuse	Landfill	Reuse	Landfill	Reuse
Farm	28%	28%	27%	27%	18%	18%	16%	16%
Processing	12%	12%	11%	11%	8%	8%	7%	7%
Warehouse and distribution	0%	0%	0%	0%	0%	0%	0%	0%
Pick-up at retailer	0%	0%	3%	3%	0%	0%	13%	13%
Use	60%	60%	59%	58%	74%	74%	64%	64%
End of life	0%	0%	0%	0%	0%	0%	0%	0%
Total[a]	100%	100%	100%	100%	100%	100%	100%	100%

[a]The total may not be exactly 100% due to rounding errors.

Table 10.12 **Water consumption, garment (L as a percentage of total life cycle consumption)**

	EU scenario for garment				US scenario for garment			
Distribution	**Online**		**Retail**		**Online**		**Retail**	
End of life	**Landfill**	**Reuse**	**Landfill**	**Reuse**	**Landfill**	**Reuse**	**Landfill**	**Reuse**
Farm	38%	38%	37%	37%	27%	27%	26%	26%
Processing	18%	18%	18%	18%	13%	13%	12%	12%
Warehouse and distribution	0%	0%	0%	0%	0%	0%	0%	0%
Pick-up at retailer	0%	0%	1%	1%	0%	0%	4%	4%
Use	44%	44%	44%	44%	60%	60%	58%	58%
End of life	0%	0%	0%	0%	0%	0%	0%	0%
Total[a]	100%	100%	100%	100%	100%	100%	100%	100%

[a]The total may not be exactly 100% due to rounding errors.

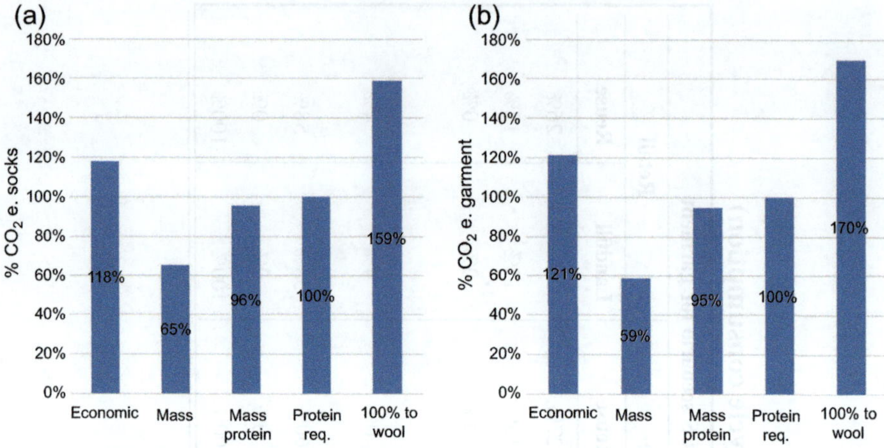

Figure 10.7 Impact of allocation method for the cradle-to-farm gate phase on GWP relative to biophysical allocation (protein required), shown as 100%. Results are for the products ((a) Socks and (b) Garment) at the point of a distribution centre.

10.3.5 *Interpretation and conclusions*

The following conclusions were drawn from the case study for each major life cycle phase.

10.3.5.1 *Merino sheep farming*

- Methane emissions from enteric fermentation and nitrous oxide emissions from soils were the main drivers of GWP.
- Primary energy demand was predominantly associated with fertiliser supply and water use.
- Water consumption was driven by on-farm water sources used as livestock drinking water.
- The choice of co-product allocation method significantly affected the results.

10.3.5.2 *Apparel processing*

- GWP and PED were predominantly associated with energy use.
- Water consumption was predominantly associated with water used for cooling systems and dyeing processes.

10.3.5.3 *Warehousing and distribution*

- The impacts were minimal for some aspects of this phase. However, online and retail scenarios showed a significant difference in impacts with retail having a higher impact than online distribution.

10.3.5.4 *Use phase*

- Washing and drying caused approximately half of the water consumption with the conservative assumptions applied here and one-third of the primary energy consumption during the life cycle of the product. Considering the uncertainty regarding consumptive water use in this phase, further research of water system supply and wastewater treatment is required to refine this estimate.

10.3.5.5 End of life

- The end-of-life scenario contributed 1–4% of the GWP of the Socks and Garment. Breakdown in anaerobic landfill resulted in higher emissions due to some carbon being released as methane.

10.3.5.6 Limitations of the study

- The cradle-to-grave results are very sensitive to assumptions in the use phase since the number of lifetime washes, the impacts of those washes and the background water systems can vary widely in practice. More consumer behaviour data, information from appliance manufacturers and information regarding background water systems could improve the accuracy of the use phase model, but the range of consumer behaviour (and therefore large variations in use phase burdens) would likely remain.

10.4 Summary and future developments in wool LCA studies

Apparel and textiles made from the natural fibre, wool, have characteristics throughout their life cycle from farming of sheep to the end-of-life stage that are both highly diverse and complex. Quantification of the environmental performance of wool textiles using a LCA approach can only be interpreted in the context of this complexity and the resultant methodological challenges. Substantial challenges relate to choice of the method of allocation of impacts and resource use between co-products of sheep production, notably meat and fibre, but also other economically and socially significant products or services. Sensitivity analysis in the case study presented in Section 10.3 demonstrates the sensitivity of results to the choice of allocation method and the importance of recent research that has provided greater consensus on methods for handling co-products of wool supply chains. The relatively large contribution from the cradle-to-farm gate stage indicates potential for increased sheep production efficiency and the importance of research on options for improving resource use efficiency and reduced environmental emissions. LCA studies can help to define the magnitude of benefits from these farm system options and encourage their adoption.

The case study for two merino wool apparel items showed for the first time the relative contributions of each phase in the full life cycle across multiple impact or resource use categories. In this assessment, socks and a long-sleeved T-shirt garment were produced from NZ merino wool, with processing and manufacture involving facilities in China, Mexico, Vietnam and the US for retail in the US or Europe. For the GWP impact using biophysical allocation, production on farm represented one-third to half of the life cycle GHG emissions, while processing was 30–35% of the total.

The majority of published wool LCA studies have assessed only GHG emissions, but the case study presented here showed that the relative contribution of life cycle stages can vary between resource or impact categories. For primary energy demand, the processing and use phases contributed most, and for consumptive water use, the use phase had the highest contribution. Understanding these differences in 'hotspots' and potential for trade-offs between environmental categories is important in targeting mitigation actions but also in conclusions on overall environmental performance as stress and priorities vary regionally. The case study results for the post-use or end-of-life phase showed a small contribution to the primary apparel products (1–4%) that use 100% virgin wool. However, the analysis of reuse and recycling showed the importance of these processes through effects on inputs to secondary products. The use of recyclate as an input has generally not been quantified but is being recognised as a knowledge gap in assessment of environmental impacts of wool textiles (BSI, 2014).

For a balanced perspective of the environmental impacts of any textile product, including those composed of wool, cradle-to-cradle thinking is required. Hence, studies should take account of the durability of different textile products, the frequency of use versus storage time, as well as the specific route that might be adopted for reuse and recycling (e.g. Laitala and Boks, 2012). These processes cannot be easily generalized and more accurate quantification of the environmental impacts and resource use of wool textiles and clothing using LCA requires new data across full, representative value chains.

References

AATCC (American Association of Textile Chemists and Colorists, 2011. Technical Manual, TM150-2010, p. 255. Standardization of Home Laundry Test Conditions, p. 402. Reference Liquid Laundry Detergent, "2003 detergent", p. 396.

Barber, A., Pellow, G., 2006. Life Cycle Assessment: New Zealand Merino Industry, Merino Wool Total Energy Use and Carbon Dioxide Emissions. Report by The Agribusiness Group, New Zealand.

Basset-Mens, C., Rankin, D.A., Boyes, M., Ledgard, S., 2007. Life Cycle Assessment of New Zealand Wool Carpet – Preliminary Study. Report by AgResearch, Hamilton, New Zealand.

Bevilacqua, M., Ciarapica, F.E., Giacchetta, G., Marchetti, B., 2011. A carbon footprint analysis in the textile supply chain. International Journal of Sustainable Engineering 4, 24–36.

Bowyer, J.L., 2009. Life Cycle Assessment of Flooring Materials: A Guide to Intelligent Selection. Dovetail Partners. Incorporated. http://www.dovetailinc.org.

Brent, A.C., Hietkamp, S., 2003. Comparative evaluation of life cycle impact assessment methods with a South African case study. International Journal of Life Cycle Assessment 8, 27–38.

Bretagnolle, V., Gauffre, B., Meiss, H., Badenhausser, I., 2011. The role of grassland areas within arable cropping systems for the conservation of biodiversity at the regional level. In: Lemaire, G., Hodgson, J., Chabbi, A. (Eds.), Grassland Productivity and Ecosystem Services. CABI International, Wallingford, UK, pp. 251–260.

Brock, P.M., Graham, P., Madden, P., Alcock, D.J., 2013. Greenhouse gas emissions profile for 1 kg of wool produced in the Yass Region, New South Wales: a life cycle assessment approach. Animal Production Science 53, 495–508.

BSI, 2011. PAS 2050:2011—Specification for the Assessment of the Life Cycle Greenhouse Gas Emissions of Goods and Services. BSI, London, UK.

BSI, 2014. PAS 2395:2014—Specification for the Assessment of the Life Cycle Greenhouse Gas (GHG) Emissions from the Whole Life Cycle of Textile Products. BSI Standards Limited, London, UK, 42 pp.

Chapman, A., July 2010. Review of Life Cycle Assessments of Clothing. Oakdene Hollins, 32 pp.

Cotton Inc, 2014. Life Cycle Assessment of Cotton Fiber & Fabric. Full Report. Prepared for VISION 21. Carz, NC, United States of America.

CSIRO, 2007. In: Freer, M., Dove, H., Nolan, J.V. (Eds.), Nutrient Requirements of Domesticated Ruminants. CSIRO Publishing, Clayton South, Australia.

CSIRO, 2008. Flame Resistance of Wool. http://www.csiro.au/Outcomes/Food-and-Agriculture/WoolFlameResistance.aspx.

Drycleaning Institute of Australia Ltd 2014. http://www.drycleanersweb.com.au/sites/default/files/user-content/resources/file/fair_claims_guide_0.pdf.

Eady, S., Carre, A., Grant, T., 2012. Life cycle assessment modelling of complex agricultural systems with multiple food and fibre co-products. Journal of Cleaner Production 28, 143—149.

Ekvall, T., Weidema, B.P., 2004. System boundaries and input data in consequential life cycle inventory analysis. International Journal of Life Cycle Assessment 9, 161—171.

EPA, 2014. Recycling. http://www.epa.gov/climatechange/wycd/waste/downloads/recycling-chapter10-28-10.pdf (accessed 21.10.14.).

FAOSTAT, 2014. Food and Agriculture Organisation of the United Nations: Statistics. FAO. http://faostat.fao.org/site/613/default.aspx; http://faostat.fao.org/site/613/default.aspx#ancor (accessed 06.10.2014).

Finnveden, G., Hauschild, M.Z., Ekvall, T., Guinee, J., Heijungs, R., Hellweg, S., Koehler, A., Pennington, D., Suh, S., 2009. Recent developments in life cycle assessment. Journal of Environmental Management 91, 1—21.

Goodall, C., 2012. How to Live a Low-carbon Life: The Individual's Guide to Stopping Climate Change. Routledge.

Henry, B.K., 2011. Understanding the Environmental Impacts of Wool: A Review of Life Cycle Assessment Studies. http://www.iwto.org/uploaded/publications/Understanding_Wool_LCA2_20120513.pdf.

ILCD, 2011. ILCD Handbook. Recommendations for Life Cycle Impact Assessment in the European Context. European Commission — Joint Research Centre. EUR 24571 EN. Publications Office of the European Union, Luxemburg, 159 pp.

IPCC, 2007. In: Core Writing Team, Pachauri, R.K., Reisinger, A. (Eds.), Climate Change 2007: Synthesis Report: Contribution of Working Groups I, II and III to the Fourth Assessment Report of the Intergovernmental Panel on Climate Change. IPCC, Geneva, Switzerland, 104 pp.

ISO, 2006. ISO 14044. Environmental Management — Life Cycle Assessment — Requirements and Guidelines. International Organisation for Standardisation, Geneva, Switzerland.

ISO, 2013. ISO/TS 14067. Carbon Footprint of Products — Requirements and Guidelines for Quantification and Communication. ISO/TC 207/SC 7. International Organization for Standardization, 52 pp.

ISO, 2014. Environmental Management — Water Footprint — Principles, Requirements and Guidelines. International Organisation for Standardisation, Geneva, Switzerland.

IWTO, 2013. IWTO Wool Market Information 2013 Edition: Statistics for the Global Wool Production and Textile Industry.

IWTO, 2014. Wool Fact Sheets. http://www.iwto.org/wool/factsheet/ (Accessed October 2014).

Josephy, 2010. Cold Wash Do Prejudices Impede High Energy Savings? Topten International Services TIS. Mibelle Group, Federation of Migros Cooperatives.

Laitala, K., Boks, C., 2012. Sustainable clothing design: use matters. Journal of Design Research 10, 121–139.

LEAP, 2014a. Greenhouse Gas Emissions and Fossil Energy Demand from Small Ruminant Supply Chains: Guidelines for Quantification. LEAP, FAO, Rome, Italy. http://www.fao.org/fileadmin/user_upload/benchmarking/docs/LEAP_Small_Ruminants_DRAFT.pdf.

LEAP, 2014b. Environmental Performance of Animal Feeds Supply Chains. LEAP, FAO, Rome, Italy. http://www.fao.org/fileadmin/user_upload/benchmarking/docs/LEAP_Anima_feeds_DRAFT.pdf.

Ledgard, S.F., Lieffering, M., Coup, D., O'Brien, B., 2011. Carbon footprinting of New Zealand lamb from an exporting nations perspective. Animal Frontiers 1, 27–32.

M&S (Marks and Spencer), 2014. Shwopping. Available from: http://www.ecouterre.com/ms-to-sell-first-shwop-coat-made-from-secondhand-garments/ http://corporate.marksandspencer.com/plan-a/about-plan-a/shwopping (cited 21.10.14.).

McNeil, S.J., Sunderland, M.R., Zaitseva, L.I., 2007. Closed-loop wool carpet recycling. Resources Conservation and Recycling 51, 220–224.

Murphy, Norton, 2008. Life Cycle Assessment of Natural Fibre Insulation Materials. Final report Imperial College London, Prepared for the NNFCC. http://eiha.org/media/attach/372/lca_fibre.pdf.

Nolan, E., Farrell, T., Ryan, M., Gibbon, C., Ahmadi-Esfahani, F.Z., 2013. Valuing quality attributes of Australian merino wool. Australian Journal of Agricultural and Resource Economics 58, 314–335.

NRC, 2007. Nutrient Requirements of Small Ruminants: Sheep, Goats, Cervids, and New World Camelids. National Research Council. The National Academies Press, Washington, USA, 362 pp.

Patagonia, 2014. Recycled Wool. Available at: http://www.patagonia.com/us/patagonia.go?assetid=93863 (cited 21.10.14.).

PE (PE International), November 2013. GaBi Database & Modelling Principles 2013. Version 6.0. http://documentation.gabi-software.com.

Petersen, A.K., Solberg, B., 2004. Greenhouse gas emissions and costs over the life cycle of wood and alternative flooring materials. Climatic Change 64, 143–167.

Potting, L., Blok, K., 1995. Life-cycle assessment of four types of floor covering. Journal of Cleaner Production 3, 201–213.

Radetić, M.M., Jocić, D.M., Jovančić, P.M., Petrović, Z.L., Thomas, H.F., 2003. Recycled wool-based nonwoven material as an oil sorbent. Environmental Science and Technology 37, 1008–1012.

Radetić, M., Ilic, V., Radojevic, D., Miladinovic, R., Jocić, D., Jovancic, P., 2008. Efficiency of recycled wool-based nonwoven material for the removal of oils from water. Chemosphere 70, 525–530.

Rebitzer, G., Ekvall, T., Frischknecht, R., Hunkeler, D., Norris, G., Rydberg, T., Schmidt, W.-P., Suh, S., Weidema, B.P., Pennington, D.W., 2004. Life cycle assessment: Part 1: framework, goal and scope definition, inventory analysis, and applications. Environment International 30, 701–720.

Ripoll-Bosch, R., de Boer, I.J.M., Bernués, A., Vellinga, T.V., 2013. Accounting for multi-functionality of sheep farming in the carbon footprint of lamb: a comparison of three contrasting mediterranean systems. Agricultural Systems 116, 60–68.

Rockström, J., Lannerstad, M., Falkenmark, M., 2007. Assessing the water challenge of a new green revolution in developing countries. Proceedings of the National Academy of Sciences 104, 6253–6260.

Russell, S.J., Swan, P., Trebowicz, M., Ireland, A., 2015. Review of wool recycling and reuse. Proceedings of the 2nd International Conference on Natural Fibres. Azores Portugal, 27th–29th April 2015. http://www.icnf2015.fibrenamics.com.

Schmidt, A.C., Jensen, A.A., Clausen, A.U., Kamstrup, O., Postlethwaite, D., 2004. A comparative life cycle assessment of building insulation products made of stone wool, paper wool and flax. The International Journal of Life Cycle Assessment 9, 53–66.

Shiklomanov, I.A., 2000. Appraisal and assessment of world water resources. Water International 25, 11–32.

Steinfeld, H., Gerber, P., Wassenaar, T., Castel, V., de Haan, C., 2006. Livestock's Long Shadow: Environmental Issues and Options. Food and Agriculture Organisation of the United Nations, Rome.

USA, 2010. Energy Conservation Program for Consumer Products: Test Procedure for Residential Clothes Washers. Department of Energy. Notice of Proposed Rulemaking, 75 Federal Register 182 (21 September 2010), pp. 57556–57595.

US Environmental Protection Agency (EPA) and Department of Energy (DOE). (2010) USEPA, DOE, 2010. Life Cycle Cost Estimate for 1 Energy Star Qualified Residential Clothes Washer(s). Retrieved from: http://www.energystar.gov/index.cfm?fuseaction=find_a_product.showProductGroupandpgw_code=CW.

Verikios, G., 2009. Modelling the world wool market: a hybrid approach. Economic Modelling 26, 418–431.

Ward, G.D., Hewitt, A.D., Russell, S.J., 2013. Fibre composition of donated post-consumer clothing in the UK. Proceedings of the ICE-Waste and Resource Management 166, 29–37.

Wheeler, D.M., Ledgard, S.F., de Klein, C.A.M., 2007. Using the OVERSEER nutrient budget model to estimate on-farm greenhouse gas emissions. Australian Journal of Experimental Agriculture 48 (2), 99–103.

WHO, 2009. Water Sanitation and Health. World Health Organisation. http://www.who.int/water_sanitation_health/hygiene/en (accessed 1.03.09.).

Wiedemann, S.G., Ledgard, S.F., Henry, B.K., Yan, M-J., Mao, N., Russell, S.J., 2015. Application of life cycle assessment to sheep production systems: Investigating co-production of wool and meat using case studies from major global producers. The International Journal of Life Cycle Assessment 20, 463–476.

Report EUR 26316 EN. In: Wolf, O., Kougoulis, J., Cordella, M., Dodd, N. (Eds.), January 2014. Environmental Improvement Potential of Textiles (IMPRO Textile). European Commission, Joint Research Centre, Institute for Prospective Technological Studies (IPTS).

Woolridge, A.C., Ward, G.D., Phillips, P.S., Collins, M., Gandy, S., 2006. Life cycle assessment for reuse/recycling of donated waste textiles compared to use of virgin material: an UK energy saving perspective. Resources Conservation and Recycling 46, 94–103.

WRAP (Waste & Resources Action Programme), 2012a. A Carbon Footprint for UK Clothing and Opportunities for Saving. http://www.wrap.org.uk/sites/files/wrap/Appendix%20IV%20-%20Carbon%20footprint%20report.pdf.

WRAP (Waste & Resources Action Programme), 2012b. Textiles Flow and Market Development Opportunities in the UK.

WRAP (Waste & Resources Action Programme), 2014. Textiles Collection Guide. Available at: http://www.wrap.org.uk/content/textile-collection-guide (cited 21.10.14.).

WRI, 2011. Greenhouse Gas Protocol – Product Life Cycle Accounting and Reporting Standard, 2011. World Resources Institute and World Business Council for Sustainable Development, Washington DC, USA.

Life cycle assessment of silk production – a case study from India

11

M.F. Astudillo, G. Thalwitz, F. Vollrath
University of Oxford, Oxford, UK

11.1 Introduction

Assessing the environmental impact of textiles allows both informed consumer choice and useful comparison of different materials as well as processing methods in engineering applications. Applying life cycle assessment (LCA) to production processes can identify 'hotspots' contributing disproportionally to environmental impact, elucidate trade-offs between production systems and serve as a decision support tool to producers and policymakers.

LCA has been extensively employed to analyse fibre and textile production and forms the basis for 'best practice' environmental guidelines such as ecolabelling. It is also the preferred methodology for assessing disparate production systems of functionally comparable man-made and natural fibres (Van der Velden et al., 2013). This latter point is particularly relevant for silks as will become apparent.

Despite its long history and unique properties as a biomaterial, to our knowledge, the LCA literature on silk production is limited to the studies presented in this chapter.

The chapter starts with a description of what silk is, where and how it is produced, focusing on South Indian sericulture. We then review the existing silk LCAs and other environmental assessments that underpin this study. Following an overview scope and overview of the life cycle inventory (LCI), and results, the discussion compares the environmental impacts of silk with those of other natural fibres. Major contributors to the high observed impact ('hotspots') are described and opportunities and barriers improvement outlined.

11.1.1 Silk production

Silk is a proteinous animal fibre – the only natural long filament fibre used in textile manufacture. Silks have been used in textiles for approximately 5000 years (Xia et al., 2009) from a number of species both wild and domesticated.

All silks are extrusion spun into a protective cocoon before the larva undergoes metamorphosis. This cocoon consists of a continuous double strand of fibroin protein, conglutinated by the sericin protein, which acts as a binder to glue the threads into the cocoon shell.

Handbook of Life Cycle Assessment (LCA) of Textiles and Clothing. http://dx.doi.org/10.1016/B978-0-08-100169-1.00011-3

Over 90% of commercially produced silk is by the domesticated silkworm *Bombyx mori*. This is a monophagous insect with a diet that is restricted to the leaves of white mulberries, *Morus alba* and *Morus indica*. The production of silk requires the maintenance of mulberry plantations (moriculture), the rearing of silkworms to the cocoon spinning stage (sericulture) and finally the unravelling of cocoons (reeling). Each subsection of the production will be outlined below.

Multivoltine (MV, many cohorts raised per year) produce final silks that do not match with those of bivoltine silks (BV, two cohorts per year) in quality. Crosses between multi- and bivoltine silkworms are not much better than pure MV silks. As BV silks are internationally traded and comparable across regions, we focus our analysis on Indian silk production from bivoltine hybrids capable of producing high-grade raw silk.

11.1.2 Silk regions

Over 90% of all *B. mori* raw silk production is in India and China. Lesser producers are Thailand, Vietnam and Brazil. There are significant regional differences between temperate sericulture areas such as Eastern China and tropical/subtropical ones such as South India in agricultural practices, matching varieties of mulberry and silkworms employed as well as in silk processing. The only published LCA of cocoon production and reeling is based on data from India, and it will be the focus of this overview. Our results and conclusions are thus highly specific, and while the insights gained can be extended to silks produced in other regions and climatic regimes, this should be done with caution and discretion.

The great majority of Indian silk is produced in three southern states: Karnataka, Andhra Pradesh and Tamil Nadu. In tropical climates such as these, mulberry can grow year-round if provided with sufficient irrigation and nutrients, although high summer temperatures can pose limits to silkworm rearing.

11.1.3 Mulberry leaf production

Production of leaves from the deciduous genus *Morus*, most commonly *M. alba* or *M. indica*, is the starting point of sericulture. In subtropical conditions such as found in the study area, this fast-growing tree starts to produce commercial quantities of leaves within one year of planting.

The plant is pruned regularly to obtain the leaves, the frequency depending on the length of the growing period of the region. If sufficient water and fertilization are made available, plants can be pruned up to six times in southern India. In temperate regions, plants are harvested twice a year. The majority of high-quality Indian mulberry production for rearing bivoltine silkworms is irrigated. Efforts have been made to introduce drip irrigation, but to date most of the farms are irrigated using furrows (Ganga, 2003a). Plants are fertilized regularly with a combination of synthetic fertilizers and farmyard manure (FYM) (Table 11.2). Pesticide application is limited by the sensitivity of silkworms to pesticides, many of which are typically lethal to these highly sensitive lepidopterans.

Leaves are the primary output of moriculture, with woody biomass from shoot and trunk growth being the principal coproduct. Leaves unsuitable for silkworm rearing are secondary coproduct and are generally used as fodder for livestock (Madhu Prasad et al., 2010).

11.1.4 Silkworm rearing

Broadly there are two varieties of silkworms, distinguished by voltinism (number of generations/cohorts per year). Bivoltines (BV) originate from the temperate regions of China and Japan, while local Indian varieties are multivoltine. Crossbreeds (CB) between bivoltines and multivoltines are the most commonly used. Bivoltines produce silk of high quality but are more susceptible to environmental stresses and disease than CB silkworms. The silkworms go through five instars before spinning their cocoons, which takes about 28 days.

Silkworms are provided by suppliers (commercial or governmental) to farmers either as eggs, or in the third instar, when silkworms are less susceptible to disease and feeding requirements increase significantly. The insects are raised indoors, in well-vented rooms where temperature and humidity should be kept closely controlled to maximize quality and yields. In tropical India, heating requirements are minimal but high summer temperatures can pose limits to silkworm rearing (Ganga, 2003b). The rearing environment should be regularly disinfected to avoid disease outbreak. A dilute solution of formaldehyde is commonly used as disinfectant (Ganga, 2003b; Dandin et al., 2003).

11.1.5 Cocoon drying and reeling

In temperate silk regions, storage of cocoons prior to further processing is necessary as cocoons are harvested only in spring and late summer. Cocoons are dried with hot air, killing the pupae and preventing eclosion of the moth. Importantly, correct drying increases both silk yield and quality and is a recommended practice even in tropical areas with year-round availability (Yong-woo, 1999). While this is a recommended practice in India, a consistent supply of cocoons from year-round rearing means cocoons are often reeled 'fresh', after spinning but before the moth emerges.

Reeling requires immersing the cocoons in water, in order to soften the sericin binder. Brushes are then used to find the end of the continuous filament. The free silk ends of several cocoons are attached to a reeling machine and unravelled onto spools. Finally, the silk is dried and transferred (re-reeled) to standardized spools. The resulting consolidated fibre is 'raw silk'. Coproducts generated in reeling are unreelable silk, sericin and pupae. Raw silk is an internationally traded commodity (United Nations, 2013) and with the exception of degumming, further processing steps are broadly comparable to those of other textiles.

11.1.6 Literature review

This section summarizes previous work on LCA of silk production and underpinning sustainability assessments, from both an agricultural and processing point of view.

LCA is commonly performed for textiles – however, the relatively small proportion of silk in the global textile trade, and possibly the complexity of the agricultural production chain, has so far precluded the generation of a large body of silk LCA suitable for meta-analysis.

Underpinning assessments (Table 11.1) include the study on health and safety issues (documented by Akter et al., 1998), waterborne emissions from degumming (Fabiani et al., 1996) and assessments of firewood consumption in reeling (Mande et al., 2000).

Vollrath et al. (2013) compiled a first LCI of farm practices based on a pilot survey in Karnataka, focusing on energy use. The results showed that energy requirements for silk were above other natural fibres and farm practices diverged significantly from guideline values. A full LCA incorporating coproducts was conducted by Astudillo et al. (2014). This study did not include the degumming and dying of the fibres, which are steps covered by the detailed work of Sara and Tarantini (2003, 2004) on the technical aspects of these steps in silk processing.

Table 11.1 Underpinning sustainability assessments: scope and main results

Scope	Conclusions	Source
Screening of health and safety issues of sericulture in Bangladesh	Inadequate health and safety practices, with lack of protective measures and uncontrolled disposal of waste.	Akter et al. (1998)
Waterborne emissions from silk degumming	Pollution from degumming can be effectively reduced recovering sericin through ultrafiltration.	Fabiani et al. (1996)
Firewood and water consumption in silk reeling	Unravelling cocoons requires between 3660 and 5440 KJ of wood per kg fresh cocoon.	Mande et al. (2000)
LCA of silk degumming, dyeing and finishing	Silk degumming is responsible for most of the environmental impact of raw silk processing.	Sara and Tarantini (2003, 2004)
LCA of raw silk based on pilot survey focusing on cumulative energy demand (CED)	Silk has a high CED when compared with other natural fibres. Electricity for irrigation is responsible for most of the nonrenewable energy requirements.	Vollrath et al. (2013)
LCA of raw silk comparing guideline recommendations and farm practices	On a mass basis, environmental impacts are above those of other fibres. Farm practices tend to have a higher environmental impact.	Astudillo et al. (2014)

We are not aware of comparative LCAs of use or end-of-life scenarios for silk and silk products. It should also be noted that, although the final application for the vast majority of silk is indeed in textiles, applications in medical devices are increasingly common.

11.2 LCI of Indian silk

Indian silk production is dominated by low capital-intensity smallholder production, with support from government agencies at national and state level. Despite significant extension work, farm practices often differ from recommended guidelines (Vollrath et al., 2013; Astudillo et al., 2014). Here we present a LCA based on guidelines and government recommendations for irrigated sericulture of BV silkworms (Government of Andhra Pradesh, 2013a,b). The assessment follows ISO 14044 guidelines (2006). Silk-specific inventory data was obtained from sericulture manuals and government recommendations (Ganga, 2003a,b; Dandin et al., 2003; Government of Andhra Pradesh, 2013a,b) as well as survey data from Karnataka (Vollrath et al., 2013). Several datasets available from Ecoinvent database (Frischknecht et al., 2005) were adapted to represent Indian conditions. Background data such as electricity provision are based on existing inventories developed by Itten et al. (2012).

As outlined above, several methods of silk production are possible. They require differing intensity of production and input use, and result in differing quality of cocoon output. Functional units need to be compared thus with caution, as low-quality silks must be processed differently, are unsuitable for machine weaving, and are less valuable.

11.2.1 Goal and scope

The goal of the study was first characterization of the potential environmental impact of Indian raw silk production on the assumption that production follows official guidelines and government recommendations and then identification of possible areas of improvement. The functional unit is one kg of bivoltine raw silk, as produced in Karnataka during the period 2000–2010. Within system boundaries are: mulberry production, silkworm rearing, cocoon drying and cooking, and silk reeling. Production of inputs (e.g. eggs, fertilizers, fuel) and capital goods are also within system boundaries (Figure 11.1). Silk weaving, degumming and dyeing were excluded from the analysis but partially covered in previous studies (Sara and Tarantini, 2003, 2004).

To cover a wider spectrum of potential environmental impacts, several indicators from different impact assessment methods were employed, as implemented in Simapro v8.0.2: global warming potential over 100 years (GWP_{100}) (IPCC, 2006), renewable and nonrenewable cumulative energy demand (CED) [v 1.08] (Hischier et al., 2010), ecotoxicity [USEtox v1.03] (Rosenbaum et al., 2008), agricultural land occupation (ALO) and freshwater eutrophication (FE) [ReCIPe v.1.09] (Goedkoop et al., 2013).

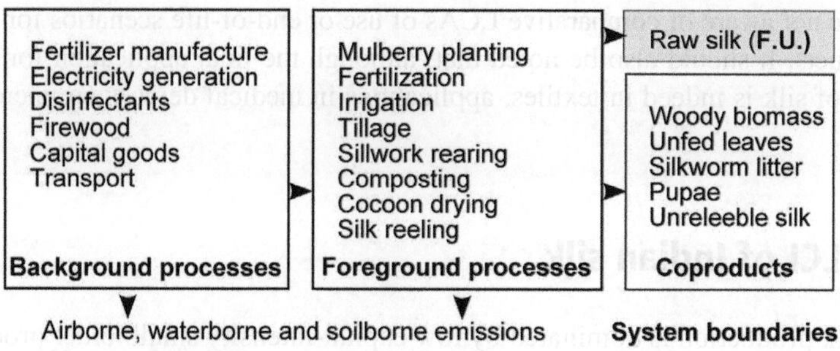

Figure 11.1 System boundaries.

Blue water footprint (BWF) was calculated following Hoekstra et al. (2011). Assumptions are detailed in the relevant subsections.

11.2.2 Mulberry production

To ensure both yields and leaf matter are of sufficient protein content, comparatively high fertilizer doses are applied to mulberry, often combining synthetic fertilizer and FYM (Table 11.2). It is assumed that macronutrients are provided in the form of urea, single superphosphate and potassium chloride, which are the most common fertilizers used in India (FAO, 2005). Burdens from imported FYM production are attributed to livestock, but transport and emissions derived from FYM applications on-farm are attributed to mulberry production. On-farm emissions of N_2O from fertilization were modelled following the Tier 1 IPCC approach (2007). NH_3 and phosphate emissions were estimated according to methods proposed by Ecoinvent developers (Nemecek and Kagi, 2007). Due to lack of reliable data regarding phosphorous emissions to freshwater from erosion, we assumed losses per hectare as specified in the Ecoinvent module of Indian kenaf. Emissions of heavy metals to soil from fertilization were estimated based on composition of the fertilizers (Nemecek and Kägi, 2007).

The other main limiting factor for mulberry production is availability of water. In the absence of crop coefficients for mulberry (Allen et al., 1998), crop water requirements were estimated based on irrigation recommendations. Following the recommendations of Dandin et al. (2003) for furrow irrigation, we estimate approximately 16,000 m^3/ha of water as annual use. Irrigation requirements are then estimated as the difference between crop water requirements and effective rainfall, assuming optimum scheduling. Although its use is widespread in India, furrow irrigation is not particularly water efficient, and 40% of applied water is estimated to be lost by seepage (FAO, 1989), albeit remaining within the watershed. Evaporation losses at dams may play an important role on the water footprint of silk and other agricultural products that do not rely on groundwater extraction (Mekonnen and Hoekstra, 2012) but are not yet commonly assessed in LCA and are not considered here.

In silk-producing regions of India, water is sourced either through groundwater extraction or surface irrigation, depending on the infrastructure available. We estimate 63% of irrigation from surface irrigation based on an average data from Karnataka

Table 11.2 Main silk production parameters

Indicator	Recommended practices
Fertilizer use (kg/ha) (N, P_2O_5, K_2O)	300:120:120 (Government of Andhra Pradesh, 2013a)
FYM (tonnes/ha)	20 (Government of Andhra Pradesh, 2013a)
Mulberry leaf (tonnes/ha)	50 (Government of Andhra Pradesh, 2013a)
Cocoon yield (kg/100DFL)	58 (Government of Andhra Pradesh, 2013a)
Cocoon yield (kg/ha)	1810 (Government of Andhra Pradesh, 2013a)
Feed requirements (kg leaf/kg cocoons)	26.2 (Government of Andhra Pradesh, 2013a)
Capacity of utilization (% of mulberry used for rearing)	95% (Government of Andhra Pradesh, 2013a)
Electricity in irrigation	0.25 kWh/m^3
Irrigation water requirements	8590 m^3/ha (Vollrath et al., 2013)
Field operations	Ploughing – 5 times year (after harvest) (Vollrath et al., 2013)
Rearing requirements (per 100 DFL)	5 l of formalin, 6.2 kg bleaching powder, 5 g detergent, 3.5 kg of newspaper (Dandin et al., 2003)
Pesticide	2.96 kg/ha/y of dichlorvos (Dandin et al., 2003)
Transport	30 km per 100 DFL (by coach) (Vollrath et al., 2013)
Water use in reeling	19.6 L/kg fresh cocoon (Venkataraman et al., 2010)
Firewood use in reeling	2.6 kg/kg fresh cocoon (Kathari et al., 2011)
Emissions from firewood use	CH_4 and N_2O (Astudillo et al., 2014) PM and others (Saud et al., 2013)
Electricity use in cocoon drying	1 kWh/kg fresh cocoon (Singh, 2011)

(Dinesh Kumar et al., 2013). For the case of surface irrigation, we follow (FAO, 1989) and assume 5% evaporation losses in conveyance canals. Seepage losses remain within the watershed and do not contribute to the BWF.

Electricity for groundwater irrigation is estimated from pump efficiency and total head (Tyson et al., 2012) (Eqn (11.1)). Water table depths vary greatly by location, and we assume an average gross head of 55 m based on Central Groundwater Board data from Karnataka (2004, 2012).

$$E_{gi}\left[kwh/m^3\right] = \frac{\rho g h}{\eta_e} \tag{11.1}$$

where E_{gi} electricity use in groundwater irrigation; ρ: water density, g: gravity constant, h: gross working head, $\eta_{e:}$ pump efficiency.

Tilling is recommended after pruning to control weeds. In less-capitalized operations, tilling is done by draft animals. Due to the lack of specific data, we have based the impact assessment on mechanized tilling.

Survey data indicate that dichlorvos is a commonly used pesticide (Vollrath et al., 2013). For this, LCI applications are based on recommended doses (Dandin et al., 2003). A fraction of the pesticides volatilizes before reaching the soil, and proportions emitted to air, soil and water were estimated as in Berthoud et al. (2011).

The area under study is not experiencing conversion of forestland into agricultural land use (FSI, 2011), and the total area dedicated to moriculture is decreasing (Central Silk Board, 2012a). It is reasonable then to assume that mulberry fields are not causing direct land use change and associated impacts.

Coproducts are often used as firewood and livestock fodder (Madhu Prasad et al., 2010); the model is detailed in the firewood and allocation sections. Inputs required for the establishment of mulberry plantations were considered and distributed uniformly across the 20-year expected life cycle of the plantation.

11.2.3 Rearing silkworms

The main physical inputs in the rearing process are mulberry leaves, silkworm eggs and disinfectants to prevent disease outbreak. Egg production is inventoried in a separate module, based on the model proposed in Dandin et al. (2003). Disinfectant use is detailed in Table 11.2 and assumed to be disposed to soil. It is known that *B. mori* does not produce methane (Hackstein and Stumm, 1994). However, as some other insects, it may produce N_2O. As a precautionary assessment, we assumed 0.6 g CO_2eq/kg biomass/day, based on emissions from other insects (Oonincx et al., 2010). Waste from rearing is composed of uneaten leaves and stems, and silkworm excreta. It is typically composted on-farm (Vollrath et al., 2013). Resulting greenhouse gas emissions were assumed to be equivalent to backyard composting of organic waste (Amlinger et al., 2008).

11.2.4 Cocoon drying, cooking and reeling

Cocoon drying and silk reeling can be a significant expenditure of fuel and an onerous expense to small operators. This partially explains why a large proportion of Indian cocoons are reeled ‘fresh’ despite reductions in silk quality. The most common dryers are electric although wood-fired models exist, and multiend reeling units use wood-fired boilers to generate steam. Fuel, water use and emissions from firewood combustion are referenced in Table 11.2. Reeling waste, as well as unreelable cocoons, are generally processed into short-fibre ‘waste silk’ and have been modelled as such.

11.2.5 Firewood

Firewood is both a product and an input to the system, although wood purchased for use in reeling operations generally does not derive directly from mulberry plantations. In the absence of datasets specific to India, we adapted the existing Ecoinvent module for *eucalyptus* produced in Thailand (Frischknecht et al., 2005), as it is commonly used in reeling clusters (Mande et al., 2000). It is assumed that wood is left to dry until it reaches equilibrium moisture content. Considering a higher heating value (HHV) of 19.55 MJ/kg (dry basis) (Jenkins and Ebeling, 1985), changes in density are estimated with the formulae proposed by Frischknecht et al. (2007). Because forest cover in the studied area has been constant since the last decade (FSI, 2011), no burden related to land use change from firewood production is attributed to Indian sericulture.

Mulberry woody biomass production is based on leaf/stem ratios (Babu et al., 2012). Wood characteristics were obtained from literature data (Guha and Reddy, 2013; Lu et al., 2009). As for the case of *eucalyptus*, we assumed wood is left to dry until equilibrium moisture content is reached.

11.2.6 Capital goods

Compared to synthetic polymer production, silk production requires significantly less capital investment; the average annual silk output per surveyed reeling facility is under 5 t (Vollrath et al., 2013). Rearing and reeling building space required was modelled according to production capacity (Government of Andhra Pradesh, 2013b; Yong-woo, 1999). Reeling machinery was approximated by the Ecoinvent dataset '*industrial machinery, heavy*' based on their weight and capacity (Central Silk Board, 2012b).

11.2.7 Allocation of coproducts

Accounting for multifunctionality is particularly important for animal-derived fibres due to the larger amounts of coproducts generated. In the case of silk, total dry matter conversion from mulberry biomass to raw silk is less than 5% (Dai-Gang, 2013). ISO standards recommend using the 'system expansion' or 'avoided burden' approach, with allocation a less preferred option. In this study we employed both approaches and provide both results for comparison.

Leaves unsuitable for silkworm rearing are generally fed to livestock. We have modelled them as a substitute for 'intensive hay' (Ecoinvent), as mulberry leaves have nutritional properties similar to alfalfa hay (Doran et al., 2007). Mulberry woody biomass is treated as a substitute for *eucalyptus* firewood, with corrections for lower HHV. Rearing waste is assumed to reduce the need of imported FYM, as they have a similar composition (Dandin et al., 2003) and composted waste makes effective fertilizer for mulberry. There are few products directly comparable to pupae, silk, and silk waste from reeling that are exclusively allocated based on their economic value.

Alternatively, results of economic allocation using constant 2005 prices are presented. As for many other agricultural products, producer prices for sericulture products exhibit high volatility. Relative prices of final products have however been consistent over time. Where available market price data has been used. Cocoons and silk products are traded on regulated exchanges and price data is available. Direct data for firewood from mulberry is not directly available, the value of close substitutes such as locally collected firewood has been employed where necessary.

11.2.8 Sensitivity analysis

To quantify the impact of uncertainty of the LCI data on the results, we performed a one-at-a-time sensitivity analysis for the following parameters: mulberry yield, cocoon yield, boiler efficiency, drier efficiency, rendita, pump efficiency and required feed. We changed the value of each parameter by 10% 'one by one' and assessed the percentage change in the impact assessment indicator. For simplicity, we used a baseline based on system expansion approach.

11.3 Results and discussion

11.3.1 Environmental impact of silk production

Results from impact assessment models for silk and other fibres are summarized in Table 11.3. Where datasets and assessments have followed a comparable approach, the environmental impact of silk on a mass basis is notably higher than that of other fibres. It should be noted that silk has very different mechanical properties from other natural textiles (Porter and Vollrath, 2009) and has no close substitute among natural fibres. Silk is also much more valuable both as fibre, yarn, and final textile.

Several factors contribute to higher observed impact. Firstly, deficiencies in infrastructure, specifically electricity supply (Itten et al., 2012) and irrigation, play an important role, affecting not just silk but all Indian agricultural production. Secondly, animal fibres generally require higher inputs compared to plant-derived fibres, resulting also in higher emissions on a mass basis.

A large amount of coproducts is generated but for the case of silk these are currently not valorized. Pupae and sericin constitute over 50% of dry weight of the final output, and we did not find evidence of these currently being utilized in Southern India (personal comm. GK Rajesh). Lastly, while silk research in India has been supported by significant funding, the primary focus has been on increasing yields of mulberry and quality of silk, with mitigation of environmental impact a secondary concern.

The majority of environmental burden stems from the agricultural phase of moriculture and cocoon production. GWP_{100}, FE and ecotoxicity is dominated by on-farm emissions (N_2O, P and pesticides, respectively). ALO, CED and BWF are mostly related to mulberry production (Table 11.4). Egg production, mulberry establishment and capital goods play a minor role on overall impact.

Table 11.3 **Environmental impact of silk production: comparison with other fibres**

Category	GWP_{100} (kg CO_{2eq}/kg)	CED R. (MJ/kg)	CED NR. (MJ/kg)	Ecotoxicity (CTU_e/kg)	ALO (m^2a/kg)	BWF (m^3/kg)	FE (g P_{eq}/kg)
Raw silk – Karnataka[a]	51.5	1349.9	110.1	522.8	19.7	24.6	4.8
Cotton fibres (China)[b]	3.4	19.7	0.1	71.2	7.8	7	0.8
Nylon 66[b]	8	1.3	$7.0*10^{-4}$	$6.0*10^{-4}$	$2.0*10^{-4}$	0.2	0.3
Sheep wool at farm (US)[b]	18.5	81.7	0.1	3.4	53.5	0.2	0.5

[a]This study.
[b]Ecoinvent database (Frischknecht et al., 2005).

Table 11.4 **Share of environmental impact by process**

	GWP	CED R	CED NR	ALO	FE	BWF	Ecotoxicity
Mulberry field	28%[a]	74%[b]		62%	50%[a]		97%
Composting	17%						
Irrigation	11%[c]		34%[c]		9%	37%[d]; 62%[e]	
Fertilizer manufacture	8%				18%		
Field operations	5%						
Cocoon drying	18%[c]		58%[c]		14%		
Cocoon reeling		25%[f]	5%	37%[f]			
Other	13%	1%	4%	1%	11%		3%

[a]Field emissions from fertilization.
[b]Solar energy requirement.
[c]Electricity.
[d]Groundwater.
[e]Surface.
[f]Firewood provision.

It is, however, worth noting that particulate matter and emissions of heavy metals were not yet modelled in USEtox at the time of analysis. In addition, the employed USEtox version is not designed to assess indoor exposure to toxics such as formalin. NO_3 emissions from reeling have no assessed eutrophication potential in freshwater as phosphorous is considered to be the limiting factor for FE (Goedkoop et al., 2013).

11.3.2 Comparison with Chinese silk

China is by far the largest silk producer. Unfortunately, complete LCIs of Chinese silk are not yet available. Here we highlight some of the similarities and differences between the system described above and Chinese production. Chinese sericulture is increasingly concentrated in subtropical areas such as Guangxi province, following concerted government efforts to shift industry location. Traditional regions still produce significant amounts of silk, at lower yields due to shorter growing seasons. As in India, cocoon production is dominated by smallholders, however, contract farming arrangements are increasingly common (Ni and Hisano, 2013), potentially affecting productivity levels through improved input provision, quality control and extension. Nevertheless, surveys at farm level suggest unbalanced fertilizer applications are not uncommon (Chen et al., 2009; Chun-yan et al., 2010). In contrast to India, Chinese farmers exclusively rear BV silkworms, producing cocoons of higher quality suitable for industrial processing and ultimately export. The reeling industry is also structured differently, with individual enterprises significantly larger and automatic reeling machinery prevalent. Overall, while the environmental impact of silk reeling is likely to differ significantly in China, agricultural processes in subtropical areas exhibit a number of similarities.

11.3.3 Coproduct use

To account for the importance of coproduct utilisation, system expansion and economic allocation were estimated and compared with the case where all the impacts are exclusively attributed to raw silk (Figure 11.2). For certain impact categories, valorization of coproducts can reduce the environmental burden by up to 60%.

In the presented case, the avoided burden is dominated by firewood from mulberry stems. This is probably not exclusive to India. In China, for example, mulberry prunings exceed household consumption of firewood where overall mulberry productivity is comparable to that of fast-growing trees (Lu et al., 2009). Using a system expansion approach, reductions in environmental impact are concentrated in categories relevant for production of firewood: land and solar energy (Figure 11.2).

11.3.4 Sensitivity analysis

Results from the sensitivity analysis are presented in Figure 11.3. Changes in the efficiency of machinery (e.g. dryers, reeling units and pumps) have a lower effect than changes in agricultural parameters (e.g. yields and reeling efficiency, 'rendita').

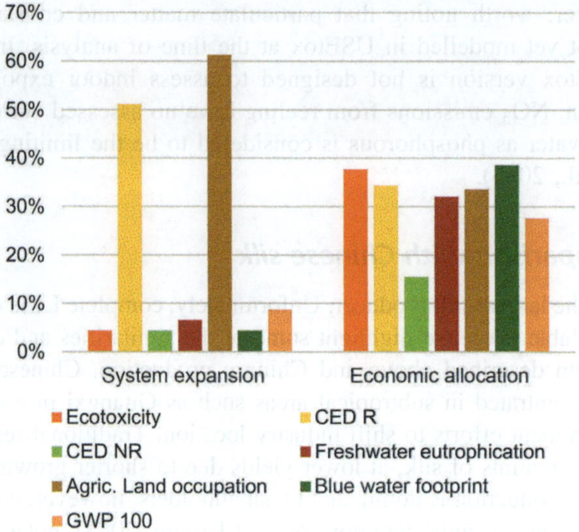

Figure 11.2 Changes in environmental impact due to coproduct utilization: system expansion versus economic allocation.

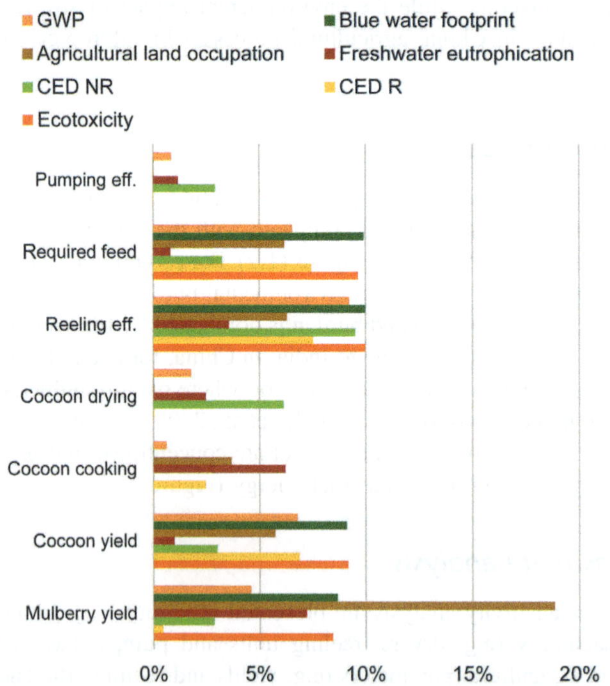

Figure 11.3 Sensitivity analysis: 10% change in production parameters.

ALO is particularly sensitive to changes in yields because coproducts of mulberry production (firewood and fodder) both require considerable amounts of land. These results can help identify effective ways of reducing environmental burden.

11.3.5 Opportunities for improvement

Fertilizer and FYM production and use are responsible for the majority of GWP_{100} and FE. Optimization of fertilizer used based on, for example, soil testing can thus help to reduce burden by improving efficiency. However, given the small size of the farms, implementation may be uneconomical at current costs. Additionally, processing of organic waste could be significantly improved. Anaerobic composting could be substituted by vermicomposting or anaerobic digestion, the latter commonly used in India to produce biogas. Adequate manure management will minimize the loss of nutrients and emissions of CH_4, effectively reducing the need for synthetic fertilizer (Babu et al., 2012). Ultimately, there is inevitably a trade-off between fertilizer and land use. Less-intensive sericulture may be limited by land availability, given high population density in India's silk-producing regions.

Energy requirements for cocoon drying could be reduced by promoting solar dryers. Compared to electric driers, dryers based on solar energy have been shown to reduce electricity consumption ten-fold (Singh, 2011).

Water use can be reduced by up to 66% without compromising yields by substituting drip irrigation systems for furrows (Siddalingaswamy et al., 2007). This will have additional benefits, reducing electricity to run the pumps. A number of silk-producing areas of India (e.g. Kolar district) are under severe water stress and electricity subsidies pose a large fiscal burden (Kimmich, 2013; World Bank, 2010).

Most ecotoxicity is linked to the use of pesticides. The use of formalin in disinfection plays a smaller role, although as discussed above exposure in enclosed environments is not modelled. In industries dominated by smallholders with low capital, it is crucial to train farmers to use these effectively and safely, as incorrect practices can incur serious health risks. For a more complete assessment of occupational health risks in sericulture and reeling, USEtox should be used in conjunction with models to assess indoor exposure (see, e.g. Rosenbaum, 2014).

Environmental sustainability can also be improved by utilizing coproducts that currently are discarded, thereby improving overall productivity and distributing incurred burdens. Pupae are often discarded (Vollrath et al., 2013) despite being an effective substitution for fish meal (Rangacharyulu et al., 2003), to the extent that they have been traditionally used in integrated agriculture–aquaculture systems. With growing interest in the use of insects as both animal feed substitutes and for direct human consumption, the only fully domesticated insect can serve as a model organism. As shown in the sensitivity analysis, the improvement of yields through, for example, improved varieties will also reduce overall environmental impact.

11.3.6 Constraints for improvement

Inevitably, there are socioeconomic barriers to the improvement of production practices, and it is crucial that recommendations take these constraints into account.

Indian sericulture is dominated by microenterprises that operate with very low margins and have little investment capacity (Mande and Kishore, 2007). Even when inputs are subsidized, their availability and correct use are not assured. There is a lack of incentives to optimize their use; for example, including a premium price for 'sustainable' silk. Additionally, improvements are limited by the capacity and quality of available infrastructure.

Technical barriers would also need to be overcome. All silks are protein-based fibres. Mulberry plants do not fix nitrogen, making the process dependent on external fertilization. It is open to question whether lower levels of fertilization can be employed without compromising the quality of cocoons (Kumar and Kumar, 2011).

11.4 Conclusions

Even following recommended practices, on a mass basis the environmental impact of Indian silk is higher than it is for other fibres – and this happens for most of the studied impact categories. Evidence suggests that the impact of present-day farm practices may well be higher. To reduce burden, several opportunities such as better coproduct management have been identified, but appropriate strategies need to be designed to overcome existing technical and socioeconomic barriers. Sericulture R&D has in the past decades focused on maximizing yields. We suggest that an emphasis on resource efficiency would be an important direction for new research. Government institutions have a prominent role in modernizing sericulture, as they are already heavily involved in most phases of the production process.

References

Akter, N., Rahman, A., Chowdhury, M., 1998. Environmental Investigation and Evaluation of Sericulture Programme and Ayesha Abed Foundation. http://research.brac.net/reports/sericulture.pdf (accessed October 2014).

Allen, R.G., Pereira, L.S., Raes, D., Smith, M., 1998. Crop Evapotranspiration e Guidelines for Computing Crop Water Requirements. FAO Irrigation and Drainage Paper 56, Rome. http://www.fao.org/docrep/X0490E/X0490E00.htm (accessed October 2014).

Amlinger, F., Peyr, S., Cuhls, C., 2008. Green house gas emissions from composting and mechanical biological treatment. Waste Management and Research 26, 47–60.

Astudillo, M.F., Thalwitz, G., Vollrath, F., 2014. Life cycle assessment of Indian silk. Journal of Cleaner Production 81, 158–167.

Babu, C.M., Dandin, S.B., Thippeswamy, T., Qadri, S., 2012. Effect of different organic manures on growth and yield of mulberry. Indian Journal of Sericulture 51, 7–10.

Berthoud, A., Maupu, P., Huet, C., Poupart, A., 2011. Assessing freshwater ecotoxicity of agricultural products in life cycle assessment (LCA): a case study of wheat using French agricultural practices databases and USEtox model. The International Journal of Life Cycle Assessment 16, 841–847.

Central Groundwater Board, 2004. Activities 2003–2004 (www document). http://cgwbswr.kar.nic.in/aboutus.htm (accessed March 2014).

Central Ground Water Board, 2012. Ground Water Year Book – India. Faridabad. http://cgwb.gov.in/documents/Ground%20Water%20Year%20Book%20-%202011-12.pdf (accessed October 2014).

Central Silk Board, 2012a. Annual Report 2011–2012. Bangalore, India. http://www.csb.gov.in/assets/Uploads/pdf-files/AR-1112-EN.pdf (accessed October 2014).

Central Silk Board, 2012b. CDP XI Plan Schemes e Post Cocoon Sector (www document). Bangalore, India. http://www.indiansilk.kar.nic.in/XIPLAN-SCHEMES/CDPXIPLAN.htm (accessed October 2013).

Chen, F., Lu, J., Zhang, M., Wan, K., Liu, D., 2009. Mulberry nutrient management for silk production in Hubei Province of China. Journal of Plant Nutrition and Soil Science 172, 245–253.

Chun-yan, L., Chao-wen, L., Shi-hua, T., Liang-yu, P., Gang, L., Cai-qun, H., Deng-wei, F., 2010. Nutrient management status in mulberry plantations of hilly regions of Sichuan Province.pdf. Southwest China Journal of Agricultural Sciences 23, 1155–1159.

Dai-gang, W., 2013. Construction and empirical analysis of eco-sericulture technology system in Qin-Ba Mountain Area. Chinese Journal of Eco-Agriculture 21, 251–260.

Dandin, S.B., Jayaswal, J., Giridhar, K., 2003. Handbook of Sericulture Technologies, third ed. Central Silk Board, Bangalore, India.

Dinesh Kumar, M., Sivamohan, M.V.K., Bassi, N., 2013. Water Management, Food Security and Sustainable Agriculture in Developing Countries. Routledge, London.

Doran, M.P., Laca, E.A., Sainz, R.D., 2007. Total tract and rumen digestibility of mulberry foliage (*Morus alba*), alfalfa hay and oat hay in sheep. Animal Feed Science and Technology 138, 239–253.

Fabiani, C., Pizzichini, M., Spadoni, M., Zeddita, G., 1996. Treatment of waste water from silk degumming processes for protein recovery and water reuse. Desalination 105, 1–9.

FAO, 1989. Annex 1: irrigation efficiencies. In: FAO (Ed.), Irrigation Water Management: Irrigation Scheduling. http://www.fao.org/docrep/T7202E/T7202E00.htm (accessed October 2014).

FAO, 2005. Fertilizer Use by Crop in India. Rome, Italy. ftp://ftp.fao.org/docrep/fao/009/a0257e/ (accessed October 2014).

Frischknecht, R., Jungbluth, N., Althaus, H.-J., Doka, G., Dones, R., Heck, T., Hellweg, S., Hischier, R., Nemecek, T., Rebitzer, G., Spielmann, M., 2005. The ecoinvent database: overview and methodological framework. International Journal of Life Cycle Assessment 10, 3–9.

Frischknecht, R., Jungbluth, N., Althaus, H., Doka, G., Heck, T., Hellweg, S., Hischier, R., Nemecek, T., Rebitzer, G., Spielmann, M., Wernet, G., 2007. Overview and Methodology. Dubendorf. http://www.ecoinvent.org/support/old-doc/rep/ (accessed October 2014).

FSI, 2011. State of Forest Report 2011. Ministry of Environment & Forests, Dehradun. http://fsi.nic.in/details.php?pgID=sb_16 (accessed October 2014).

Ganga, G., 2003a. Comprehensive Sericulture, first ed., vol. 1. Science Publishers, Inc., Enfield, USA.

Ganga, G., 2003b. Comprehensive Sericulture, first ed., vol. 2. Science Publishers, Inc., Enfield, USA.

Goedkoop, M., Heijungs, R., Huijbregts, M., De Schryver, A., Struijs, J., Van Zelm, R., 2013. ReCiPe 2008, Report 1: Characterization. http://www.lcia-recipe.net/file-cabinet (accessed October 2014).

Government of Andhra Pradesh, 2013a. Farm Model with One Acre Mulberry (V-1 Variety) Garden and Rearing of CSR Hybrid (5 Crops in a Year). http://www.seri.ap.gov.in/pm/4_2_CSRModelOne_acre_5_crops.pdf (accessed October 2014).

Government of Andhra Pradesh, 2013b. Economics of a Revived Multi-end Reeling Unit (10 Basin Capacity). http://www.seri.ap.gov.in/pm/5_C_3_10%20B_MERU.pdf (accessed October 2014).

Guha, A., Reddy, A.R., 2013. Leaf functional traits and stem wood characteristics influencing biomass productivity of mulberry (*Morus* spp. L) genotypes grown in short-rotation coppice system. BioEnergy Research 6, 547–563.

Hackstein, J.H., Stumm, C.K., 1994. Methane production in terrestrial arthropods. Proceedings of the National Academy of Sciences of the United States of America 91, 5441–5445.

Hischier, R., Weidema, B., Althaus, H., Bauer, C., Doka, G., Dones, R., Frischknecht, R., Hellweg, S., Humbert, S., Jungbluth, N., Köllner, T., Loerincik, Y., Margni, M., Nemecek, T., 2010. Implementation of Life Cycle Impact Assessment Methods. Ecoinvent Report No. 3. St. Gallen. http://www.ecoinvent.org/support/old-doc/rep/ (accessed October 2014).

Hoekstra, A.Y., Chapagain, A.K., Aldaya, M.M., Mekonnen, M.M., 2011. The Water Footprint Assessment Manual. Earthscan, London. http://www.waterfootprint.org/downloads/TheWaterFootprintAssessmentManual.pdf (accessed October 2014).

IPCC, 2006. Intergovernmental Panel on Climate Change Guidelines for National Greenhouse Gas Inventories. http://www.ipcc-nggip.iges.or.jp/public/2006gl/index.html (accessed October 2014).

ISO, 2006. ISO 14044 Environmental Management — Life Cycle Assessment — Requirements and Guidelines.

Itten, R., Frischknecht, R., Stucki, M., Scherrer, P., Psi, I., July 2012. Life Cycle Inventories of Electricity Mixes and Grid. Uster. http://www.esu-services.ch/publications/energy/ (accessed October 2014).

Jenkins, B.M., Ebeling, J.M., 1985. Thermochemical properties of biomass fuels. California Agriculture 14–16.

Kathari, V.P., Patil, B.G., Das, S., 2011. An energy efficient re-reeling process for silk reeling industry to reduce deforestation. Indian Journal of Fibre & Textile Research 36, 96–98.

Kimmich, C., 2013. Incentives for energy-efficient irrigation: empirical evidence of technology adoption in Andhra Pradesh, India. Energy for Sustainable Development 17, 261–269.

Kumar, J.S., Kumar, N.S., 2011. Production efficiency of cocoon shell of silkworm, *Bombyx mori* L. (Bombycidae: Lepidoptera), as an index for evaluating the nutritive value of mulberry, *morus* sp. (Moraceae), varieties. Psyche: A Journal of Entomology 2011, 1–3.

Lu, L., Tang, Y., Xie, J., Yuan, Y., 2009. The role of marginal agricultural land-based mulberry planting in biomass energy production. Renewable Energy 34, 1789–1794.

Madhu Prasad, V.L., Naika, R., Ravindra, U., Gokul Raj, M.P., 2010. Analysis of utilization pattern of seri-byproducts among sericulturist in Kolar district. Agriculture Update 5, 274–276.

Mande, S., Kishore, V.V.N., 2007. Towards Cleaner Technologies: A Process Story on Biomass Gasifiers for Heat Applications in Small and Micro Enterprises. TERI Press, New Delhi. http://bookstore.teriin.org/docs/books/SDC-Gasifier%20full.pdf (accessed October 2014).

Mande, S., Pai, B., Kishore, V.V.N., 2000. Study of stoves used in the silk-reeling industry. Biomass and Bioenergy 19, 51–61.

Mekonnen, M.M., Hoekstra, A.Y., 2012. The blue water footprint of electricity from hydropower. Hydrol. Earth Syst. Sci. 16, 179–187.

Nemecek, T., Kägi, T., 2007. Life Cycle Inventories of Agricultural Production Systems. Final report Ecoinvent v.2.0. Zurich and Dubendorf.

Ni, H., Hisano, S., 2013. Development of contract farming in Chinese sericulture and the silk industry. In: The Political Economy of Agro-food Markets in China, pp. 236–256.

Oonincx, D.G.A.B., van Itterbeeck, J., Heetkamp, M.J.W., Van den Brand, H., van Loon, J.J.A., van Huis, A., 2010. An exploration on greenhouse gas and ammonia production by insect species suitable for animal or human consumption. PLoS One 5.

Porter, D., Vollrath, F., 2009. Silk as a biomimetic ideal for structural polymers. Advanced Materials 21, 487–492.

Rangacharyulu, P.V., Giri, S.S., Paul, B.N., Yashoda, K.P., Rao, R.J., Mahendrakar, N.S., Mohanty, S.N., Mukhopadhyay, P.K., 2003. Utilization of fermented silkworm pupae silage in feed for carps. Bioresource Technology 86, 29–32.

Rosenbaum, R.K., Bachmann, T.M., Gold, L.S., Huijbregts, M.A.J., Jolliet, O., Juraske, R., Koehler, A., Larsen, H.F., MacLeod, M., Margni, M., McKone, T.E., Payet, J., Schuhmacher, M., Meent, D., Hauschild, M.Z., 2008. USEtox—the UNEP-SETAC toxicity model: recommended characterisation factors for human toxicity and freshwater ecotoxicity in life cycle impact assessment. The International Journal of Life Cycle Assessment 13, 532–546.

Rosenbaum, R.K., 2014. Human Health Impact Assessment of Indoor Pollutants with USEtox in LCA, SETAC Europe 24th Annual Meeting, 11–14 May, Basel, Switzerland.

Sara, B., Tarantini, M., 2003. Life Cycle Assessment of Silk- and Charged Silk Yarn in I09 Company. Ravenna. http://spring.bologna.enea.it/towefo/results_public_area/D20%20LCA/D20%20LCA_I15.pdf (accessed October 2014).

Sara, B., Tarantini, M., 2004. Life Cycle Assessment of Silk Fabric in I15 Company. http://spring.bologna.enea.it/towefo/results_public_area/D20%20LCA/D20%20LCA_I15.pdf (accessed October 2014).

Saud, T., Saxena, M., Singh, D.P., Dahiya, M., Sharma, S.K., Datta, A., Gadi, R., Mandal, T.K., 2013. Spatial variation of chemical constituents from the burning of commonly used biomass fuels in rural areas of the Indo-Gangetic Plain (IGP), India. Atmospheric Environment 71, 158–169.

Siddalingaswamy, N., Basavaiah, U.D., Dandin, S.B., Narayanagowda, S.N., Shivaprakash, R.M., 2007. A study on the efficiency of micro-irrigation systems on growth yield and quality of mulberry. Indian Journal of Sericulture 46, 76–79.

Singh, P.L., 2011. Silk cocoon drying in forced convection type solar dryer. Applied Energy 88, 1720–1726.

Tyson, A., George, B., Aye, L., Nawarathna, B., Malano, H., 2012. Energy and greenhouse gas emission accounting framework for groundwater use in agriculture. Irrigation and Drainage 61, 542–554.

United Nations, 2013. UN Comtrade (www document). Commod. Trade Stat. http://comtrade.un.org/.

Van der Velden, N.M., Patel, M.K., Vogtländer, J.G., 2013. LCA benchmarking study on textiles made of cotton, polyester, nylon, acryl, or elastane. The International Journal of Life Cycle Assessment 2, 331–356.

Venkataraman, C., Sagar, A.D., Habib, G., Lam, N., Smith, K.R., 2010. The Indian National Initiative for Advanced Biomass Cookstoves: the benefits of clean combustion. Energy for Sustainable Development 14, 63–72.

Vollrath, F., Carter, R., Rajesh, G.N., Thalwitz, G., Astudillo, M.F., 2013. Life cycle analysis of cumulative energy demand on sericulture in India. In: 6th BACSA International Conference, pp. 1–13. http://users.ox.ac.uk/~abrg/spider_site/pdfs/Silk%20LCA%20Vollrath%20et%20al%202013.pdf (accessed October 2014).

World Bank, 2010. Deep Wells and Prudence. Washington DC. http://siteresources.worldbank.org/INDIAEXTN/Resources/295583-1268190137195/DeepWellsGroundWaterMarch2010.pdf (accessed October 2014).

Xia, Q., Guo, Y., Zhang, Z., Li, D., Xuan, Z., Li, Z., Dai, F., Li, Y., Cheng, D., Li, R., Cheng, T., Jiang, T., Becquet, C., Xu, X., Liu, C., Zha, X., Fan, W., Lin, Y., Shen, Y., Jiang, L., Jensen, J., Hellmann, I., Tang, S., Zhao, P., Xu, H., Yu, C., Zhang, G., Li, J., Cao, J., Liu, S., He, N., Zhou, Y., Liu, H., Zhao, J., Ye, C., Du, Z., Pan, G., Zhao, A., Shao, H., Zeng, W., Wu, P., Li, C., Pan, M., Li, J., Yin, X., Li, D., Wang, J., Zheng, H., Wang, W., Zhang, X., Li, S., Yang, H., Lu, C., Nielsen, R., Zhou, Z., Wang, J., Xiang, Z., Wang, J., 2009. Complete resequencing of 40 genomes reveals domestication events and genes in silkworm (*Bombyx*). Science 326, 433–436.

Yong-woo, L., 1999. Silk Reeling and Testing Manual. FAO, Rome. http://www.fao.org/docrep/x2099e/x2099e00.HTM (accessed October 2014).

Comparative life cycle assessment of natural and man-made textiles

12

S.S. Muthu
Global Sustainability Services, SGS Hong Kong Limited, Hong Kong

12.1 Introduction

As a systems analysis tool, life cycle assessment (LCA) is a wonderful technique available for making scientific comparisons amongst various products and/or processes/technologies/methods. LCAs are used for single-product systems as well as comparative product systems, apart from being used to analyse the environmental impacts of processes and services. For a single-product system, LCAs can be used to identify the hotspots in the different life cycle phases of the studied system, and in case of comparative assertions, LCAs are used to compare the environmental impacts. Later case is very important and predominantly used in many sectors, and this type of comparative LCA is used for various decision-making processes (getting rid of one product system completely if the environmental impacts are found to be more compared to the rival studied, purchasing decisions – e.g. choosing an ecofriendly raw material), marketing purposes, etc. Since the results of such LCA studies decide the fate of a business/products entirely, extra caution needs to be exercised in terms of such studies. There are a lot of things need to be discussed and settled in terms of such comparative studies, and then only the results can be compared. For instance, the following factors in a study need to be the same if one intends to compare the products from the outcome of an LCA study:

- Scope
- Assessment method/(s)
- System boundaries
- Assumptions
- Functional unit
- Data collection
- Analysis techniques, etc.

It is a long-standing dilemma in textiles when trying to conclude which one is better for the environment – a natural textile product or a man-made textile product? The industry is full of myths and general perceptions, which may or may not be true. As mentioned in one of the previous books, a scientific LCA study conducted with all the above-mentioned elements consistently is the way to find out the answer

Handbook of Life Cycle Assessment (LCA) of Textiles and Clothing. http://dx.doi.org/10.1016/B978-0-08-100169-1.00012-5

for this question (Muthu, 2014). People commonly perceive that natural fibre textiles are good since they are from renewable resources and can biodegrade, etc.; and they perceive man-made textiles as environmentally unfriendly since they are manufactured from nonrenewable products, cannot biodegrade, etc. Certainly, this should not be the way this industry should be looked at. Although natural fibres have many good points, they do lack some environmental indicators, and man-made textiles surpass natural products in some indicators. One has to weigh all the parameters across in a common manner. Just from studying one issue, for example water consumption for the production of textiles, one cannot conclude whether or not a fibre is good or bad, sustainable or not. Hence, a scientific LCA study is crucial to help sort out the answers.

Unfortunately, such LCA studies are very limited in textiles and there are scant studies reported in the literature to help with such comparisons. This chapter covers the discussions of such limited studies and highlights the results from such studies.

12.2 Natural and man-made textiles

In the world of textiles, there are three major fibre types:

1. Natural – As the name implies, they arise from natural resources including both cellulosic (cotton, flax, etc.) and protein bases (wool, silk from animal origins)
2. Regenerated – These principally originate from natural polymers but they demand various treatments and processing sequences (viscose, bamboo)
3. Synthetic – These are principally derived from nonrenewable resources such as petrochemical products (polyesters, nylon).

The second and third types of fibres, regenerated and synthetic, are commonly grouped as 'man-made' fibres. Hence, this chapter focuses on a comparative look at these natural and man-made textiles with an LCA lens. The detailed aspects of production processes, pertinent environmental impacts, etc. have already been reviewed and presented by the author in a previous book (Muthu, 2014).

12.3 LCA studies

There are a limited number of studies in the literature that have compared man-made and natural textiles in the same study under similar conditions. Of course, there is a great deal of available information from the Internet and also from several consultancy studies/research papers that provide comparative figures of natural and man-made textiles. This kind of information on the environmental impact of various textile fibres, processes and technologies under natural and man-made categories of textiles have already been thoroughly reviewed and discussed in the author's previous book (Muthu, 2014). Hence, the same information is not repeated in this chapter; this chapter is concerned with studies that have directly compared both natural and man-made textiles under similar conditions/system.

12.3.1 Cotton versus polyester – a T-shirt case study (Institute for Sustainable Resources, 2009)

A case study was conducted on the LCA of a 100% Australian-cotton T-shirt (titled this way, however, the assessment included the LCA of both cotton and polyester T-shirts) by the Institute for Sustainable Resources, Queensland University of Technology, published in 2009. This is a full LCA (cradle-to-grave) study, which included all the stages of the life cycle, namely production of raw materials, product assembly, use and disposal phases. The study's boundary included environmental impact from producing inputs (petrol, buildings, fertilisers, etc.) as well, and it was aimed at identifying the major environmental hotspots in the life cycle stages of a T-shirt. The focus was primarily on two indicators, namely carbon footprint and fossil fuel depletion, and two additional indicators, ozone layer depletion and mineral resources depletion. From the results, it was determined that cotton in its production stage created 3.3 kg CO_2 e/kg cotton, and polyester in its production stage created 20 kg CO_2 e/kg polyester. In the production of cotton, the production of fertilisers was a major issue, and in case of polyester, it was the production of purified terephthalic acid. When the final results are looked at for a complete product level (T-shirt), cotton was found to be better for the environment than polyester in this study that assessed four indicators. Final carbon footprint results of these two T-shirts are given below:

- Cotton: 4 kg CO_2 e/T-shirt
- Polyester: 5 kg CO_2 e/T-shirt (Institute for Sustainable Resources, 2009).

12.3.2 Cotton versus polyester – environmental profile of cotton and polyester-cotton fabrics – bed sheets

A now relatively old study, but an important study, was published in 1999 on analysing the LCA of hotel textiles. This study analysed various functional and ecological aspects of textile fabrics made out of cotton and polyester. What is interesting and worthwhile to discuss here is the life cycle inventory (LCI) results of organic, conventional cottons and polyester. Discussion of fabrication is not necessary here since this is not a direct comparison of polyester and cotton textiles. However, at the fibre stage these three types of fibres were directly compared. The inventory assessment included many elements taken into consideration, namely energy production and consumption, chemical consumption and emissions to air and water. Energy and water results are discussed below. From the LCI results, it was revealed that the production of cotton fibres is less energy consuming than polyester production. As reported in the study, cotton fibres consumed about 40% less energy than the production of polyester fibres (even after feedstock values of nonrenewable resources, such as crude oil and natural gas used as polyester raw materials were included in the energy consumption calculation).

For 1 kg of production, organic cotton needed 53.6 MJ and conventional cotton needed 59.8 MJ, whereas polyester required 97.4 MJ of energy. In the case of water consumption, the results were reversed and supported polyester as the best choice.

Per kilogram of organic cotton demanded 24,000 kg of water and in case of conventional cotton it was 22,200 kg, where polyester demanded only 17.2 kg (Kalliala and Nousiainen, 1999).

12.3.3 Biodiversity study of various textile fibres

In 2014, a study was published on incorporating land use impacts on biodiversity into LCA for the apparel industry. This research study was done by students as part of their master's degree work (Egorova et al., 2014). The apparel industry is one of the reasons for the global decline in biodiversity, due to the considerable amount of land used by this industry. Four Patagonia T-shirts were utilised in this study to quantify each product system's biodiversity impacts. Using a previous model along with the appropriate regional biodiversity characterisation factors, this study evaluated the biodiversity impacts of four Patagonia T-shirts made from cotton, wool, polyester and lyocell, comparing the biodiversity impacts of the unique land use requirements of each of the four materials assessed.

This was basically a cradle-to-gate study, and the LCI was compiled with the aid of primary data collected from Patagonia's suppliers, which included the following parameters:

- Property size
- Location
- Land use type
- Input requirements
- Total output for each process used in manufacturing a particular T-shirt.

LCI quantified the amount of land occupied in m^2∗years per functional unit for each unit process from raw material extraction and production through manufacturing, and the functional unit assumed for this study was 1 million T-shirts.

The basic approach of calculating land occupation is the same across textiles. The calculation was based on quantifying how much land occupation a particular unit process needed to make 1 million T-shirts by multiplying the output per m^2 of each unit process by the inverse yield of every subsequent unit process. Followed by this, each unit process is assigned with a characterisation factor based on its location and land use type. Further, the land occupation and characterisation factor were multiplied together to convert land occupation into an absolute measure of biodiversity loss. The final result derived was a biodiversity impact (measured in potentially lost nonendemic species per functional unit) of each unit process, which was finally summed up to quantify the impact of the entire product system from cradle to factory gate of the selected T-shirts.

As evidenced from this study, the total land occupation of the cotton T-shirt was 1.59×10^7 m^2∗years per functional unit and the total biodiversity impact of the cotton T-shirt was 2.29×10^{-3} potentially lost nonendemic species per functional unit. For wool, the total land occupation was 1.77×10^9 m^2∗years per functional unit and the total biodiversity impact for the wool T-shirt was 4.75×10^{-2} potentially lost nonendemic species per functional unit. Land occupation of the polyester T-shirt was

1.92×10^4 m^2∗years per functional unit and the biodiversity impact was 5.02×10^{-6} potentially lost nonendemic species per functional unit. For lyocell, the land occupation was 9.77×10^5 m^2∗years per functional unit and the total biodiversity impact was 3.02×10^{-4} potentially lost nonendemic species per functional unit. The ratio of each type of textile made out of cotton, lyocell and polyester to wool, the textile with the greatest biodiversity impact, was also calculated and it worked out to 1:21 for cotton, 1:157 for lyocell and 1:9461 for polyester.

From the results of this study, it was revealed that the biodiversity impact of wool is approximately 20 times greater than cotton, 120 times greater than lyocell and more than 9000 times greater than polyester. It was found that the production of 1 million wool T-shirts led to the potential loss of roughly 4.8% of a nonendemic species regionally, while all other textiles produced a loss of less than 1%. Low yield, which eventually demands a higher amount of land for wool grazing, is the reason for these results. It was also noticed from the results that the raw material production was the major hotspot, and it was responsible for 99% of the total biodiversity impact and land occupation for the cotton and wool T-shirts, and 92% of the lyocell T-shirt. These results are attributed to the fact that these types of textiles demand a huge amount of agricultural-, forest- and pasture-based land units with considerably lower yields (thus they need more land per unit than their counterparts, i.e. urban manufacturing processes such as polyester). It was also understood from the results that the biodiversity impacts of cotton and wool manufacturing were primarily dominated by the land occupation required for the manufacturing processes, whereas the impacts of polyester and lyocell were chiefly influenced by the characterisation factors used in the model applied in this study. This model has many uncertainties and hypotheses and limitations, which are well explained in this study. A sensitivity study and a list of recommendations were also part of this study. This is really a good comparative study in terms of a single factor measured across the uniform platforms of various fibres of natural and man-made origin.

12.3.4 LCA benchmarking study on textiles made of cotton, polyester, nylon, acryl and elastane (van der Velden et al., 2014)

This study was published in 2013, focusing on furnishing improved insight into the environmental burden of textiles made out of the well-known textile materials, namely cotton, polyester (PET), nylon, acryl and elastane. This study was aimed at answering the research question formed for this research, which was: Which base material and which life cycle stage (cradle-to-gate as well as cradle-to-grave) have the biggest impact on the environment? To answer this question, LCA methodology was followed. LCI data were collected from various sources such as the literature, (LCA) databases and emission registration database of the Dutch government, along with the communications of production equipment and textile companies. Four single indicators were earmarked to express the life cycle impact assessment results: Ecocosts, 2012 (a prevention-based indicator), CO_2 equivalent (carbon footprint), cumulative energy demand and ReCiPe (a damage-based indicator).

The LCA results of cradle-to-grave analysis from raw material extraction to discarded textiles revealed that the textiles made out of acryl and PET have the least impact on the environment, followed by elastane, nylon and cotton. Additionally, it was also found out from this study that the use phase has less relative impact than is suggested in the classical literature (van der Velden et al., 2014). There were many other conclusions, which will not be discussed here.

12.4 Concluding remarks

A long-standing question remains unsolved in the textile industry: Which material is sustainable – natural or man-made? This is a long-standing debate and people have their own perceptions and conclusions about both of these textiles. Additionally, there are many myths as well in concluding which one is least environmentally friendly. The best answer can be given by a full LCA study that is conducted scientifically by including all the life cycle phases of both products made out of natural and man-made textiles; the methodological aspects (functional unit, assumptions, scope, boundary, assessment method, models) should be robust and consistent. Such consistent and scientific studies will be the way to provide answers to these questions. Reports/papers conducted on one material and comparing the other with the data obtained from other sources may not be able to lead us in the right direction, due to the inconsistencies in the methodological aspects discussed above.

Unfortunately, there are scant LCA studies on many textile products or product categories and again, there is a dearth of such comparative LCA studies on textiles. There are only a few textile studies that have compared both natural and man-made textiles on a single platform and measured different environmental indicators. The very tiny number of studies available in the literature were referred to and discussed in this chapter.

LCA researchers in textiles should notice this literature gap and attempt to fill this by means of conducting more research studies in various geographical locations, on various product categories and again for measuring many different environmental indicators (such as carbon footprint, ecological footprint, water and energy footprints, eutrophication, water consumption, acidification, human toxicity, etc.) to augment the limited knowledge in this sphere. Certainly such studies are needed and we hope that this literature gap will be reduced in the very near future.

References

Egorova, Elena, Perry, Heather, Smythe, Louisa, Song, Runsheng, Sorensen, Sarah, April 2014. Incorporating Land Use Impacts on Biodiversity into Life Cycle Assessment for the Apparel Industry, Master of Environmental Science and Management for the Bren School of Environmental Science & Management. University of California, Santa Barbara.

Institute for Sustainable Resources, 2009. Life Cycle Assessment of a 100% Australian-Cotton T-Shirt. Queensland University of Technology. Obtained from: http://cottonaustralia.com.au/uploads/publications/Case_Study_-_Life_Cycle_Assessment,_Peter_Grace.pdf.

Kalliala, Eija M., Nousiainen, Pertti, 1999. Environmental profile of cotton and polyester-cotton fabrics. Autex Research Journal 1 (1), 8–20.

Muthu, S.S., March 2014. Assessing the Environmental Impacts of Textiles and the Clothing Supply Chain. Woodhead Publishing, UK.

van der Velden, Natascha M., Patel, Martin K., Vogtländer, Joost G., September 2014. LCA benchmarking study on textiles made of cotton, polyester, nylon, acryl, or elastane. International Journal of Life Cycle Assessment 19 (2), 331–356.

Further reading

Baan, Laura de, Alkemade, Rob, Koellner, Thomas, 2013. Land use impacts on biodiversity in LCA: a global approach. The International Journal of Life Cycle Assessment 18 (6), 1216–1230.

BSR, 2009. Apparel Industry Life Cycle Carbon Mapping. Business for Social Responsibility.

BTTG (British Textile Technology Group), 1999. Report 4: Textile Mass Balance and Product Life Cycles. BTTG, Manchester, UK.

Cartwright, J., Cheng, J., et al., 2011. Assessing the Environmental Impacts of Industrial Laundering: Life Cycle Assessment of Polyester/cotton Shirts (M.Sc. thesis). Bren School of Environmental Science & Management, University of California.

Cotton Incorporated, 2011. Life Cycle Assessment of Cotton Fibre and Fabric Executive Summary (15 Pages) and Brochure (4 Pages). Prepared for Vision 21 (A Project of the Cotton Foundation).

Cherrett, N., Barrett, J., Clemett, A., Chadwick, M., Chadwick, M.J., 2005. Ecological Footprint and Water Analysis of Cotton, Hemp and Polyester. Stockholm Environment Institute can be obtained from: http://www.sei-international.org/mediamanager/documents/Publications/SEI-Report-EcologicalFootprintAndWaterAnalysisOfCottonHempAndPolyester-2005.pdf.

Dahllöf, L., 2004. LCA Methodology Issues for Textile Products. Technical report no. 2004:8. (Licentiate thesis). Chalmers University of Technology, Göteborg.

Laursen, S.E., Hansen, J., 1997. Environmental Assessment of Textiles: Life Cycle Screening of Textiles Containing Cotton, Wool, Viscose, Polyester or Acrylic Fibres. Danish Environment Protection Agency, Copenhagen, Denmark.

Laursen, S.E., Hansen, J., et al., 1997. Environmental Assessment of Textiles. Environmental Project No. 369. Danish Environmental Protection Agency.

Smith, G.G., Barker, R.H., 1995. Life cycle analysis of a polyester garment. Resources, Conservation and Recycling 14, 233–249.

Steinberger, J.K., Friot, D., Jolliet, O., Erkman, S., 2009. A spatially explicit life cycle inventory of the global textile chain. International Journal of Life Cycle Assessment 14, 443–455.

Tobler-Rohr, M.I., 2011. Handbook of Sustainable Textile Production. Woodhead Publishing Series in Textiles. Woodhead Publishing Limited, Cambridge.

Walser, T., Demou, E., Lang, D.J., Hellweg, S., 2011. Prospective environmental life cycle assessment of nanosilver T-shirts. Environmental Science & Technology 45 (10), 4570–4578.

LCA of cotton shopping bags

13

S.S. Muthu
Global Sustainability Services, SGS Hong Kong Limited, Hong Kong

13.1 Introduction

In this environmentally conscious era, it is quite important to quantify the environmental impacts of any produced product. Environmental impact assessments of produced products are slowly moving from a voluntary basis to mandatory basis. Though it is not mandatory in all countries to assess and label the environmental impacts of produced/sold products, the day is not far where all countries will demand this and make labelling of products with environmental impacts a must. This holds true for all products and it is not luxury but rather a necessity for products that are perceived to be a symbol of a throwaway society such as shopping bags and diapers. It is not urgent, but emergent, to assess the environmental impacts of these products and act wisely to reduce them to preserve our living planet.

Assessment of the environmental impacts of products on their entire life cycle phases by the concept of life cycle assessment (LCA) is becoming popular and widespread. The International Organization for Standardization (ISO) has earmarked two important standards pertaining to LCA study, ISO 14040 and ISO 14044 (ISO, 2006; ISO, 2006). According to these standards, an LCA study should include the following four interconnected steps/phases:

- Goal and scope definition
- Inventory analysis
- Impact assessment
- Interpretation

Details pertaining to the concept of LCA and the details of these above-mentioned phases have been discussed at a sufficient length and a plenty of information is available on this subject.

Shopping bags appear to be small entities by their physical dimensions and look, and pose greater threats to the environment in their life cycle phases, especially at the disposal phase. Although this statement cannot be generalized and it differs for various types of shopping bags, it is a standing statement when it comes to the environmental impacts of shopping bags. The reason behind the importance of the disposal phase of these products, which have a short life span, is that they reach their end-of-life point very quickly and their daily usage quantities are very high by an individual. Moreover, unfortunately if they are destined to be disposed at landfills, it becomes a nightmare to everyone because of many issues such as their nonbiodegradability if they are made out of synthetic materials and the landfill spaces are universally getting

Handbook of Life Cycle Assessment (LCA) of Textiles and Clothing. http://dx.doi.org/10.1016/B978-0-08-100169-1.00013-7

scarcer every day. Shopping bags made out of different materials uniquely impact the environment in different phases of their life cycle, which cannot be generalized. Hence, it is pertinent and very important to quantify and study the environmental impacts created by different types of shopping bags across the different life cycle phases by LCA technique.

13.2 Shopping bags and their environmental impacts

Shopping bags can be distinguished in terms of different criteria. The two well-known criteria that are used to classify them are their number of intended uses and the type of material used to manufacture a bag. Based on the former classification, shopping bags can be divided into two types, namely single use and multiple use, and based on the later classification, the well-known types are plastic, paper, nonwoven (polypropylene) and cotton shopping bags. Detailed discussions pertaining to the types and manufacturing processes are explained in detail by the author in Muthu and Li, 2014a,b; Muthu et al., 2011, 2012a,b and in the references therein.

Assessment of environmental impacts of shopping bags is currently an important subject of discussion. This subject is of keen interest not only to academicians and scientists but also to governments and even consumers. LCA is one of the techniques widely used to assess the environmental impact of products/processes/services, as stated earlier, and many LCA studies that were conducted on the subject of shopping bags used for grocery bags are discussed in detail in previous publications by the author (Muthu and Li, 2014c). The author has made notable contributions in this subject in terms of evaluating the eco-impacts of various shopping bags and correlating these eco-impacts by consumer behaviour and governmental policies (Muthu et al., 2011, 2012a,b,c). The only work available in the literature on this very peculiar and an important subject has been contributed so far only by the author (Muthu et al., 2011, 2012a,b; Muthu and Li, 2014d; Li et al., 2010) and acknowledged by other researchers (Musa et al., 2013).

In the context of environmental assessment of products, a very new concept called 'eco-functional assessment' was recently developed by the author as a result of his doctoral research, which is many steps further ahead of LCA (Muthu, 2011b; Muthu et al., 2013a). The author has defined the concept of eco-functional assessment for the first time and developed eco-functional assessment methodology, a theoretical framework and eco-functional models (Muthu and Li, 2014e). Different types of shopping bags used for grocery bags were assessed in terms of their eco-functionality (Muthu et al., 2012d); see references provided (Muthu, 2011b; Muthu et al., 2012d, 2013a; Muthu and Li, 2014).

As I have mentioned in previous publications (Muthu et al., 2012c), there have been many publications on the subject of LCA of plastic and paper bags and a very little focus on reusable bags. No publication so far has discussed in the literature that earmarked only for cotton bags, hence this chapter focusses only on the LCA studies of cotton bags. Another reason for focussing only on cotton bags is that, being a handbook of LCA in textiles, it is very appropriate to have a chapter on LCA of cotton bags in this book.

13.3 Cotton shopping bags

13.3.1 Manufacturing of cotton shopping bags

Cotton bags occupy a significant position in the category of reusable bags and they have gained significant popularity amongst other reusable bags. Conventional cotton fibre is generally employed to produce these bags. Due to the environmental pollution concerns pertaining to cotton's cultivation stage, which are well known, organic cotton is also alternatively employed. These cotton shopping bags are predominantly produced by weaving process (woven products).

The production process of these cotton bags involves a lengthy supply chain and begins with cultivation of conventional/organic cotton. Produced seed cotton from cultivation follows regular spinning, weaving, textile refinement and garmenting processes (ginning, yarn formation, winding, weaving, chemical processing, garmenting) as outlined in Figure 13.1 (Muthu and Li, 2014b).

13.3.2 Environmental concerns of cotton shopping bags

Cotton bags are produced from either conventional or organic cotton and hence they are subjected to the environmental impacts pertaining to the cultivation and growth of cotton fibres (especially conventional cotton). Some of the important issues include the usage of higher amounts of water and land usage, apart from the employment of huge quantities of pesticides, herbicides and their associated environmental, health issues (Muthu, 2014). Apart from the fibre stage, cotton bags are subjected to the environmental impacts pertaining to the manufacturing stage, which consist of lengthy processes that deploy a huge amount of energy and quite a lot of consumables (Muthu, 2014).

Apart from these two life cycle stages, further stages of the life cycle of a cotton shopping bag also demand various resources that again produce different environmental impacts (Muthu, 2014).

13.3.3 Merits of cotton shopping bags vis-à-vis their rivals

LCA studies assess the environmental impacts of a product in its entire life cycle phases and every product creates a different magnitude of impacts in various life cycle stages. The crux of LCA is to evaluate the environmental impacts of a product from its cradle-to-grave stages rather than focussing on one particular stage, and LCA helps to transfer impacts from one phase to another phase of the life cycle (bringing a balance in life cycle impacts from one life cycle phase to the other/s).

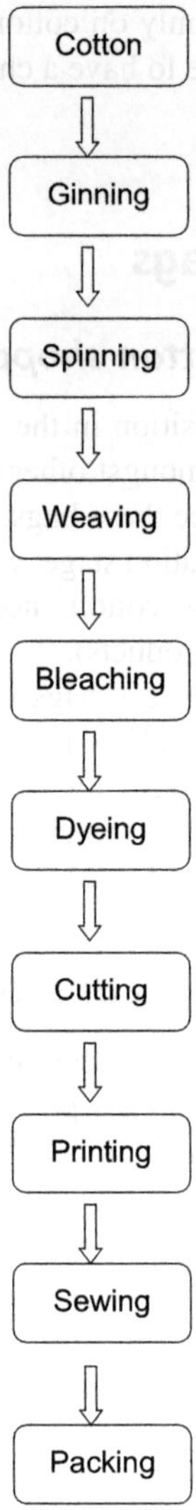

Figure 13.1 Manufacturing process of cotton shopping bags (Muthu and Li, 2014b).

But back to the case of the shopping bags – it is evident from previous studies (Muthu and Li, 2014a,b,c,d,e; Muthu et al., 2011a, 2012a,b,c,d, 2013a, 2014; Li et al., 2010) which reported that the cotton shopping bags demand huge energy requirements in their life cycle phases apart from consuming other resources to manufacture them (owing to a lengthy supply chain a cotton bag is comprised of). If the life cycle phase is divided into two stages, the first phase from cradle-to-gate would be highly detrimental for cotton bags compared to its rivals (plastic-LDPE/HDPE, paper and nonwoven bags made out of polypropylene/polyester) and the second stage would be highly beneficial for cotton woven bags compared to its above mentioned rivals and because of this second stage, it will diminish or balance the overall impacts considering the entire life cycle phases. Only because of its very good reusability, which is an important asset of these bags, its overall life cycle impacts would be significantly reduced when compared to its counterparts. Cotton woven bags would pose a major threat to the environment if they are not reused for a sufficient number of times and are discarded soon before reaching their end of life. The merit of one stage and its influence in compensating for the overall impacts can be evaluated with the aid of LCA, as mentioned earlier.

13.4 LCA studies on cotton shopping bags

Table 13.1 summarizes the studies reviewed for the discussions pertaining to the subject of this chapter. As already noted, only a few studies have been published so far on

Table 13.1 LCA studies conducted on cotton shopping bags

S. No.	Name of the study	Authors and details of affiliation	Year	Country
1.	Nolan ITU study – plastic shopping bags – analysis of levy and environmental impacts	Environment Australia, Department of the Environment and Heritage, Australia	2002	Australia
2.	The impacts of degradable plastic bags in Australia	ExcelPlas Australia, Centre for Design at RMIT& Nolan-ITU	2004	Australia
3.	Evaluation des impacts environnementaux des sacs de caisse	Carrefour	2004	France
4.	LCA of shopping bag alternatives	Zero waste South Australia	2009	Australia
5.	Life cycle assessment of supermarket carrier bags	Environment agency	2011	UK
6.	Uncertainty and sensitivity in the carbon footprint of shopping bags	Mattila et al. (Finnish study)	2011	Finland
7.	Carbon footprint of shopping (grocery) bags in China, Hong Kong and India	Muthu et al.	2011	China, HK and India

the subject of LCA of cotton bags. Again, Table 13.1 shows that the studies include cotton bags along with other bags, with no unique study/studies dealing only with cotton bags. As is evident from Table 13.1 and from the references in this author's previous publications (Muthu and Li, 2014a,b,c,d,e; Muthu et al., 2011a, 2012a,b,c,d, 2013a, 2014; Li et al., 2010), there is a scant number of journal publications available on the subject (only a few are available apart from the publications of the author) and most of the literature on this subject relies on the publications from different consultancies.

The following sections present the details of these above-mentioned studies in detail in terms of life cycle indicators studied, scope and boundaries of each study and the LCA results pertaining to cotton bags and the major conclusions from each study. All the above-mentioned studies are discussed below except for the Carrefour study, which was published in French.

13.4.1 Nolan ITU study – plastic shopping bags – analysis of levy and environmental impacts

As mentioned earlier, this Australian study was done in 2012 (Nolan-ITU Pty Ltd, 2002) and its main aim was to evaluate the environmental impacts of plastic bags and their potential alternatives by a detailed LCA study. This study evaluated 11 shopping bag alternatives from cradle-to-grave stages, including the following major aspects:

- Resource extraction and processing and other related issues of raw material preparation stage
- The manufacturing process and its detailed implications
- Distribution and transportation impacts from the manufacturer to the wholesaler/retailer
- Consumer-use stage including the patterns of use
- End-of-life or disposal options.

In this study, calico/woven cotton bags that were evaluated were mainly imported to Australia from Asia. This study has earmarked the merits of cotton bags compared to its counterparts, including the considerable resource savings due to the reduction in the materials employed for the production of single-use shopping bags (depending on the expected service life of a reusable bag) and also the reduction in the amount of litter, since the frequency of a cotton bag's reaching landfill is quite low compared to single-use bags. On the other hand, this study also mentioned the noticeable consequences of these cotton bags pertaining to the consumption of synthetic fertilizers and pesticides in the cotton growth stage and the labour issues pertaining to the cotton industry. Apart from this, the life cycle environmental impacts during the use stage of these reusable bags related to washing (water and detergent consumption, energy consumption and waterborne wastes) and ironing (some consumers would love to iron their bags, which does employ energy consumption). Again, it will be cumbersome to recycle the damaged cotton bags after their useful life.

A streamlined LCA study was done to assess the life cycle impacts of these chosen packaging systems with the aid of Simapro software, clubbed with Australian and International data. The following functional unit was used in this study: **a household**

Table 13.2 **Nolan ITU-2002 study's result – cotton bags (Nolan-ITU Pty Ltd, 2002)**

Alternative	Material Consumption (kg)	Litter (g)	Litter (m^2)	Litter (m^2/y)	Greenhouse gas emissions (CO_2 eq)	Primary energy use (MJ)
Calico	1.14	5.7	0.0041	0.0819	2.52	160

carrying approximately 70 grocery items home from a supermarket each week for 52 weeks. The relative carrying capacity and expected life of various shopping bags included in this assessment were taken into account in the assessment. For cotton woven bags, it was assumed that a bag weighs 125.4 g with a relative capacity of 1.1 (depending on the number of items it can carry, seven items used as an average). Further, it was assumed that a cotton woven bag can be capable of completing 52 trips in a year, and in total 9.1 bags were required to fulfil the assumed functional unit in adjustment to the expected lifetime of shopping bags. It was also estimated that on an average a calico bag will stay in the litter stream for 2 years. It was chosen to represent the litter with the aid of three indicators, which follow:

- Mass of material finding its way into the litter stream – representing mass of resources lost from recovery options
- The area of ground covered by litter – measured in metres squared (m^2) and based on the maximum area that a bag could cover if lost to the litter stream
- The persistence of litter measured in metres squared per annum (m^2a) to represent the area covered by litter over time.

Alternatives were assessed via LCA models assuming 52 shopping trips in one year with 10 average plastic shopping bag loads in each trip; the results pertaining to cotton bags are shown in Table 13.2.

In this assessment, woven High Density Polyethylene (HDPE) was found to be the preferred option in terms of resource efficiency, embodied energy and global warming potential, primary energy used and lowest contribution to litter. The calico bag scored well in the litter, followed by woven HDPE and PP green fibre bag (nonwoven bag). In the reusable bags category, calico's environmental footprints in terms of primary energy and green house gas emissions are higher compared to its rivals as far as this study is concerned. This is owing to the lengthy supply chain and the greater number of processes involved on a comparative scale, apart from the environmental issues of cotton bags already mentioned in the earlier paragraphs (Nolan-ITU Pty Ltd, 2002).

13.4.2 ExcelPlas Australia, Centre for Design at RMIT& Nolan-ITU – The impacts of degradable plastic bags in Australia

This is an updated version of previous study, which was discussed earlier. In this study (ExcelPlas Australia Centre for Design at RMIT& Nolan ITU, 2004), a streamlined

LCA was undertaken for bags manufactured from degradable polymers, and these bags were compared with conventional HDPE bags, paper bags, reusable plastic bags and calico bags. This study was based on the previous study as far as the streamlined LCA was concerned. Streamlined LCA modelling in this study included the following aspects:

- Laid a foundation from the previous report and discussed the assumptions regarding the rates of degradation of different shopping bag materials that were modelled in the streamlined LCA reported in Nolan ITU et al. (2002) and how they were modified in this study to suit its needs.
- Compared the environmental profile of different degradable resins for shopping bag applications pertaining to Australia.
- Importantly, made the comparative assertion of the environmental profile of the degradable bags with other single-use bags (Low Density Polyethylene (LDPE), HDPE, paper and cotton bags).

Although calico bags must be treated as reusable bags (which goes without saying), at some of the places calico bags are treated and regarded as single-use bags, and in some instances, they were mentioned as a reusable one. This dilemma prevails in this report, which is a shortfall of this study.

The goal of this study was to determine the life cycle environmental profile of degradable plastics in the application of shopping bags and to provide a comparative view against its rivals such as HDPE, LDPE, paper and calico in terms of life cycle impacts. A similar functional unit as the previous study, that is, **a household carrying approximately 70-grocery items home from a supermarket each week for 52 weeks**, was assumed in this study too. The number of single and reusable bags required to fulfil this functional unit was assessed in this study as well. Publicly available life cycle inventories were used as data sources for this study. Bags at retail outlets after the production of the bags and the consumer use were the main areas outside of the scope of this study. The same criteria as the previous study (125.4 g of bag weight; relative capacity of 1.1; 52 trips a year and in total 9.1 bags were required to fulfil the assumed functional unit) was used in this study also for calico bags. The following end-of-life scenario was assumed for cotton bags: 99.5% landfill and 0.5% litter. The final results obtained from the comparative assertions from this study are shown in Table 13.3.

In the comparative environmental profile, calico bags outscored all the other alternatives' considered in this study in terms of Litter Marine Biodiversity and Litter Aesthetics and they were the best choice in terms of these two impact categories. However, their material consumption, abiotic depletion and eutrophication scores are higher than other synthetic fibre counterparts (PP fibre reusable, woven HDPE reusable and LDPE reusable); greenhouse gas emissions are higher than PP fibre reusable and woven HDPE reusable, but lower than the LDPE reusable bag. The environmental profile of calico bags in all the categories studied in this work is significantly lower than the Kraft paper-reusable category (for that matter, even all synthetic counterparts scored well over Kraft paper). Greenhouse gas emissions of calico bags are higher than its counterparts of reusable category due to the methane emissions during landfill degradation and N_2O emissions from the fertilizing crops.

Table 13.3 **ExcelPlas Australia, Centre for Design at RMIT & Nolan ITU, 2004 study's result — cotton bags (ExcelPlas Australia Centre for Design at RMIT & Nolan ITU, 2004)**

Alternative	Material consumption (kg)	Greenhouse gas emissions (kg CO_2)	Abiotic depletion (kg Sb eq)	Eutrophication (kg PO_4 eq)	Litter marine biodiversity (kg·y)	Litter aesthetics (m^2·y)
Calico	1.141	2.56	0.059	0.01	0.003	0.03

Similarly, the eutrophication level of cotton bags is higher compared to its rivals due to the application of fertilizers to land.

Similar conclusions as in the previous study were reached here in this study as well, that is reusable bags have lower environmental impacts than all of the single-use bags, including both conventional HDPE bags and degradable bags.

13.4.3 Zero Waste South Australia, LCA of shopping bag alternatives study, 2009

This is another streamlined LCA study of different shopping bag alternatives. The reusable cotton bag considered in this study was reported to weigh 85 g, with a relative capacity of 1.4 (to HDPE) and expected to have 104 trips in its life cycle. The functional unit was similar as in previous Australian studies, and it was reported that four bags per year were needed to fulfil the desired functional unit.

The reusable cotton bag used in this study was reported to be imported 100% from Pakistan and travelled a distance of 12,300 km. This cotton bag was assumed to be landfilled to 99.5% and 0.5% to be littered. Different LCA indicators, namely material consumption, climate change, energy consumption, water use, litter marine impacts and litter aesthetics were quantified in this study for various shopping bag alternatives considered in this study.

The results were expressed in terms of a five-point scale rating. A rating scale of 1–5 was utilized to show the diversity of impacts for each impact category considered, with 1 being the lowest impact. Detailed assumptions and the calculation base used to arrive at the scaling system are in the Appendix of this report (O'Farrell, 2009). Reusable cotton bags were reported to have one point in the five impact categories of material consumption, climate change, energy consumption, litter marine impacts and litter aesthetics, and five points in the water use category. Numerical LCA modelling results (also reported in the Appendix) are indicated in Table 13.4. From the above results it can be understood that the reusable cotton bags scored well in all the impact categories except for water use.

It was concluded in this study that reusable bags were found to be more environmentally beneficial than all of the bags with only 1–3 uses. It was reported that a

Table 13.4 Numerical LCA modelling results – Zero Waste South Australia, LCA of shopping bag alternatives study, 2009 (O'Farrell, 2009)

Bag type	Climate change (kg CO_2 eq)	Cumulative energy demand (MJ LHV)	Water use (litres)	Litter marine biodiversity (g·y)	Litter aesthetics (m^2a)
Reusable calico	3.20	30.05	7600.36	0.76	0.01

considerable change in switching over to more durable bags would reduce the environmental impacts significantly mainly through reductions in greenhouse gases, energy and water use, resource depletion and litter.

13.4.4 Life cycle assessment of supermarket carrier bags – environment agency (UK)'s study, 2011

13.4.4.1 Goal, scope and methodology

This British study (Edwards and Fry, 2011) assessed the life cycle environmental impacts of the production, use and disposal of different carrier bags for the UK. HDPE bag with its some of the variants, LDPE, PP, paper and cotton bags were assessed in this study. The primary goal of this study was assessing the potential life cycle environmental impacts of various current and potential supermarket carrier bags in the UK. A supermarket carrier bag was defined in this study as one with a capacity of over 15 L that could be used at a supermarket checkout. Only bags falling under this category were assessed.

The functional unit assumed in this study was, **Carrying one month's shopping (483 items) from the supermarket to the home in the UK in 2006/07**. The reference flow for this kind of shopping bag case study is related to many factors, namely the volume of the bag, its strength and consumer behaviour when filling and using the bags; in this particular study, it was quantified with the aid of a consumer survey. The assumed volume, weight, items per bag and required reference flow for cotton bag (excluding primary reuse) were: 28.65 L of volume; 183.11 g of weight/bag; 10.59 items per bag and 45.59 number of bags were found to be needed to fulfil the assumed functional unit of this study. Compared to all the bags included in this study, the reference flow of cotton bags was found to be lower.

This study was a cradle-to-grave assessment, and hence all significant life cycle stages, namely raw material extraction, process of manufacture, distribution, use and reuse to the final management of the carrier bag as waste were included in the scope of assessment. Recycled content and recycling and composting at end of life were not included in the scope of assessment. Land filling and incineration were included in the end-of-life management scenario pertaining to cotton shopping bags in this study.

With the aid of IPCC 2007 characterization factors over a 100-year time horizon, the global warming potential (GWP) of shopping bag alternatives were assessed, and using CML 2001 method, the different impact categories were quantified, namely depletion of abiotic resources, photo-oxidant formation, eutrophication, acidification, human toxicity and aquatic and terrestrial toxicity.

13.4.4.2 Inventory analysis

Manufacturing process

Weight and volume of a cotton bag to fulfil the required functional unit of a cotton bag have already been mentioned. It was reported that a cotton bag has 181.81 g of cotton textile and 1.3 g of cotton thread to make a total weight of 183.11 g. Cotton bags studied in this report contained corrugated box 1000 g/50 bags as primary packaging. Data for the cotton bags were modelled with the aid of the ecoinvent database. Regarding the production process data of cotton bag production (bag production from fabric stage), no primary data on cotton bag production were obtained by the producers contacted by the authors of this study, and hence the data on conversion of cotton fabric to a carrier bag were only estimated. It was also assumed that the cotton bags in this study were produced in China with the aid of electric sewing machines. Therefore, electricity usage for those machines was obtained from previous projects and subsequently modelled using data from the ecoinvent database. Eleven kWh (39.6 MJ; 0.06 kWh/kg) electricity was estimated to produce 1000 cotton bags. All production waste (1800 g for 1000 cotton bags) from the manufacturing process was assumed to be landfilled (Edwards and Fry, 2011).

Transportation scenario

The transportation scenario used in this study is given in Table 13.5.

End-of-life scenario

Eighty-six percent of cotton bags were assumed to be landfilled and 14% were incinerated, and the end-of-life scenarios were modelled with the aid of WRATE (LCA software developed by the Environment Agency).

13.4.4.3 Impact assessment

The life cycle environmental impacts of the cotton bags, which were assumed to be used 173 times, are listed in Table 13.6.

From Table 13.5, it is evident that if the cotton bags are not reused, this will pose a very severe threat to the environment, and if they are used 173 times (as determined by this particular study), the environmental impacts are significantly reduced. From the life cycle impact assessment results, it was noticed that the material extraction and production contributed to almost more than 98% to all the impact categories (this might be

Table 13.5 Transportation scenario used in UK's supermarket bags study (Edwards and Fry, 2011)

Bag type	From	To	Transport modes	Distance
Cotton bag	Textile producer in China	Bag producer in China	Lorry	100 km
	Bag producer in China	Bag importer in UK	Lorry	100 km
			Sea freight	15,000 km
			Lorry	280 km
	Bag importer	Supermarket	Lorry	200 km

Table 13.6 Life cycle environmental impacts of a cotton bag as per UK's super markets carrier bag study (Edwards and Fry, 2011)

Assessment method	Impact category	Unit	Total (no reuse)	Total (used 173 times)
IPCC 2007	Global warming	Potential kg CO_2 eq	271.533	1.570
CML 2 baseline	Abiotic depletion	g Sb eq	1519.838	8.785
	Acidification	g SO_2 eq	2787.681	16.114
	Eutrophication	g PO_4 eq	304.486	1.760
	Human toxicity	kg 1,4-DB eq	66.254	0.383
	Freshwater aquatic ecotoxicity	g 1,4-DB eq	23,477.073	135.706
	Marine aquatic ecotoxicity	kg 1,4-DB eq	44,716.601	258.477
	Terrestrial ecotoxicity	g 1,4-DB eq	3208.855	18.548
	Photochemical oxidation	g C_2H_4	95.114	0.550

due to the fact that the cotton material was assumed to be a woven textile). Impacts pertaining to abiotic depletion of resources, acidification, human toxicity, freshwater and marine ecotoxicity, and photochemical oxidation categories were mainly attributed to the energy needs for the conversion of cotton fibre to cotton yarn, and eutrophication is mainly due to the use and production of fertilizers used in the cotton growth stage. Both of the above processes, namely cotton growing and the energy used for cotton processing, contributed equally to terrestrial ecotoxicity (Edwards and Fry, 2011).

From this study, the cotton bag was found to be environmentally worse than the conventional HDPE bag in seven of the nine impact categories even when used 173 times, and acidification and aquatic and terrestrial ecotoxicity impacts were found to be very high for cotton bags, which might be owing to the higher amount of energy used in producing cotton yarn and the usage of fertilizers in the cotton growth stage. Cotton bags need to be reused for at least 173 times to have a lower carbon footprint that conventional HDPE bags.

13.4.5 Finnish study – carbon footprint of shopping bags

In this Finnish study (Matilla et al., 2011), the carbon footprint of several shopping bag alternatives (five types) was studied using LCA method. The functional unit assumed was 'carrying devices for the transportation of goods from grocery stores for 1 year in Finland'. It was assumed that the cotton bag would replace 50–150 single-use bags in the entire life cycle and hence the reference flow was set at 0.5 to 1.5 for cotton bags. The ecoinvent 2.0 database was used to obtain the primary data for the production of cotton bags. This study discussed the uncertainty and sensitivity of the different shipping bags, which is not discussed here. Carbon footprints of different shopping bag alternatives were assessed for various waste management scenarios, namely:

- Current Finnish situation (when the study was made), in which a small share of municipal solid waste is combusted and about half is landfilled
- Waste incineration without energy compensation
- Waste incineration with energy compensation
- Efficient sorting and recovery of waste

The carbon footprint results of cotton bags (in grams of CO_2 equivalent per bag) for the above assumed scenarios are as follows:

- Current Finnish situation – 1800
- Waste incineration without energy compensation – 1760
- Waste incineration with energy compensation – 1720 (Mattilla et al., 2011)
- Efficient sorting and recovery of waste – 1710

In this comparative assessment, cotton bags were found to be the least preferred option in terms of climate change impacts.

13.4.6 Carbon footprint of different grocery shopping bags in China, Hong Kong and India

This study (Muthu et al., 2011a) reported the carbon footprint of various types of shopping bags (plastic, paper, nonwoven and woven) using life cycle impact technique in two stages. The first stage, which was considered as a baseline study, comprised the study of the impact of different types of shopping bags in the manufacturing phase, without considering their usage and disposal phases (cradle-to-gate stage). The next stage was the study of the carbon footprint of these bags including their usage and disposal phases (cradle-to-grave stage) and the results derived were compared with the results derived from the baseline study, which was the major focus of this research work. The GWP values were calculated by the IPCC 2007 method for 100 years with the aid of Simapro version 7.2 of LCA software. The crux of this work was the usage and disposal values that were derived by actual users of shopping bags in China, Hong Kong and India. Cradle-to-gate data were obtained from the Australian studies of LCA discussed in this chapter (Nolan-ITU Pty Ltd, 2002; ExcelPlas Australia Centre for Design at RMIT& Nolan ITU, 2004).

The functional unit assumed in this study was the 'number of shopping bags used for grocery shopping per year by an average Chinese/Indian/HK resident'. It was reported in this study that two woven cotton bags will replace 100 single-use plastic and paper bags and hence 21.9 woven bags are required to fulfil the functional unit assumed in this study for average Chinese and HK residents and three woven bags are needed to fulfil the functional unit for Indian citizens (based on the consumption behaviour of statistics reported in this study). Detailed results reported for different scenarios for cotton shopping bags can be inferred from this study, and it was noticed that reusable bags, namely nonwoven and woven shopping bags in the order of hierarchy, had lower carbon footprints compared to single-use plastic and paper bags.

13.5 Conclusions

This chapter has included discussions pertaining to the LCA of cotton shopping bags. After discussing the general aspects of LCA of cotton bags, this chapter included discussions related to the various LCA studies reported to date in the literature for cotton bags. As mentioned earlier, there is scant literature pertaining to cotton shopping bags in terms of environmental impacts. Very importantly, only a few studies can be found in the academic literature on the subject and the others (which are also limited in number) are consultancy studies. This scarcity of research needs to be overcome by more research in this field. Different conclusions were derived from various studies reported in this chapter based on the scope, assumptions and boundaries of each study.

As is evident from various LCA studies conducted, the cotton bags were intended to be reused many times and they necessarily have to be reused until the end of their functional life. At this juncture, the actual reusability of cotton shopping bags, or any shopping bag for that matter, needs to be quantified with the aid of scientific instruments such as the eco-functional tester developed by this author (Muthu et al., 2013). Results in terms of actual reusability potential of shopping bags have to be tested objectively and used in LCA studies in assessing the reference flow of shopping bags, which makes a study much more sensible and scientific.

As reported by various LCA studies discussed herein, it was noticed that the environmental impacts of the cotton bags are higher in the production stage (including raw materials stage) and this necessarily has to be looked at in terms of employing alternative sustainable raw materials such as organic or recycled cotton to gain much better environmental benefits from cotton shopping bags. Similarly, various sustainable production and processing technologies need to be used to produce cotton shopping bags to reduce the environmental impacts at the manufacturing stage (for instance, open-end spun yarns can be used vis-à-vis ring spun yarns to reduce the energy consumption and hence climate change and other impacts at yarn manufacturing and hence the total manufacturing impacts). Cotton bags are supposed to reduce the total life cycle impacts by virtue of their reusability; however, their production stage impacts are higher with the current raw materials (conventional cotton) and manufacturing technologies (as seen from various studies in this chapter). In the future, sustainable technologies with eco-friendly raw materials and increased functional limits of cotton bags will eventually produce greater reuse capacities and significantly reduce the life cycle impacts of cotton shopping bags.

References

Edwards, C., Fry, J.M., February 2011. Life Cycle Assessment of Supermarket Carrier Bags. Environment Agency, UK.

ExcelPlas Australia, Centre for Design at RMIT& Nolan-ITU, April 2004. The Impacts of Degradable Plastic Bags in Australia.

ISO 14040, 2006. Environmental Management-life Cycle Assessment-principles and Frame Work. International Organisation for Standardisation, Geneva.

ISO 14044, 2006. Environmental Management -Life Cycle Assessment. International Organisation for Standardisation, Geneva.

Li, Y., Muthu, S.S., Hu, J.Y., Mok, P.Y., Ding, X., Wang, L., Chen, W., 2010. Eco-impact of shopping bags: consumer attitude and governmental policies. Journal of Sustainable Development 3 (2), 71–83.

Mattilla, T., et al., 2011. Uncertainty and sensitivity in the carbon footprint of shopping bags. Journal of Industrial Ecology 15 (2), 217–227.

Musa, H.M., Haye, C., Bradley, M.J., Clayson, A., Gillibrand, G., 2013. Measures aimed at reducing plastic carrier bag use: a consumer behaviour focused study. Natural Environment 1 (1), 17–23.

Muthu, S.S., 2011. Eco-functional Assessment of Grocery Shopping Bags, Ph.D. thesis. The Hong Kong Polytechnic University, Hong Kong.

Muthu, S.S., 2014. The textile supply chain and its environmental impacts. In: Assessing the Environmental Impacts of the Textiles and Clothing Supply Chain. Woodhead Publishing, UK, pp. 1–31.

Muthu, S.S., Li, Y., 2014a. Basic introduction to shopping bags and eco-functional assessment of shopping bags. In: Assessment of Environmental Impact by Grocery Shopping Bags: An Eco-functional Approach, Eco Production Series. Springer, Singapore, pp. 1–6.

Muthu, S.S., Li, Y., 2014b. Manufacturing processes of grocery shopping bags. In: Assessment of Environmental Impact by Grocery Shopping Bags: An Eco-functional Approach, Eco Production Series. Springer, Singapore, pp. 7–14.

Muthu, S.S., Li, Y., 2014c. Life cycle assessment of grocery shopping bags. In: Assessment of Environmental Impact by Grocery Shopping Bags: An Eco-functional Approach, Eco Production Series. Springer, Singapore, pp. 15–54.

Muthu, S.S., Li, Y., 2014d. Consumption behaviour of shopping bags and eco-impact. In: Assessment of Environmental Impact by Grocery Shopping Bags: An Eco-functional Approach, Eco Production Series. Springer, Singapore, pp. 77–88.

Muthu, S.S., Li, Y., 2014e. Life cycle assessment of grocery shopping bags. In: Assessment of Environmental Impact by Grocery Shopping Bags: An Eco-functional Approach, Eco Production Series. Springer, Singapore, pp. 99–113.

Muthu, S.S., Li, Y., Hu, J.Y., Mok, P.Y., January 2011. Carbon footprint of shopping (grocery) bags in China, Hong Kong and India. Atmospheric Environment 45 (2), 469–475.

Muthu, S.S., Li, Y., Hu, J.Y., Mok, P.Y., Ding, X., 2012a. An eco-impact assessment study of nonwoven and woven shopping bags: influence of consumer attitude and governmental policies in China, Hong Kong and India. Energy Education Science and Technology 28 (2), 1131–1144. Carbon footprint of shopping (grocery) bags in China, Hong Kong and India.

Muthu, S.S., Li, Y., Hu, J.Y., Mok, P.Y., Ding, X., 2012b. Eco-impact of plastic and paper shopping bags. Journal of Engineered Fibres and Fabrics 7 (1), 26–37.

Muthu, S.S., Li, Y., Hu, J.Y., Mok, P.Y., Ding, X., 2012c. Influence of consumer behaviour and governmental policies in China, Hong Kong and India: an eco-impact assessment study of reusable shopping bags EEST Part A. Energy Science Research 28, 1131–1144.

Muthu, S.S., Li, Y., Hu, J.Y., Mok, P.Y., 2012d. Eco-functional assessment combined with life cycle analysis: concept and applications. Energy Education Science and Technology Part A 29 (1), 435–450.

Muthu, S.S., Li, Y., Hu, J.-Y., Mok, P.-Y., Lin, M., 2013a. Modelling and quantification of eco-functional index: the concept and applications of eco-functional assessment. Ecological Indicators 26, 33–43.

Muthu, S.S., Li, Y., Hu, J.Y., Mok, P.Y., Mao, Y.F., Wu, X.X., Li, Q.H., 2013b. Assessment of eco-functional properties of shopping bags: development of a novel eco-functional tester. International Journal of Clothing Science and Technology 25 (3), 208–225.

Nolan-ITU Pty Ltd, December 2002. Plastic Shopping Bags – Analysis of Levies and Environmental Impacts. Final Report.

O'Farrell, K., August 2009. LCA of Shopping Bag Alternatives. Final Report, Zero Waste South Australia, Hyder Consulting, Australia.

LCA of fibre-reinforced composites

14

A.D. La Rosa, G. Cicala
University of Catania, Catania, Italy

14.1 Introduction

14.1.1 Towards sustainable development: when technology meets the environment

It is recognised worldwide that we are facing an era of ecological changes. Paul Crutzen, Nobel laureate in chemistry for his studies on the atmosphere, proposed using the term *Anthropocene* for the current geological epoch. The term was coined in the year 2000 to describe the influence of human behaviour on the Earth in recent centuries as so significant as to constitute a new geological era. On this respect, the geologist Zalasiewicz and his colleagues claimed that, 'since the start of the Industrial Revolution, Earth has endured changes sufficient to leave a global stratigraphic signature distinct from that of the Holocene or of previous Pleistocene interglacial phases. These changes so much established for suggestions of a Holocene–Anthropocene boundary in the recent historical past to be geologically reasonable' (Zalasiewicz et al., 2008).

The concept of the Anthropocene may serve the purpose of focusing our minds on what we are doing to the planet. Crutzen continues, saying that: 'To develop a world-wide accepted strategy leading to sustainability of ecosystems against human induced stresses will be one of the great future tasks of mankind' (Crutzen and Stoermer, 2000).

Appropriate and feasible methods to quantify material and energy flows of ecological-socioeconomic systems are required because it is the responsibility of the technology developer to guarantee the environmental protection. Some attempts to evaluate the sustainability of materials have been carried out. In Table 14.1, Ljungberg (2007) gives an interesting classification of materials reporting advantages, disadvantages and sustainability. This classification of sustainability, though it is useful as a preliminary screening, requires a more comprehensive and very rigorous analysis of the data in order to understand, for example if a new technology is more environmentally convenient or not compared with the old one.

Life cycle assessment (LCA) seems to be one of the best methods that makes such analysis possible. The term *life cycle* refers to the notion that for a fair, holistic assessment, the raw material production, manufacture, distribution, use and disposal (including all intervening transportation steps) need to be assessed.

Handbook of Life Cycle Assessment (LCA) of Textiles and Clothing. http://dx.doi.org/10.1016/B978-0-08-100169-1.00014-9

Table 14.1 Material groups: advantages, disadvantages and classification of sustainability

Material group	Examples on materials	Typical advantages	Typical disadvantages	Classification of the sustainability[a]
Metals	• Steel (Fe + C) • Aluminium • Bronze (e.g. Cu + Sn)	• Durable and strong • Often plastic formable • Often cheal	• High cost for machining • Mostly corrosion sensitive	• Easy recyclable (remeltable) 2–3
Ceramics	Synthetic materials like: • Porcelain (clay) • Mineral glass • Al_2O_3, Si_3N_4,Sic, etc.	• Non toxic • Light • Hard and durable • Corrosion resistant • High temp. resistant	• Brittle • High cost for machining when burnt • Not suitable for load in tension	• Easy to deposit (nontoxic) • Possible but expensive to remelt 2–3
Synthetic polymers	• Thermoplastics (e.g. PE, PS, PC, PP) • Two component polymers (e.g. epoxy) • Rubber (e.g. isopren)	• Nontoxic • Light • Cheap and easy forming • Often easy to recycle (e.g. by remelting or burning)	• Sometimes very toxic when burnt • Sensitive to high temperatures	• Typically non-renewable • Often easy to remelt or burn 1–3
Natural organic materials	• Wood • Cotton • Silk	• Renewable • Light • Cheap and easy forming • Recyclable by, e.g. burning	• Decomposes easily • Not durable • Toxic when impregnated	• Recyclable by, e.g. burning • Renewable 2–3
Natural inorganic materials	• Stone • Minerals	See ceramics above!	• Brittle • High cost for machining • Not suitable for load in tension	See ceramics above! 3
Composites	• Mixed materials, e.g. PS + glass fibres, Cu + W-fibres, rubber + textile fibres, asphalt (oil + stone), wood polymer composites (WPC)	• Optimised use of the materials • Often very strong and light	• Often expensive to produce • Very various properties for various composites	• Typically low sustainability due to separation problems for the mixed materials 1–2

Note that the advantages/disadvantage and the 'sustainability' can change due to the specific material type, and the examples are not valid for all materials in a certain material group.
[a]The sustainability is estimated from a scale of 1 to 3, where 3 indicates the highest (or best) sustainability and 1 the lowest.

14.1.2 Materials selection and eco-design

When designing a product, the engineer has to consider a list of properties (e.g. stiffness, strength, density, working temperature, etc.) to choose the material that best suits the intended purpose and the respective production technology (Simões et al., 2012). Nowadays, environmental issues, such as low energy consumption, low carbon emissions and recyclability, are generally added to this list of requirements (Ashby, 2009). The concept of eco-design, also known as design for environment, has gained increasing attention. The main strategies for eco-design are: using materials with low environmental impact, choosing clean production processes, avoiding hazardous and toxic materials, maximising the efficiency of the energy used for production and for the product in use and designing for waste management and recycling (Ljungberg, 2007). LCA is a useful tool for eco-design as it is a very comprehensive way of determining the total environmental impacts. One important advantage of LCA is that it involves a 'holistic' approach, bringing the environmental impacts into one consistent framework. Another advantage is the 'cradle-to-grave' approach of LCA meaning that when an environmental problem is remarked in one stage of the product's life cycle, it cannot be shifted to another stage, but it must be solved.

14.2 Types of fibre-reinforced composites

14.2.1 Traditional fibre-reinforced composites

Composite materials consist of three main constituents: matrix, reinforcement and interface. The matrix in composite materials has key functions. It transfers load to the reinforcement and protects the reinforcement from environmental degradation. When polymeric matrix materials are used, composites called polymeric matrix composites (PMCs). Composite can be classified according to the reinforcement used as shown in Figure 14.1 (Cicala and Lo Faro, 2012).

Polymeric matrices are divided into thermoplastics and thermosets. Thermoplastics can be melted by heating and solidified by cooling. This property is of high interest for recycling because thermoplastics can be reshaped. Thermoset materials once polymerised cannot be remelted or reformed. The polymerisation process is commonly called curing. Thermosets present good thermal stability and chemical resistance but cannot be reshaped after curing.

14.2.1.1 Traditional fibres

Traditional fibres used as reinforcement in polymer composites are generally either polymers or ceramics: the polymer aramids, glass, carbon, boron, aluminium oxide and silicon carbide. Carbon is a high-performance fibre material that is the most commonly used reinforcement in advanced polymer-matrix composites. Glass fibre is readily available and may be fabricated into a glass-reinforced plastic economically using a wide variety of composite-manufacturing techniques.

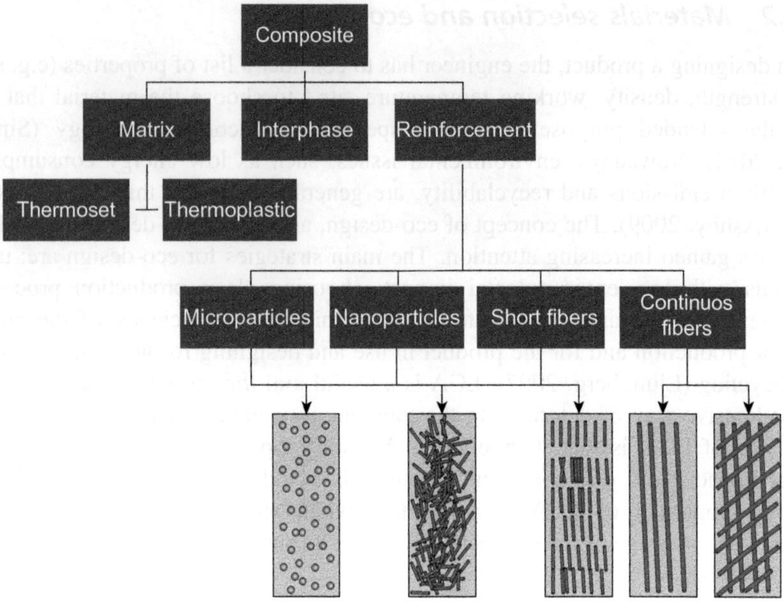

Figure 14.1 Composites classified according to reinforcement types.

The aramid fibres (Kevlar and Nomex are the trade names for two of the most common) are especially desirable for their strength-to-weight ratios being superior to metals. They are most often used in composites having polymer matrices (generally epoxies and polyesters). Since the fibres are relatively flexible and somewhat ductile, they may be processed by most common textile operations.

14.2.2 Eco-composites

Polymer composites reinforced with natural fibres have received increasing attention during the last decades. These composites have been referred to as bio-composites or eco-composites and have found a wide array of applications in the building and construction industry, the automobile industry and other sectors (Joshi et al., 2004; Dornburg et al., 2004; Mohanty et al., 2002). Natural fibres such as hemp, sisal, flax, jute, kenaf, bamboo and coir possess good reinforcing capability when properly compounded with polymers. Chemical–physical and mechanical properties of natural fibres have been studied and compared to traditional glass fibre as reported in Table 14.2 (Dittenber and GangaRao, 2012). Tensile strength and thermal resistance are lower in natural fibres, but cost reduction and low density represent good opportunities.

The main advantage of using natural fibre composites is the reduction of the environmental impact due to the use of nonrenewable energy/material sources, lower pollutant emissions, lower greenhouse gas emissions, enhanced energy recovery and end-of-life biodegradability of components as reported by Corbière-Nicollier et al. (2001). However, Dissanayake et al. (2009a,b) have presented data that suggest that flax fibres may require equivalent or higher energy consumption due to heavy reliance

Table 14.2 **Properties of natural fibers**

Fiber type	Density (g/cm^3)	length (mm)	Diameter (μm)	Tensile strength (MPa)	Tensile modulus (GPa)	Specific modulus (approx)	Elongation (%)	Cellulose (wt%)	Hemi-cellulose (wt%)	Lignin (wt%)	Pectin (wt%)	Waxes (wt%)	Micro-fibrillar angel (deg)	Moisture content (wt%)
E-glass	2.5–2.59	–	<17	2000–3500	70–76	29	1.8–4.8	–	–	–	–	–	–	–
Abaca	1.5	–	–	400–980	62–20	9	1.0–10	56–63	20–25	7–13	1	3	–	5–10
Alfa	0.89	–	–	35	22	25	5.8	45.4	38.5	14.9	–	2	–	–
Bagasse	1.25	10–300	10–34	222–290	17–27.1	18	1.1	32–55.2	16.8	19–25.3	–	–	–	–
Bamboo	0.6–1.1	1.5–4	25–40	140–800	11–32	25	2.5–3.7	26–65	30	5–31	–	–	–	–
Banana	1.35	300–900	12–30	500	12	9	1.5–9	63–67.6	10–19	5	–	–	–	8.7–12
Coir	1.15–1.46	20– 150	10–460	95–230	28–6	4	15–51.4	32–43.8	0.15–20	40–45	3–4	–	30–49	8.0
Cotton	1.5–1.6	10–60	10–45	287–800	55–12.6	6	3–10	82.7–90	5.7	<2	0–1	06	_	7.85–8.5
Curaua	1.4	35	7–10	87–1150	11.8–96	39	1.3–4.9	70.7–73.6	9.9	7.5–11.1	–	–	–	–
Flax	1.4–1.5	5–900	12–600	343–2000	27.6–103	45	1.2–3.3	62–72	18.6–20.6	2–5	2.3	1.5–1.7	5–10	8–12
Hemp	1.4–15	5–55	25–500	270–900	23.5–90	40	1–3.5	68–74.4	15–22.4	3.7–10	0.9	0.8	2–6.2	6.2–12
Henequen	1.2	–	–	430–570	10.1–16.3	11	3.7–5.9	60–77.6	4–28	8–13.1	–	0.5	–	–
Isora	1.2–1.3	–	–	500–600	–	–	5–6	74	–	23	–	1.09	–	–
Jute	1.3–1.49	1.5–120	20–200	320–800	8–78	30	1–1.8	59–71.5	13.6–20.4	11.8–13	0.2–0.4	05	8.0	12.5–13.7
Kenaf	1.4	–	–	223–930	14.5–53	24	1.5–2.7	31–72	20.3–21.5	8–19	3–5	–	–	–
Nettle	–	–	–	650	38	–	1.7	86	10	–	–	4	–	11–17
Oil palm	0.7–1.55	–	150–500	80–248	0.5–3.2	2	17–25	60–65	–	11–29	–	–	42–46	–
Piassava	1.4	–	–	134–143	1.07–4.59	2	7.8–21.9	28.6	25.8	45	–	–	–	–
PALF	0.8–1.6	900–1500	20–80	180–1627	1.44–82.5	35	1.6–14.5	70–83	–	5–12.7	–	–	14.0	11.8
Ramie	1.0–1.55	900–1200	20–80	400–1000	24.5–128	60	1.2–4.0	68.6–85	13–16.7	0.5–0.7	1.9	0.3	7.5	7.5–17
Sisal	1.33–1.5	900	8–200	363–700	9.0–38	17	2.0–7.0	60–78	10.0–14.2	8.0–14	10.0	2.0	10–22	10–22

on agrochemicals (when all burdens are assigned to the primary product as recommended by Ekvall and Finnveden, 2001).

Both thermoplastic and thermoset polymers reinforced with natural fibres are being produced. Thermoplastic polymers reinforced with natural fibres have been widely studied for industrial applications (Cicala et al., 2010; La Mantia and Morreale, 2011). In the automobile industry, for instance, they are already used to make a variety of indoor parts such as seat bottoms, heat restraints, back cushions, underfloor body panels and so on (Ashori, 2008). This is partly due to legislation in the European Union (EU) and other countries and partly to the effort of automakers to contribute to the goal of sustainability by using more eco-friendly materials (Chawla and Chawla, 2012). Thermoset composites reinforced with natural fibres are widely used in the automotive, naval and building sectors. For many applications, sandwich structures are preferred because of their ideal combination of high flexural stiffness and low weight. Eco-sandwich composites have been studied for building applications (La Rosa et al., 2014a).

14.2.3 Green composites

Recently, bio-based composites from natural fibres and bio-resins are getting increased attention. Depending on their chemical structure, bio-resins can be thermoplastic or thermoset, biodegradable or not. A list of bio-derived commercially available polymers is reported in Table 14.3. Data reported in Table 14.3 were collected from technical data sheets available for each compound on the company's website. Thermoset bio-derived resins commercially available on the market are mainly epoxy resins, unsaturated polyester resins and polyurethane. These resins have a bio-content ranging between 20% and 50% and are not biodegradable. Among the thermoplastics, we find nonbiodegradable bio-resins like bio-polyolefin, bio-nylons, cellulose and lignin-based polymers, and bio-degradable bio-resins like poly(lactic acid) PLA, poly(butylene succinate) PBS and polyhydroxyalkanoates (PHA). A new class of totally biodegradable 'green' composites obtained by combining natural fibres with biodegradable resins is being developing (Netravali and Chabba, 2003). Furthermore, several applications are being carried out using recycled petro-based thermoplastics like polyethylene and polypropylene reinforced with natural fibres.

Figure 14.2 is an SWOT (strengths, weaknesses, opportunities and threats) analysis that summarises benefits and drawbacks of the use of green composites produced with natural fibres and bio-derived matrices to replace traditional composites. The benefits appear superior to the disadvantages and the market opportunities are increasing for many industrial sectors.

14.3 LCA method and models

14.3.1 Method description

The LCA methodology is considered the most widespread technique for evaluating the environmental impacts associated with material products (Bovea and Vidal, 2004). LCA is a cradle-to-grave assessment of a product or service that analyses the impact

Table 14.3 **Commercial bio-derived resins available in the market**

Tradename	Producer	Type	BioContent %	Recycled content %
BiOH	Cargill	PU	5–20	–
Envirez	Ashland	UPR	8–22	0–44
Epobiox	Amroy Europe Oy	Epoxy	60–90	–
Palapreg eco	DSM	UPR	55	–
EcoFlex – Ecovio	Basf	Biodegradable copolyester	–	–
Eco Tek	AOC	UPR	0–30	0–45
Envirolite 31325-00	Reichold	UPR	25	–
SuperSap	Entropy	Epoxy	25–55	–
BioRez	TransFuranChemical	Furanic	–	–
Exaphen (Novocard/ Polycard)	Cimtec Lab	Biobased NovolacResin/ polyols	–	–
Arboform	Tecnaro/Albis	Thermoplastic	100	–
Cellidor	Albis	Thermoplastic cellulose ester	45	–

Strengths	**Weaknesses**
1. Reduction of the environmental impacts 2. Cost reduction 3. Weight reduction in composites 4. Rural development 5. Strong market potential 6. Biodegradability 7. Recyclability	1. Reduction of mechanical performance 2. Reduction of thermal resistance 3. Biomass cropping
Opportunities	**Threats**
1. New and emerging markets 2. Emerging processing technologies 3. Emerging emphasis on environmental issues 4. Growing demand for green-materials	1. Lack of production/distribution chains 2. Infancy of the value chain 3. Land use

Figure 14.2 SWOT analysis of the market of green composites compared to traditional composites.

of any product over a lifetime from the extraction of raw materials to the waste disposal of its various components. LCA is focused on studying the whole product system, as it not only studies one single process but also the complete chain of production and lifetime.

LCA includes the following important steps: materials extraction, manufacturing and waste production, packaging, transportation, product use and product disposal. In order to calculate the amount of emissions and waste created during the life cycle of a product, much fundamental information is needed on manufacturing processes, materials and energy use. A life cycle inventory (LCI) is therefore compiled to record the emissions and resources consumed that can be attributed to a specific product. Databases have been developed that enable the user to input new information or data specific to their products and processes. Ecoinvent v3 (http://www.ecoinvent.org/database/) is probably the most comprehensive international LCI database. The ecoinvent database is included in most LCA software models. The following are some of the software suitable for LCA of plastics (O'Neill, 2003):

1. SimaPro, Pre Product Ecology Consultants, Amersfoort, NL, www.pre.nl/simapro/
2. Boustead, Boustead Consulting Ltd, West Grinstead, UK, www.boustead-consulting.co.uk
3. Gabi, Polymer Institute University of Stuttgart www.peproduct.de/GABI/htdoc/home_englisch.htm
4. Umberto, Institute for Environmental Informatics, Hamburg, www.umberto.de/english/

LCA is governed by specific standards from the International Organisation for Standards (ISO). The LCA study consists of defining the scope and goal, inventory analysis, assessing the impact and interpreting the data. ISO 14040 (ISO 14040, 2006) and ISO 14044 (ISO 14044, 2006) Environmental Management Life Cycle Assessment provides the principles, framework, requirements and guidelines to conduct a LCA study. The impact stage involves using the inventory analysis results to come to relevant midpoint and endpoint indicators. Midpoint indicators might include acidification, radiation, climate, fossil fuel, ecotoxicity, etc. Endpoint indicators can include respiratory diseases, seawater level, resource depletion, etc.

14.3.2 Environmental impact assessment methods

A wide range of life cycle impact assessment (LCIA) methodologies has been developed, based on midpoint and/or endpoint indicators (ILCD Handbook, 2010):

- *CML 2002* (Guinée et al., 2002)
- *Eco-Indicator 99* (Goedkoop and Spriensma, 2000)
- *EDIP* (*1997–2003*) (Wenzel et al., 1997; Hauschild and Wenzel, 1998; Hauschild and Potting, 2005; Potting and Hauschild, 2005)
- *EPS2000* (Steen, 1999a,b)
- *Impact 2002+* (Crettaz et al., 2002; Jolliet et al., 2003; Payet, 2004; Pennington et al., 2005, 2006; Rochat et al., 2006; Rosenbaum, 2006; Rosenbaum et al., 2007)
- *LIME* (Itsubo et al., 2004; Hayashi et al., 2000, 2004, 2006)
- *LUCAS* (Toffoletto et al., 2007)

- *ReCiPe* (De Schryver et al., 2009; Huijbregts et al., 2005a,b; Van Zelm et al., 2007a,b; Wegener Sleeswijk et al., 2008)
- *Swiss Ecoscarcity or Ecological scarcity* (Brand et al., 1998; Müller-Wenk, 1994; Ahbe et al., 1990; Frischknecht, 2006a,b)
- *TRACI* (Bare, 2002; Bare et al., 2003; Hertwich et al., 1997, 1998, 1999, 2001; Norris, 2002)
- *MEEuP methodology* (Kemna et al., 2005).

14.3.3 Environmental impact classification factors

The environmental issues analysed in an LCIA are called environmental impact classification factors as outlined in ISO/TR 14047 (2003). A brief description follows:

1. Acidification potential (AP): is a consequence of acids being emitted to the atmosphere and subsequently deposited in surface soils and waters. AP classification factors are mainly based on the contributions of SO_2, NO_x, HCl, NH_3 and HF and expressed as SO_2 equivalent.
2. Aquatic toxicity potential is calculated based on the maximum tolerable concentrations of different toxic substances in water by aquatic organisms.
3. Human toxicity potential is calculated in kg and takes into account the releases of materials toxic to humans in air (A), water (W) and soil (S) based on human toxicological factors.
4. Eutrophication potential (EP). Eutrophication is referred to as the pollution state of aquatic ecosystems where the overfertilisation of water and soil has turned into an increased growth of biomass. EP is calculated in kg based on a weighted sum of the emission of nitrogen and phosphorus derivatives such as N_2, NO_x, NH^{4+}, PO_4^{3-}, P and chemical oxygen demand (COD). The classification factors for EP are expressed as phosphates equivalents.
5. Global warming potential (GWP) is calculated for each different greenhouse gas, including CO_2, N_2O, CH_4 and volatile organic compounds (VOCs). GWP is expressed as CO_2 equivalent.
6. Nonrenewable/abiotic resource depletion potential is calculated for fossil fuels, metals and minerals by dividing the quantity of resource used by the estimated total world reserves of that resource.
7. Ozone depletion potential represents the potential of depletion of the ozone layer due to the emissions of chlorofluorocarbon compounds and other halogenated hydrocarbons.
8. Photochemical oxidants creation potential. Photochemical smog is caused by the degradation of VOCs and nitrogen and is expressed in kg of ethylene.

Other factors such as *land use* and *loss of biodiversity* are considered in natural fibre production.

The output of the LCIA consists of a list of such category indicators; it does not provide an evaluation of actual environmental impact. All emissions are sorted into classes according to the effect they have on the environment. For example, chemicals that contribute to the greenhouse effect or to ozone layer depletion are divided between those two classes, whereas emissions such as nitrogen oxides may simultaneously belong to several classes, such as aquatic toxicity, acid rain and eutrophication. This is called classification. The following step is called characterisation, which means to assign and convert LCI results into numerical indicator results. An example of output is shown in Figure 14.3 (La Rosa et al., 2013) that reports a comparative LCA between

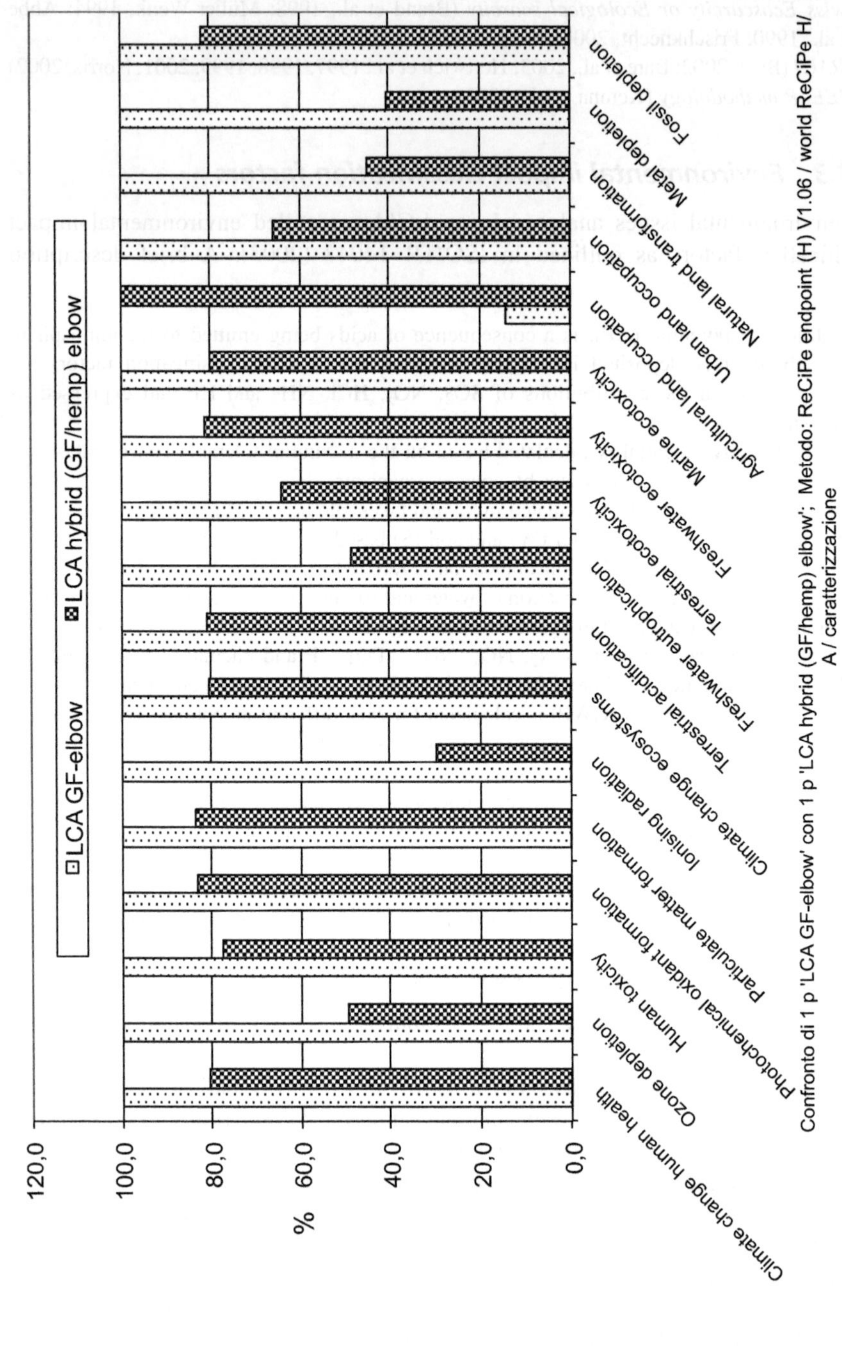

Figure 14.3 Contribution of different impact categories with ReCipe Endpoint method.

a traditional composite made with epoxy resin reinforced with glass fibres and an innovative hybrid composite structure where part of the glass fibre was replaced by hemp fibres.

In this study the Simapro software and the ReCiPe endpoint impact assessment model were used, but other software (e.g. GaBi) and other impact assessment methods (e.g, CML 2002; Impact 2002+) are suitable for composites.

14.4 LCA of composites made out of different materials: examples and case studies

During the last decade, industrial production has been driven towards more sustainable policies. Many regulations have been defined for different sectors, including the following:

- End of Life Vehicles (ELV) Directive (2000/53/EC), for automotive;
- Waste Electrical and Electronic Equipment (WEEE) Directive (2002/96/EC);
- Discussions about Nearly Zero Energy Buildings (NZEB) by 2020 in Europe.

Furthermore, for 2020 the EU has committed to cutting its emissions to 20% below 1990 levels; for 2050 EU leaders have endorsed the objective of reducing Europe's greenhouse gas emissions by 80–95% compared to 1990 levels. LCA has been developed as a useful tool to evaluate the environmental impacts of products and processes. It is very useful for material selection, representing a very important step to design more sustainable products. Some considerations on composite products must be assumed:

- thermoplastic matrix composites offer the opportunity for recycling at the end of life but they require a greater energy input in the fabrication of structures due to the melting point normally being ~200 °C above the use temperature;
- thermoset composites are difficult to recycle and may be a potential landfill problem, but large structures can be fabricated with minimal energy requirements and lower carbon emissions;
- natural fibres and bio-based matrices are more carbon neutral, but they require the use of fertilisers, herbicides, pesticides, etc.;
- petroleum-based composites display better mechanical resistance, but they require high energy for production processes.

LCA of fibre-reinforced composite materials has helped to better characterise composite products, in terms of environmental performance, and compare them with other classes of materials. Generally, LCAs of bio-based polymer composites have shown favourable results in terms of environmental impacts and energy use compared to petroleum-based products (Corbiére-Nicollier et al., 2001; Dornburg et al., 2004; González-García et al., 2010; Wool and Sun, 2005). However, calculation of these impacts always depends on the system and boundary conditions considered during the study.

14.4.1 LCA of automotive components

Polymer composites are widely used in the automotive sector. As reported in Table 14.4 (Mallick, 2010), the amount of polymer composites used in typical automobiles is about 9%.

According to the legislation End of Life Vehicles (ELV) Directive (2000/53/EC), vehicles must be designed, manufactured and assembled giving consideration to their disposal. Recycling has become a very important issue in terms of avoiding virgin material production. LCA of polymer composite products has pointed out that in the production phase the standard polymeric matrices have higher environmental impact compared to the reinforcement benefits for the development of green composites. In a comparative LCA study, La Rosa et al. (2014b) evaluate the environmental impacts of two different elbow fittings made of glass-fibre/epoxy vinyl ester and hybrid (glass-hemp)/epoxy vinyl ester, respectively.

The functional unit of the study was defined as the function of one elbow fitting used in the seawater-cooling pipeline of a Sicilian chemical plant, with an estimated life of 20 years. In both cases, the major impacts were associated with the epoxy resin as shown in Figures 14.4 and 14.5 that report, respectively, the global warming (GW) impact associated with the glass/epoxy composite and with the hemp/epoxy composite production. Figure 14.4 reports an impact of 66.9% associated with the matrix and an impact of 27.9% associated with the glass fibres, while the impact associated with the hemp (Figure 14.5) was only 7.29%.

The study concluded that the hybrid (hemp/glass)-reinforced thermoset was more eco-efficient than the conventional glass fibre alternative. The main advantage of the hybrid material was the reduction of glass fibre and resin content compared to standard

Table 14.4 Material distributions in typical automobiles

Material	Percentage of vehicle weight	Major areas of application
Steel	55	Body structure, body panels, engine and transmission components, suspension components, driveline components
Cast iron	9	Engine components, brakes, suspension
Aluminum	8.5	Engine block, wheel
Copper	1.5	Wiring, electrical components
Polymers (plastics) and polymer-matrix composites	9	Interior components, electrical and electronic components, under-the-hood components, fuel line components
Elastomers	4	Tires, trims, gaskets
Glass	3	Glazing
Other	10	Carpets, fluids, lubricants, etc.

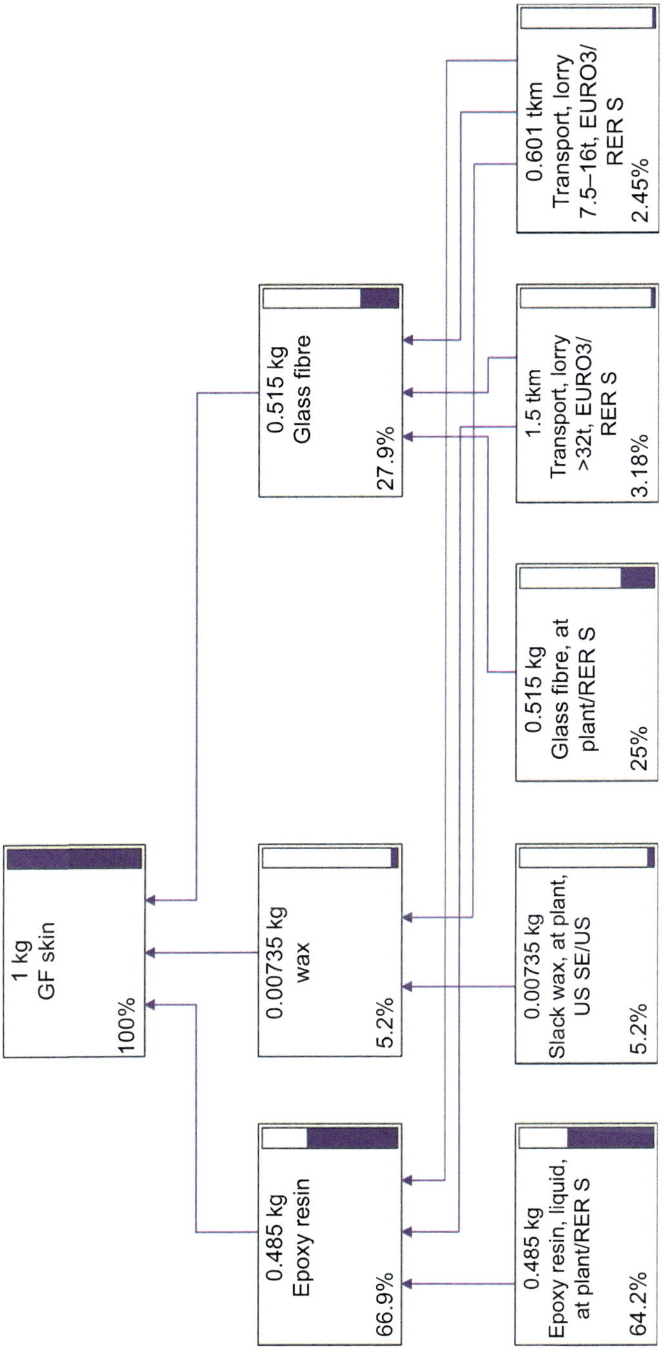

Figure 14.4 Percentage of GW impact contribution evaluated for each component of the flowchart referring to the production of 1 kg of glass fibre/epoxy resin composite.

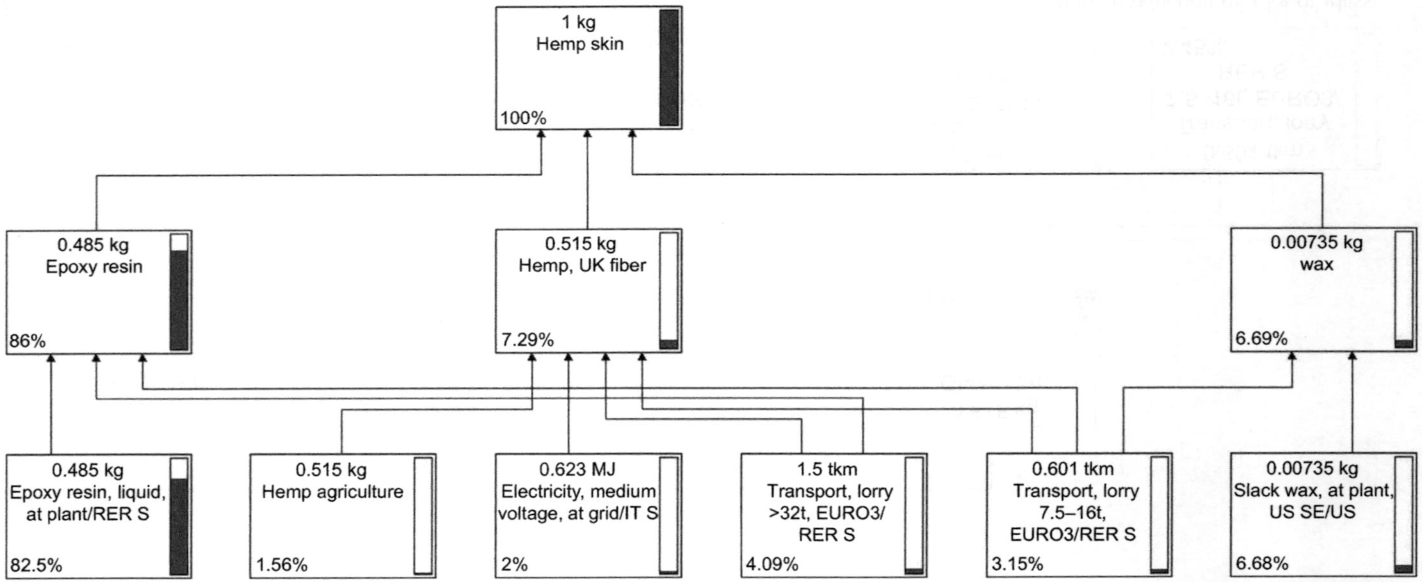

Figure 14.5 Percentage of GW impact contribution evaluated for each component of the flowchart referring to the production of 1 kg of hemp/epoxy resin composite.

glass fibre/composites. A significant environmental advantage was found during the production phase, mainly attributable to the 'green' origin of the hemp mat. It was also reported that renewable raw materials generally score better than petrochemical polymers with regard to fossil energy use and greenhouse gas emissions while they score worse with regard to land use, ecotoxicity and eutrophication. This limit could be overcome by organic cropping that avoids the use of fertilisation and pesticide. The magnitude of the environmental advantage also depends on the kind of application and obviously on the distance between the country of production of the materials and the country where they are used. Finally, material cost reduction was significant in the hybrid elbow formulation, according to the LCC results.

Though the overall impact result of the hemp/epoxy composite was lower than the impact of the glass/epoxy composite, the impact associated with the matrix was very high. There are two ways to reduce the impacts associated with the matrix: one is using bio-based matrix, the other is using a recycled matrix. Many studies involve the use of bio-based epoxy resin in thermoset composites, while recycled polypropylene and recycled polyethylene have been studied in new thermoplastic formulations.

Another important aspect to consider in the automotive application is the use of lightweight materials.

In 1999, Wötzel et al. developed an LCA study for the interior side panel of an Audi A3 made of hemp/epoxy resin composite. They studied the environmental impacts of using hemp/epoxy resin in the composite formulation and compared the results with a conventional side panel manufactured by injection moulding of virgin acrylonitrile-butadiene-styrene. The results of the LCAs indicated the ecological dominance of the vehicle's use phase compared to its production and recycling phase. Weight reduction for a single basic component amounts to 300 g using hemp fibre composites, and this leads to a reduction in fuel consumption. For the automotive industry, where weight reduction always is an issue, this was said to be the original reason for the development of interior parts with natural fibres. Furthermore, natural fibres based on cellulose have a relatively low density and are relatively stiff and strong. Therefore, their specific properties, that is the properties divided by the density, or the 'properties per kilo', are rather high, and actually comparable to those of glass fibres. In order to combine both lightweight with recyclability properties, recently there has been increasing interest in researching thermoplastic mixed matrix by compounding polypropylene and lignin (Sahoo et al., 2011). The reason is mainly due to economic and environmental benefits: lignin is an abundant resource and is a natural thermoplastic polymer.

14.4.2 LCA in the building sector

The LCA of a building requires a lot of time for quantifying building materials, energy use, etc. Therefore, specific software to facilitate the use of LCA in the building sector has been developed, as reported in Table 14.5 (Zabalza Bribián et al., 2009).

Several studies have been carried out on the LCA methodology applied to entire buildings (Sartori and Hestness, 2007; Oeriz et al., 2009).

Table 14.5 **Specific building life cycle assessment tools**

ECO-QUANTUM	www.ecoquantum.nl
LEGEP	www.legep.de
EQUER	www.izuba.fr
ATHENA	www.athenaSMI.ca
OGIP	www.ogip.ch/
ECO-SOFT	www.ibo.at/de/ecosoft.htm
ENVEST 2.0	envestv2.bre.co.uk
BECOST	www.vtt.fi/rte/esitteet/ymparisto/lcahouse.html
BEES	www.bfrl.nist.gov/oae/software/bees.html
GREENCALC	www.greencalc.com
ECOEFFECT	www.ecoeffect.se
ECO-QUANTUM	www.ecoquantum.nl
LEGEP	www.legep.de
EQUER	www.izuba.fr

Most references show that a big part of a building's environmental impact comes from its use phase, primarily its water and energy use. Orientation, insulation, building operation, lighting, etc. are very important issues. Rossi et al. (2012) reports that the phase with the highest environmental impact is the operation phase, representing approximately 62–98% of the life cycle total impacts, while the construction phase accounts for a total of 1–20% and the dismantling phase represents less than about 0.2–5%. Concerning the choice of materials for buildings applications, an optimum material solution should have the essential structural properties of concrete but with lower thermal conductivity. The reduced thermal conductivity is required to provide better thermal insulation and reduce energy consumed for cooling and heating in the use phase. Structural insulated panels (SIPs) are an example of energy-efficient building technology that is increasingly used for walls, partitions and flooring (Mullens and Arif, 2006). In a recent study, an eco-sandwich containing cork, plant-derived epoxy resin and flax fibre was proposed as a structural insulated solution for buildings (La Rosa et al., 2014a). Thermal conductivity was evaluated for cork (0.045–0.057 W/m K) and for the eco-sandwich (0.074–0.081 W/m K). The interesting result was that the thermal conductivity value obtained for the eco-sandwich was much lower than for cement (0.9 W/m K). In the same work, a comparative LCA method was applied to the manufacturing phase of four composite systems for building walls, as described in Figure 14.6.

The functional unit of the LCA study was defined as the material assembly used for the production of 1 m^2 of a building's wall with thermal transmittance $U \leq 0.46\ W/m^2\ K$ (according to the Italian law).

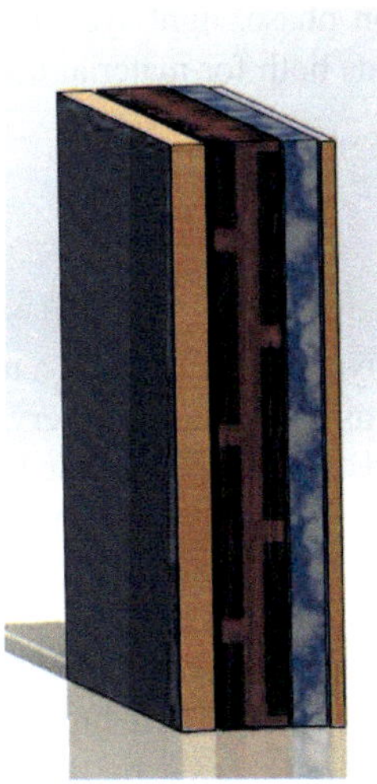

System A. From right to left: sandwich panel; air interspace; perforated brick 12; bioverd.

System B. From right to left: cement; perforated brick 8; air interspace; perforated brick 12; bioverd.

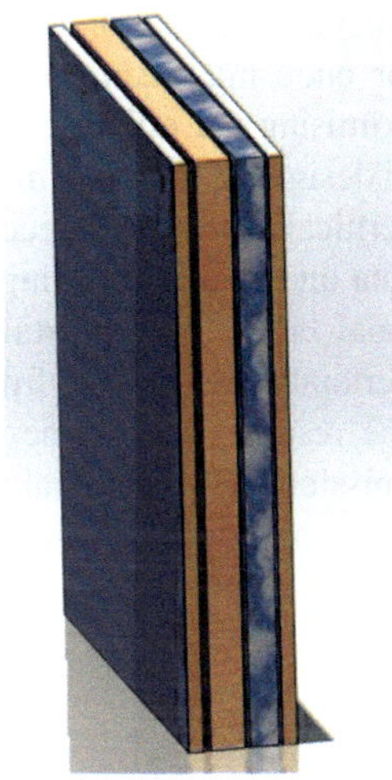

System C. From right to left: sandwich panel; air interspace; cork panel; sandwich panel.

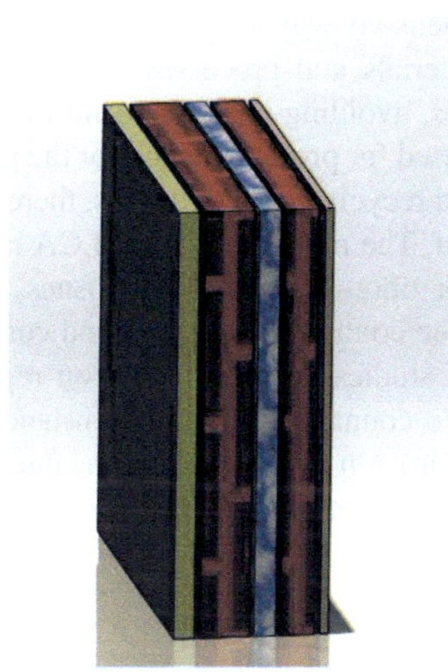

System D. From right to left: cement; perforated brick 8; air interspace; perforated brick 12; PVC foam.

Figure 14.6 Graphical representation of the four systems.

The conclusion of the study was that materials with low thermal conductivity (e.g. the eco-sandwich) enables the application of relatively thin building envelopes, but the environmental impacts evaluated for the manufacture phase depend more on the type of materials used than on their weight. The system containing the eco-sandwich was the lightest, but the environmental impact due to the epoxy resin was very high. Nevertheless, light walls and roofs have the advantage that they are easy to erect.

Furthermore, when considering the building construction phase, light systems could perform with lower impacts due to the minor energy needs both for material transportation and installation.

14.5 Conclusions

LCA methodology applied in the development of new polymer composites is a relative new field, but it has already been widely used in many industrial sectors as a very useful tool to drive the choice of materials and processes towards a more sustainable production system. Several LCA applications to both traditional composites and new green composites have pointed out that the benefits of using green composites produced with natural fibres and bio-derived matrices, compared to traditional composites, appear superior, and their market opportunities are increasing for many industrial sectors.

Traditional composites are still very important in those sectors where high thermomechanical performance is required (e.g. aerospace, sport races). In these cases though, nanotechnology is becoming more and more effective. LCA of nanocomposite is still at a very early stage but already in progress.

The main benefit of using LCA in composite production is that it provides a comprehensive way of determining the total environmental impact for designing new materials and processes. LCA is a useful tool for choosing clean production processes, avoiding hazardous and toxic materials, maximising the efficiency of the energy used for production and for the product in use and designing for waste management and recycling. Nevertheless, there are still several critical aspects that need to be looked at. The main limit of the LCA methodology is data uncertainty. Existing databases for fibre-reinforced composites, green or traditional based, are uncertain and need to be continually updated and corrected. Data uncertainty also derives from the fact that studies are carried out on regional site and the results are to some extent subject to country-specific circumstances (e.g. GHG emissions from national power production). On the other hand, the uncertainties related to conclusions can be reduced if several independent analyses for different countries arrive at similar conclusions.

Acknowledgment

The authors are very grateful to the Italian Ministry of University and Research for funding the project PON 2239 'MATRECO'. Parts of these funds have been used to evaluate the environmental impacts of eco-composites by means of the LCA methodology.

References

Ahbe, S., Braunschweig, A., Müller-Wenk, R., 1990. Methodology for Ecobalances Based on Ecological Optimization. BUWAL (SAFEL). Environment Series No. 133, Bern.

Ashby, M.F., 2009. Materials and the Environment: Eco-informed Material Choice. Heinemann, Butterworth.

Ashori, A., 2008. Wood–plastic composites as promising green-composites for automotive industries! Bio-resource Technology 99, 4661–4667.

Bare, J.C., 2002. Developing a Consistent Decision-making Framework by Using the U.S. EPA's TRACI Systems Analysis Branch, Sustainable Technology Division, National Risk Management; Research Laboratory. US Environmental Protection Agency, Cincinnati, OH.

Bare, J.C., Norris, G.A., Pennington, D.W., McKone, T., 2003. TRACI, the tool for the reduction and assessment of chemical and other environmental impacts. Journal of Industrial Ecology 6, 3–4. http://mitpress.mit.edu/journals/pdf/jiec_6_3_49_0.pdf.

Bovea, M.D., Vidal, R., 2004. Increasing product value by integrating environmental impacts, costs and customer valuation. Journal of Resources Conservation Recycling 41, 133–145.

Brand, G., Braunschweig, A., Scheidegger, A., Schwank, O., 1998. Weighting in Ecobalances with the Ecoscarcity Method – Ecofactors 1997. Environment Series No. 297, Bern. BUWAL (SAFEL).

Chawla, K.K., Chawla, N., 2012. Automotive composites. In: Wiley Encyclopedia of Composites (1–6), pp. 72–77.

Cicala, G., Cristaldi, G., Recca, G., Latteri, A., 2010. Composites based on natural fibre fabrics. In: Dubrovski, P.D. (Ed.), Woven Fabric Engineering. InTech, ISBN 978-953-307-194-7. http://dx.doi.org/10.5772/10465. http://www.intechopen.com/books/woven-fabric-engineering/composites-based-on-natural-fibre-fabrics.

Cicala, G., Lo Faro, C., 2012. Material selection: polymeric composites matrix. In: Wiley Encyclopedia of Composites (1–6), pp. 1572–1610.

Corbiére-Nicollier, T., Gfeller Laban, B., Lundquist, L., Leterrier, Y., Manson, J.A.E., Jolliet, O., 2001. Life cycle assessment of biofibres replacing glass fibres as reinforcement in plastics. Resources Conservation and Recycling 33, 267–287.

Crettaz, P., Rhomberg, L., Brand, K., Pennington, D.W., Jolliet, O., 2002. Assessing human health response in life cycle assessment using ED10s and DALYs: carcinogenic effects. International Journal of Risk Analysis 22 (5), 929–994.

Crutzen, P.J., Stoermer, E.F., 2000. The "Anthropocene". Global Change Newsletter 41, 17–18.

De Schryver, A.M., Brakkee, K.W., Goedkoop, M.J., Huijbregts, M.A.J., 2009. Characterization factors for global warming in life cycle assessment based on damages to humans and ecosystems. Environmental Science and Technology 43 (6), 1689–1695.

Dissanayake, N.P.J., Summerscales, J., Grove, S.M., Singh, M.M., 2009a. Energy use in the production of flax fiber for the reinforcement of composites. Journal of Natural Fibers 6 (4), 331–346.

Dissanayake, N.P.J., Summerscales, J., Grove, S.M., Singh, M.M., 2009b. Life cycle impact assessment of flax fibre for the reinforcement of composites. Journal of Biobased Materials and Bioenergy 3 (3), 245–248.

Dittenber, D.B., GangaRao, H.V.S., 2012. Critical review of recent publications on use of natural composites in infrastructure. Composites: Part A 43, 1419–1429.

Dornburg, V., Lewandowski, I., Patel, M., 2004. An analysis and system extension of life cycle assessment studies. Comparing the land requirements, energy savings, and greenhouse gas emissions reduction of biobased polymers and bioenergy. Journal of Industrial Ecology 7 (3–4), 93–116.

Ekvall, T., Finnveden, G., 2001. Allocation in ISO 14041-a critical review. Journal of Cleaner Production 9 (3), 197–208.

Frischknecht, R., Steiner, R., Braunschweig, A., Egli, N., Hildesheimer, G., 2006b. Swiss ecological scarcity method: the new version 2006. In: Proceedings of the 7th International Conference on EcoBalance, Tsukuba, Japan, November 2006.

Frischknecht, R., Steiner, R., Jungbluth, N., 2006a. Methode der ökologischen Knappheit – Ökofaktoren 2006. ö.b.u. und Bundesamt für Umwelt, Bern.

Goedkoop, M., Spriensma, R., 2000. Eco-indicator 99, a Damage Oriented Method for Lifecycle Impact Assessment, Methodology Report (update April 2000).

González-García, S., Hospido, A., Feijoo, G., Moreira, M.T., 2010. Life cycle assessment of raw materials for non-wood pulp mills: hemp and flax, Resources. Conservation and Recycling 54, 923–930.

Guinée, J.B., Gorrée, M., Heijungs, R., Huppes, G., Kleijn, R., de Koning, A., van Oers, L., Wegener Sleeswijk, A., Suh, S., Udo de Haes, H.A., de Bruijn, J.A., van Duin, R., Huijbregts, M.A.J., 2002. Handbook on life cycle assessment: operational guide to the ISO standards. In: Series: Eco-efficiency in Industry and Science. Kluwer Academic Publishers, Dordrecht. Hardbound, ISBN 1-4020-0228-9; Paperback, ISBN 1-4020-0557-1.

Hauschild, M., Potting, J., 2005. Spatial Differentiation in Life Cycle Impact Assessment – the EDIP2003 Methodology. Environmental News no. 80. The Danish Ministry of the Environment, Environmental Protection Agency, Copenhagen.

Hauschild, M.Z., Wenzel, H., 1998. Environmental Assessment of Products. In: Scientific Background, 565, vol. 2. Chapman & Hall, United Kingdom, ISBN 0412 80810 2. Kluwer Academic Publishers, Hingham, MA, USA.

Hayashi, K., Itsubo, N., Inaba, A., 2000. Development of damage function for stratospheric ozone layer depletion – a tool towards the improvement of the quality of life cycle impact assessment. IJLCA 5 (5), 172–265.

Hayashi, K., Nakagawa, A., Itsubo, N., Inaba, A., 2006. Expanded damage function of stratospheric ozone depletion to cover major endpoints regarding life cycle impact assessment. IJLCA 11 (3), 150–161.

Hayashi, K., Okazaki, M., Itsubo, N., Inaba, A., 2004. Development of damage function of acidification for terrestrial ecosystems based on the effect of aluminum toxicity on net primary production. IJLCA 9 (1), 13–22.

Hertwich, E., McKone, T., Pease, W., 1999. Parameter uncertainty and variability in evaluative fate and exposure models. Risk Analysis 19, 1193–1204.

Hertwich, E.G., Pease, W.S., Koshland, C.P., 1997. Evaluating the environmental impact of products and production processes: a comparison of six methods. The Science of the Total Environment 196, 13–29.

Hertwich, E.G., Pease, W.S., McKone, T.E., 1998. Evaluating toxic impact assessment methods: what works best? Environmental Science and Technology 32, A138–A144.

Hertwich, E.G., Mateles, S.F., Pease, W.S., McKone, T.E., 2001. Human toxicity potentials for life cycle assessment and toxics release inventory risk screening. Environ. Toxi- col. Chem. 20, 928–939.

Huijbregts, M.A.J., Rombouts, L.J.A., Ragas, A.M.J., Van de Meent, D., 2005b. Human-toxicological effect and damage factors of carcinogenic and non-carcinogenic chemicals for life cycle impact assessment. Integrated Environmental Assessment and Management 1 (3), 181–244.

Huijbregts, M.A.J., Struijs, J., Goedkoop, M., Heijungs, R., Hendriks, A.J., Van de Meent, D., 2005a. Human population intake fractions and environmental fate factors of toxic pollutants in life cycle impact assessment. Chemosphere 61 (10), 1495–1504.

ILCD Handbook: Analysing of Existing Environmental Impact Assessment Methodologies for Use in Life Cycle Assessment, 2010. European Union.

International Organization for Standardization ISO 14040, 2006. Environmental Management – Life Cycle Assessment – Principles and Framework. ISO 14000 International Standards Compendium, Switzerland.

International Organization for Standardization ISO 14044, 2006. Environmental Management – Life Cycle Assessment – Requirements and Guidelines. ISO 14000 International Standards Compendium, Switzerland.

ISO/TR 14047, 2003. Technical Report: Environmental Management – Life Cycle Assessment – Examples of Application of ISO 14042.

Itsubo, N., Sakagami, M., Washida, T., Kokubu, K., Inaba, A., 2004. Weighting across safeguard subjects for LCIA through the application of conjoint analysis. IJLCA 9 (3), 196–2005.

Jolliet, O., Margni, M., Charles, R., Humbert, S., Payet, J., Rebitzer, G., Rosenbaum, R., 2003. IMPACT 2002+: a new life cycle impact assessment methodology. International Journal of LCA 8 (6), 324–330.

Joshi, S.V., Drzal, L.T., Mohanty, A.K., Arora, S., 2004. Are natural fiber composites environmentally superior to glass fiber reinforced composites? Composites: Part A 35, 371–376.

Kemna, R., van Elburg, M., Li, W., van Holstein, R., 2005. MEEuP – the Methodology Report. EC, Brussels (Final version, Delft 28-11-2005).

La Mantia, F.P., Morreale, M., 2011. Green composites: a brief review. Composites: Part A 42, 579–588.

La Rosa, A.D., Cozzo, G., Latteri, A., Recca, A., Björklund, A., Parrinello, E., Cicala, G., 2013. Life cycle assessment of a novel hybrid glass-hemp/thermoset composite. Journal of Cleaner Production 44, 69–76.

La Rosa, A.D., Gagliano, A., Summerscales, J., Latteri, A., Cozzo, G., Cicala, G., 2014a. Environmental impacts and thermal insulation performance of innovative composite solutions for building applications. Construction and Building Materials 55, 406–414.

La Rosa, A.D., Summerscales, J., Recca, G., Latteri, A., Cozzo, G., Cicala, G., 2014b. Bio-based versus traditional polymer composites. A life cycle assessment perspective. Journal of Cleaner Production 74, 135–144.

Ljungberg, L.Y., 2007. Materials selection and design for development of sustainable products. Materials and Design 28 (2), 466–479.

Mallick, P.K., 2010. Thermoplastics and thermoplastic–matrix composites for lightweight automotive structures. In: Materials, Design and Manufacturing for Lightweight Vehicles. Woodhead Publishing Series in Composites Science and Engineering, pp. 174–207.

Mohanty, A.K., Misra, M., Drzal, L.T., 2002. Sustainable bio-composites from renewable resources: opportunities and challenges in the green materials world. Journal of Polymers and the Environment 10 (1/2).

Mullens, M.A., Arif, M., 2006. Structural insulated panels: impact on the residential construction process. Journal of Construction Engineering and Management 132 (7), 786–794.

Müller-Wenk, R., 1994. The ecoscarcity method as a valuation instrument within the SETAC-framework. In: Udo de Haes, Jensen, Klöpffer, Lindfors (Eds.), Integrating Impact Assessment into LCA. SETAC-Europe, Brussels, pp. 115–120.

Netravali, A.N., Chabba, S., 2003. Composites get greener. Materials Today 6, 22–29.

Norris, G., 2002. Impact characterisation in the tool for the reduction and assessment of chemical and other environmental impacts: methods for acidification, eutrophication, and ozone formation. Journal of Industrial Ecology 6 (3/4), 83–105.

O'Neill, T.J., 2003. Life cycle assessment and environmental impact of polymeric products. Number 12. In: Rapra Review Report 156, Vol. 13. ISSN:0889–3144.

Ortiz, O., Castells, F., Sonnemann, G., 2009. Sustainability in the construction industry: a review of recent developments based on LCA. Construction Build Materials 23, 28–39.

Payet, J., 2004. Assessing Toxic Impacts on Aquatic Ecosystems in LCA (thesis 3112). Ecole Polytechnique Fédérale de Lausanne EPFL, Lausanne.

Pennington, D.W., Margni, M., Amman, C., Jolliet, O., 2005. Multimedia fate and human intake modeling: spatial versus non-spatial insights for chemical emissions in western Europe. ES&T 39 (4), 1119–1128.

Pennington, D.W., Margni, M., Payet, J., Jolliet, O., 2006. Risk and regulatory hazard-based toxicological effect indicators in life-cycle assessment (LCA) (June 2006: best HERA paper of year 2006 in Integrated Risk Assessment). Human and Ecological Risk Assessment 12 (3), 450–475.

Potting, J., Hauschild, M., 2005. Spatial Differentiation in Life Cycle Impact Assessment – The EDIP2003 Methodology. Environmental News no. 80. The Danish Ministry of the Environment, Environmental Protection Agency, Copenhagen.

Rochat, D., Margni, M., Jolliet, O., 2006. Continent-specific intake fractions and characterization factors for toxic emissions: does it make a difference? International Journal of LCA 11 (1), 55–63.

Rosenbaum, R., 2006. Multimedia and Food Chain Modelling of Toxics for Comparative Risk and Life Cycle Impact Assessment (thesis 3539). Chapter 5 on uncertainties, Ecole Polytechnique Fédérale de Lausanne EPFL, Lausanne.

Rosenbaum, R., Margni, M., Jolliet, O., 2007. A flexible matrix algebra framework for the multimedia multipathway modeling of emission to impacts. Environment International 33 (5), 624–634.

Rossi, B., Marique, A.F., Glaumann, M., Reiter, S., 2012. Life-cycle assessment of residential buildings in three different European locations, basic tool. Building and Environment 51, 395–401.

Sahoo, S., Misra, M., Mohanty, A.K., 2011. Enhanced properties of lignin-based biodegradable polymer composites using injection moulding process. Composites: Part A 42, 1710–1718.

Sartori, I., Hestness, A.G., 2007. Energy use in the life cycle of conventional and low energy buildings: a review article. Energ and Build 39, 249–257.

Simões, C.L., Costa Pinto, L.M., Bernardo, C.A., 2012. Modelling the environmental performance of composite products: benchmark with traditional materials. Materials and Design 39, 121–130.

Steen, B., 1999a. A Systematic Approach to Environmental Priority Strategies in Product Development (EPS). Version 2000-general System Characteristics. CPM report 1999:4. Chalmers University of Technology, Gothenburg, Sweden.

Steen, B., 1999b. A Systematic Approach to Environmental Priority Strategies in Product Development (EPS). Version 2000-Models and Data of the Default Method. CPM report 1999:5. Chalmers University of Technology, Gothenburg, Sweden.

Toffoletto, L., Bulle, C., Godin, J., Reid, C., Deschênes, L., 2007. LUCAS – a new LCIA method used for a Canadian-specific context. International Journal of Life Cycle Assessment 12 (2), 93–102.

Van Zelm, R., Huijbregts, M.A.J., Harbers, J.V., Wintersen, A., Struijs, J., Posthuma, L., Van de Meent, D., 2007a. Uncertainty in msPAF-based ecotoxicological freshwater effect factors for chemicals with a non-specific mode of action in life cycle impact assessment. Integrated Environmental Assessment and Management 3 (2), 203–210.

Van Zelm, R., Huijbregts, M.A.J., Van Jaarsveld, H.A., Reinds, G.J., De Zwart, D., Struijs, J., Van de Meent, D., 2007b. Time horizon dependent characterisation factors for acidification in life-cycle impact assessment based on the disappeared fraction of plant species in European forests. Environmental Science and Technology 41, 922–927.

Wegener Sleeswijk, A., Van Oers, L., Guinée, J.B., Struijs, J., Huijbregts, M.A.J., 2008. Normalisation in product life cycle assessment: an LCA of the global and European economic systems in the year 2000. Science of the Total Environment 390 (1), 227–240.

Wenzel, H., Hauschild, M.Z., Alting, L., 1997. Environmental Assessment of Products, 544 pp. In: Methodology, Tools and Case Studies in Product Development, vol. 1. Chapman & Hall, United Kingdom, ISBN 0 412 80800 5. Kluwer Academic Publishers, Hingham, MA. USA.

Wool, R.P., Sun, X.S., 2005. Bio-based Polymers and Composites. Elsevier Science & Technology Books, ISBN 0127639527.

Zabalza Bribián, I., Aranda Usón, A., Scarpellini, S., 2009. Life cycle assessment in buildings: state-of-the-art and simplified LCA methodology as a complement for building certification. Building and Environment 44, 2510–2520 table 3 p. 2512.

Zalasiewicz, J., et al., 2008. Are we now living in the Anthropocene? GSA (Geological Society of America) Today 18 (2), 4–8.

Life cycle assessment and the environmental and social labels in the textile and clothing industry

15

M. Koszewska
Lodz University of Technology, Lodz, Poland

15.1 Environmental and social labels: definitions

Giving environmental and social labelling a precise and unambiguous definition is not simple nor is it easy to determine how the labels are actually understood by the consumers of clothing and textile products.

This is mainly so because of the high complexity of sustainable production of textiles and clothing. The textiles and clothing industry has one of the longest, intricate, dispersed and globally stretched supply chains. Its manufacturing processes are also less transparent when compared, for instance, with the production of food. All of this means that understanding and defining the sustainability of textile and clothing products is ambiguous and fairly problematic.

Most people associate sustainability with the environment, even though two other aspects of it – economy and society – are just as important. Considered with respect to their social dimension, sustainable textile and clothing products are those made in good working conditions, without the involvement of child labour, and in conformity with the fair trade rules. In the past, labelling systems focused on the ecological aspect of raw materials or a production process (OEKO-TEX® Standard 100, Blue Angel, EU Ecolabel), but with the passing of time, labels accentuating the social and ethical aspects (Fair Trade, No Sweat label) gained in importance. The present trend in labelling stresses a possibly comprehensive assessment of textile and clothing products. The approach can be called sustainable because both environmental and social dimensions of companies' activities are investigated before a label is granted (Global Organic Textile Standard, GOTS; Sustainable Textile Production, SteP; Global Recycle Standard, GRS). The trend also points to a gradually blurring line between ecological labels and social labels.

Every stage in the life of a textile product can have some social and environmental impacts, and what seems positive at some point of the supply chain may have negative consequences at its later stages; for example, bamboo is a natural renewable material used to produce rayon, but its processing into rayon is a chemically intensive process (Gill, 2012).

Consequently, increasing attention is now given to labels granted after certification processes that assess products over their life cycle and considers both social and environmental impacts of production processes. Attempts are also being made to ensure a

Handbook of Life Cycle Assessment (LCA) of Textiles and Clothing. http://dx.doi.org/10.1016/B978-0-08-100169-1.00015-0

possible high degree of integrity of labels granted to products at particular life cycle stages, for instance by means of so-called transactional or scope certificates.

The literature on this subject offers a number of more or less precise definitions that try to explain the nature of environmental and social labelling. Very broadly, the labels can be described as any words, particulars, trademarks, brand names, pictorial matter or symbols on any packaging, document, notice, board or collar accompanying or referring to a product (Fliess et al., 2007). Social and environmental labels are examples of approaches to providing information about social and environmental aspects of the business (Zadek et al., 1998). They also tell consumers how the product they want to buy has been made and so they help consumers choose goods meeting their environmental and social responsibility standards (Fliess et al., 2007).

The above definitional approaches indicate one of the primary aims of social and environmental labelling schemes is to make sustainable products more widely available and visible, and to convince consumers that by buying a specific group of products they contribute to measurable social and environmental benefits. From this perspective, labelling schemes can be deemed crucial to more efficient development of sustainable markets.

The following section reviews the definitions and explanations of environmental and social labels used in the literature.

15.1.1 Environmental and ecological labels

Hilowitz (1997) defines ecological labels as an instrument for informing consumers about the environmentally safe mode of production or of the ecological benefits of using a specific product.

The European Union (EU) views ecological labels as the type of labels that are granted to products with properties significantly contributing to the improvement of important aspects of the environment (EU Regulation No. 1980/2000).

The authors of the pertinent OECD document have highlighted the differences between ecological and environmental labels. In their view, eco-labels are different from the more general environmental labelling schemes in that they are voluntary and provide information on a wider range of issues, thereby offering a more complete picture of the environmental impact of a product. The definition the authors propose is the following: ‘eco-labels are voluntary labels which convey information to consumers about the environmental implications associated with all elements in the product’s ‘life’, that is its production, distribution, use and disposal’ (Vitalis, 2002).

The difference between environmental and ecological labelling is also stressed in the definition developed by the Global Ecolabelling Network (GEN), which describes an ecological label as one that identifies overall environmental preference of a product within a product category based on life cycle considerations. This feature distinguishes the label from a self-styled environmental symbol or claim statement developed by a manufacturer or service provider. An ecological label is awarded by an impartial third party to products that meet established environmental leadership criteria (GEN, 2004). The International Organization for Standardization (ISO) has divided ‘environmental labelling’ into three types: Type I – environmental labelling (eco-labels); Type II – self-declaration claims; and Type III – environmental declarations (more information on this subject can be found in the section dealing with the classifications of labels).

The above implies that the narrowly understood ecological labels are (1) voluntary, (2) show the environmental impacts of a product to the consumer and (3) are preferably based on life cycle considerations and are verified by a third party (governmental and nongovernmental organisations). Because of these characteristics, they can be viewed as a subgroup of a wider category of environmental labels.

15.1.2 Social labels

Compared with environmental labels, social labels are relatively new and much fewer in number. Most of the currently used social labels were developed in the 1990s (Dickson, 2001), but origins of social labelling can be traced back to the White Label created in the US in 1899. The White Label was meant to assure consumers that women's and children's stitched cotton underwear was manufactured under decent working conditions and with no child labour. It was discontinued in 1918, when labels sponsored by trade unions guaranteeing manufacturer compliance with labour standards came into widespread use (Hilowitz, 1997). While some social labels also impose some requirements concerning the environmental protection, their focus is usually on workers' rights, security and protection of work, the development of local community and fairer trade conditions (Koszewska, 2011b).

Zadek et al. (1998) have formally defined *social labels* as 'the words and symbols on products which seek to influence the economic decisions of one set of stakeholders by providing an assurance about the social and ethical impact of a business process on another group of stakeholders'. In most cases, the existing social labels are to encourage one group of stakeholders, the consumers, to make decisions on which goods to purchase on the basis of ethical considerations, thereby affecting the circumstances under which the goods are produced and marketed towards decent working conditions and fair pricing (Steinrücken and Jaenichen, 2007).

According to Hilowitz (1997), the purpose of social labels is to inform consumers about the social aspects of the production process and to assure them that the product that they like has been made under equitable working conditions.

15.2 Classifications of social and environmental labels

Environmental and social labels can be classified using a whole range of criteria, the most important of which are:

- the range of product aspects addressed by the certification process: ecological, social and both
- the scope of life cycle assessment (LCA) – the coverage of respective life cycle stages
- the object of labelling/certification
- the scope and manner of verification and control

- the geographic scope
- industries and products covered by the label
- leaders, grants, assistance

Figure 15.1 presents the classification of environmental and social labels based on the above criteria.

Environmental and social labels can also be classified using several of the above criteria at the same time, as it has been done by the International Organization for Standardization (ISO). The ISO classification is one of the best known as far as environmental labelling is concerned.

It has already been noted that the environmental label certification schemes were established much earlier than social labelling came into being and their influence is much wider. The eco-certification rules were created jointly by a group of international organizations, including the European Community, the World Trade Organisation, the United Nations Environmental Programme (UNEP) and the ISO. Over the last 30 years, many initiatives have been undertaken to certify products and services the environmental properties of which exceeded those required by the standards in force. To ensure a consistent approach to this type of labelling, the ISO has published a series

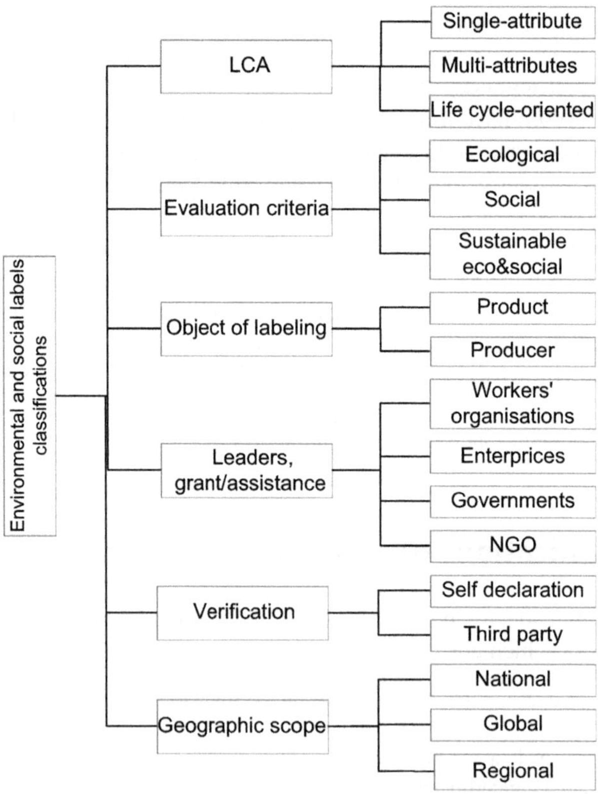

Figure 15.1 Popular classifications of environmental and social labels.

of standards applying specifically to environmental labelling (ISO 14020, ISO 14021, ISO 14024, ISO 14025) (Koszewska, 2011b).

The ISO 14020 family consists of three types of labelling schemes that describe the environmental advantages that a product should have:

- Type I (ecological label) is a voluntary, multiple-criteria-based, third-party programme that awards a licence authorising the use of environmental labels on products, indicating overall environmental preferability of a product within a product category based on life cycle considerations. (ISO 14024)
- Type II (self-declared environmental claims) is a single-attribute label developed by the producer. It is made without independent third-party certification, by manufacturers, importers, distributors, retailers or anyone else likely to benefit from such a claim. (ISO 14021)
- Type III is an eco-label whose environmental declaration reads as follows: quantified environmental data of a product under preset categories or parameters. It is awarded based on a full LCA and verified by a qualified third party. (ISO 14025)

Table 15.1 provides a comparison of labels with the applicable standards.

Given the range of criteria applied in the certification process, the classification proposed by Hartlieb and Jones (2009) seems very useful. They grouped labels into four broad categories sharing similar characteristics:

- Organic agriculture, which is one of the most important regarding sales, popularity, age and number of labelling initiatives (Mintel, 2006). As far as clothing and textiles are concerned, organic farm standards underpin many popular labels, such as the Global Organic Textile Standard or the Organic Content Standard.
- Fair trade, which is one of the oldest and most popular groups of labels as regards consumer awareness and sales (Mintel, 2006). Fair trade concentrates on issues in international trade

Table 15.1 Comparison of the types of ISO environmental labels

	Environmental labelling		
Type	**Type I – eco-labels**	**Type II**	**Type III**
Standard	ISO 14024	ISO 14021	ISO 14025
Third party involvement	Yes	No	Yes
Life cycle analysis	Simplified	No	Yes
Scope	Multi-criterial	Selected product characteristics	Parameter categories defined for the sector
Possibility of differentiating products ecologically within a group of products	Yes	No	Yes

Continued

Table 15.1 Continued

Environmental labelling			
Type	**Type I – eco-labels**	**Type II**	**Type III**
Information carrier	A label – graphic mark, logo	Graphic mark/ word/slogan	Numerical data represented by graphs, drawings, text
Voluntary	Yes	Yes	Yes
Verifiability/ reliability	High	Low	High
Prospects for development	Good	Weak (low reliability)	Average (complex procedure, analysis of a large volume of data)

Source: Koszewska (2011b).

and the related social injustice and poverty issues, usually giving a strong focus on aspects in community development. The category includes also Equitrade and Rugmark initiatives.

- Holistic approaches (sustainable labels) dealing with social and environmental issues. These labels are relatively new but have recently gained in popularity. The pertinent examples of sustainable labels applying to clothing and textiles are GOTs, Better Cotton Initiative or very recently the STeP standard proposed by OEKO-TEX®.
- Sustainable management of natural resources is also quite new, and its main focus is on the environmental side of one specific sector, such as forestry, fishery and marine aquarium organisms. Sustainable management labels are mainly created as a result of multi-stakeholder initiatives.

Figure 15.2 shows the classification of environmental and social labels applied to textiles and clothing performed using two crucial criteria: LCA and the range of issues addressed in the producer evaluation process.

From Figure 15.2 it follows that no social labels use life cycle analysis in the assessment and certification process. This is due to the fact that even though LCA is a widespread and accepted method to evaluate the global environmental impact of a product, it turns out to be inadequate if one wishes to include social and economic aspects in the product evaluation. An environmental LCA employs an input–output model, making an inventory of the flows of raw materials, energy and emissions without taking much account of what happens within the company. The actual production process is considered more or less as a black box. Since essential social criteria such as wages, working hours and discrimination are characteristic of what happens during the production process and within the organisation itself, this black box model is not applicable for social impacts (Spillemaeckers, 2007).

	Environmental labels emphasis on ecological criteria	**Sustainable labels** holistic approach ecological & social criteria	**Social labels** emphasis on social criteria
Single-attribute labels - one stage of the product's life cycle	**OEKO-TEX®** standard 100 **OCS**-Organic Content Standard	**BCI**-The Better Cotton Initiative **e^{3} TM** sustainable cotton program	**Fairtrade Standard** for cotton **CARE & FAIR RUGMARK**
Multi-attributes ecolabels almost every life cycle stages but acts on different stages separately	The EU Ecolabel The bluesign® standard	**GRS**-Global Recycled Standard **STeP**-Sustainable Textile **GOTS**-Global Organic Textile Standard	
Life cycle-oriented labels assessing the life cycle as a whole		**Cradle to Cradle** **SMaRT**-Consensus Sustainable Product	

Figure 15.2 Classification of environmental and social labels by life cycle assessment and the range of issues addressed in the producer evaluation process.

15.3 Life cycle assessment and greenwashing in textiles and clothing

In recent years, consumers have shown rising interest in ecological, or more broadly, sustainable products. An important role in this trend (also known as eco-consumption, green consumption or sustainable consumption) is played by textile and clothing products (Koszewska, 2011a, Organic Monitor, 2011). Manufacturers are always seeking new ways to differentiate their products. One way of achieving this is green marketing, whose terms have become a common feature of advertising messages such as 'eco', 'environmentally friendly', 'green', 'earth friendly' and 'sustainable'(Chen and Chang, 2013). A company's claim about the environmental quality of its products that has not been verified by an independent third party is called a 'green claim' (UNOPS, 2009). In the ISO classification, labels based on green claims are Type II labels. Many of these environmental claims pointing to ecological and/or social products' attributes are ambiguous and deceptive, leaving space for very harmful, opportunistic producer behaviour called 'greenwash' or 'greenwashing'.

Greenwashing develops at the intersection of two of a firm's behaviours: poor environmental performance and positive communication about environmental performance (Delmas and Burbano, 2011). It may occur at a firm level and then consumers are misled regarding the firm's environmental practices, or at a product level when the environmental benefits of a product are misrepresented (Parguel et al., 2011; TerraChoice, 2009).

An important issue is the asymmetry of information between the producer and consumers, arising from the fact that consumers frequently do not have enough knowledge and awareness to access the level of product sustainability or environmental performance of the firm and/or its products (Genç, 2013). In the case of textile and clothing products, the situation is more complicated because of the lack of clarity of how 'good' and 'bad' should be interpreted with respect to textile sustainability. Terms such as *organic*, *natural* and *biodegradable* are generally perceived as 'good' in contrast to *GM*, *synthetic* and *chemical* (Blackburn, 2009). Many of these terms are cleverly used to market claimed sustainability benefits. Not many consumers realise that the reliable assessment of textiles and clothing sustainability involves an investigation into the whole 'cradle-to-grave' life cycle of a product. Moreover, the complexity of sustainable production of textiles and clothing (the industry has one of the longest, most complicated, fragmented and globally expanded supply chains) makes such an assessment impossible for consumers to make. Even specialists have problems trying to understand and define sustainability of textile and clothing products.

One way of effectively reducing the asymmetry of information between the producer and consumers is third-party-certified ecological labels giving credible environmental and social characteristics of a product. Environmental standards, certification systems and consequential ecological labels have emerged as a protection against greenwashing from the 1980s' wave of green product launches. At the same time, the perceived importance of eco-labels has led to their proliferation and to the multitude of their types and degrees of meaningfulness and integrity (TerraChoice, 2010). The 'sin of worshiping false labels' is one of new sins that TerraChoice recognised in 2009. The risk of labels being misused, or of their quality and features being misinterpreted, is becoming more serious (UNOPS, 2009). According to a TerraChoice report (2010), false labels are becoming widespread. More than 32% of 'green' products found in a 2010 study carried a fake label compared with 26.8% in 2009.

The risk of misleading labels being used can be minimised by giving stronger emphasis to labels verified by a third party based on multiple criteria and life cycle considerations. The effectiveness of an environmental label as a countermeasure against greenwashing practices and a way to gain consumer trust depends on its ability to be truthful, easy to communicate and, finally, verifiable and accurate.

The most effective in this respect are Type-I labels provided in ISO 14024 (some examples of Type I labels applying to textiles and clothing products are GOTS, European Ecolabel, Nordic Swan, and Blue Angel). Much effort will be needed, though, to increase consumer awareness of what particular labels mean and of their true contents.

According to a 2010 TerraChoice report, more than 30% of products certified by an ISO 14024-based program were sin-free (compared with 4.4% of the study-wide results). This shows that while good eco-labelling helps prevent greenwashing, it cannot eliminate it.

15.4 A comparison of the selected environmental and social labels in terms of LCA

In this section, 17 popular environmental and social labels applied to textile and clothing products are compared with respect to LCA, the range of issues addressed in the producer evaluation process and the form of verification.

ISO 14044 defines LCA as a technique addressing the environmental aspects and potential environmental impacts of a product over its life cycle, from raw material acquisition through production, use, end-of-life treatment, recycling and final disposal (ISO, 2006). The LCA methodology has been used to assess the environmental impacts of the textile and clothing sector for some time now, but studies in this field have only been conducted in the last five years (Altun, 2012). What makes the methodology so important for the labelling of sustainable textile and clothing products?

As mentioned, it is not easy to assess textile and clothing products for sustainability because the industry has one of the longest, most complicated, dispersed and globally stretched supply chains. The degree to which a product can be considered 'green' is determined by a number of factors that must be analysed for each life cycle phase, starting from raw materials acquisition through use and disposal. This analysis is particularly difficult in the case of customers who usually make buying choices unable to make judgments about the sustainability of textiles and clothing products they like. They need simple information they can understand and trust. One of the most effective means of achieving this is eco-labels, which constitute a reliable and trustworthy information system on product features (Frieder and Frankl, 2005). However, it must be noted that the ever-growing number of labels makes it difficult for customers to make conscious and informed choices, leading them into confusion and distrust. The following section is an attempt at ordering ecological and social labels and dividing them with respect to LCA.

Based on LCA, social and environmental labels can be divided into three major groups (Cobut et al., 2013):

1. Single-attribute labels that cover mainly one stage of a product's life cycle, usually raw material extraction or the product-use stage, for instance Organic Content Standard (OCS) or OEKO-TEX® Standard 100.
2. Multi-attribute labels that address almost every stage of the product life cycle. This category offers a more fragmented vision of the product life cycle, though, because rather than providing a holistic evaluation it acts on different stages separately. In this sense, it is close to the ISO Type I labels.
3. Life cycle-oriented labels that are different from the multi-attribute labels in that they aim to provide an assessment of the entire life cycle. They also recommend that criteria applying to life cycle stages are considered as early as the design phase. The LCA methodology is based on the ISO 14040 series of standards or at least on a formal life cycle approach; an example of this is the Cradle2Cradle or SMaRT programme.

At the same time, efforts are being made to ensure high integrity of labels granted to various phases of product life cycle. Solutions such as the modular assessment system used for Oeko-Tex 100 (see Moore and Wentz, 2009) or the scope and transaction certificates issued with OE labels enable the integrity of labels, including single-attribute labels, across the textile supply chain.

Some of the crucial aspects of label integrity and transparency across the supply chain are the product's full life cycle and the system of verification.

As far as verification methods are concerned, it is very rare, and expensive, for one company in the textile and clothing industry to control or to own the production of fibre from raw material all the way through to the final product. Most companies have to cooperate with direct suppliers.

There are many approaches within the verification process that can assure label integrity across the supply chain (Gill, 2012):

1. Internal assurance: A textile and clothing producer using forms of internal assurance can implement systems and processes with his suppliers and monitor them to ensure that they are doing what they need to do. This method involves a lot of effort and cost. Moreover, the only assurance the customers have is a producer's word and the word of their suppliers.
2. Data collection or electronic traceability: This method allows ensuring label integrity and transparency by means of software that tracks the product from the acquisition of raw material to the finished product. The supplier feeds into the system the product information that eventually is relayed to the final brand or retailer. The technology that is frequently used in support of this method is Icebreaker Baacode. Barcodes, or baacodes, collect information along every step of the supply chain and the end consumer is able to scan the code and, for example, find out where the farm that produced their wool is.

   An example of a practical application of the method is the independent platform 'Respect-Code'. This traceability system helps brands to increase trust with their consumers, as well as allowing them to enhance collaboration with suppliers. It offers a smart risk management tool that shows what is happening across the supply chain and identifies room for improvement (www.respect-code.org). Another practical example of traceability systems is the 'MADE-BY Track&Trace' system that follows the trail of the clothes. From the MADE-BY code in the garment, the consumer can find out where the garment was made and by whom (http://www.made-by.org/).
3. Certification: This method checks the suppliers in the supply chain through third-party verification. Textile Exchange proposes two kinds of certificates that allow doing this: transaction and scope certificates.[1]

The Oeko-Tex Association has proposed a modular system of certification. A product certification process according to OEKO-TEX® Standard 100 has a modular structure: textile products can be tested and certified at any production stage (including the manufacturing of accessories). Certificates can be issued for:

- raw materials, fibres, filaments
- yarns, raw and finished

[1] A 'transaction certificate' is a document supplied by the certification body as proof that the goods being shipped from one organization to the next conform to a given standard. Transaction certificates are issued each time goods change ownership, and their details must match invoices and shipping documents. A 'scope certificate' is a document issued by the certification body as a confirmation that a given organization is qualified to produce goods to the standard. To qualify, the organization must be inspected by the certification body at least once a year (Gill, 2012).

- textile fabrics, raw and finished (dyed, printed, finished)
- ready-made products
- textile and non-textile accessories.

This modular principle generally avoids double testing and spreads the financial expense of certification across all production stages involved. The components of a finished product that have already been certified will be recognised in the framework of certification processes at subsequent processing stages upon the presentation of a valid certificate and a representative sample of materials and do not require retesting on the finished product (https://www.oeko-tex.com).

In Table 15.2, particular labels are compared by product life cycle, social and environmental criteria and forms of verification.

15.5 The utility of environmental and social labels for consumers – barriers and challenges

Environmental and social labelling seeks to inform consumers about the environmental and social impacts of the production, consumption and waste phases of the products consumed. Consequently, it has two objectives to fulfil:

- provide consumers with more information about the environmental effects of consumption with a view to bringing about a change towards more environmentally friendly consumption patterns, and
- encourage producers, governments and other agents to raise the environmental standards of products/services (Gallastegui, 2002).

On the one hand, social and environmental labels provide consumers with the information they want, so they increase market efficiency acting as information policy instrument. On the other hand, they help enterprises to mitigate the environmental and social impacts of their business, thus acting as an environmental policy instrument (see Figure 15.3).

Evaluating labelling as a factor stimulating the development of sustainable consumption and production of textiles and clothing is not simple. In some cases, eco- and social labels are a mechanism that allows effective communication with consumers, but in many cases they are a source of confusion and frustration.

Compared with other instruments that companies use to communicate with consumers (reports, codes of conduct, etc.), the main advantage of labels lies in their simplicity and visibility. Particularly in the case of consumer goods such as textiles and clothing, they offer a useful and efficient means of communicating whether the properties of the particular product meet consumers' social and environmental expectations (Koszewska, 2011b).

Studies show that in OECD countries the sustainability effects of labelling schemes are growing with heightened consumer interest in environmental and social issues (OECD, 2008). At the same time, however, as a means of communication, labelling systems have a number of weaknesses that make customers less trustful

Table 15.2 Comparison of environmental and social labels for textile and clothing products

Labelling scheme	Life cycle assessment			Emphasis on			3rd Party verification	Product labelling	Input material
	Single-attribute	Multi-attributes	Life cycle oriented	Ecological criteria	Social criteria	Both			
Global Organic Textile standard (GOTS) http://www.global-standard.org		X				X	X	Yes	Organic material – natural organic textile fibres
The EU Ecolabel http://ec.europa.eu/environment/ecolabel/		X		X			X	Yes	Limited allowable fibres
The bluesign® standard http://www.bluesign.com		X				X	X	Yes	N/A
OEKO-TEX® standard 100 (chemically safe textiles) https://www.oeko-tex.com	X[a]			X			X	Yes	N/A
Sustainable Textile Production (STeP) https://step.oeko-tex.com/en/step/step_home/home.html		X				X	X	No	N/A
OEKO-TEX® Standard 100plus (the combination OEKO-TEX® Standard 100 and STeP) https://www.oeko-tex.com								Yes	N/A

Fairtrade Standard for cotton http://www.fairtrade.net/cotton.html	X				X		X	Yes	Fairly traded raw materials
The Better Cotton Initiative (BCI) http://bettercotton.org/	X					X	Partial 2nd and some 3rd party	No	Efficient water and pesticide use
e^{3} ™ sustainable cotton program by Bayer CropScience	X					X	Self-evaluation and 3rd party audits	Yes	
Cotton made in Africa (CmiA) http://www.cotton-made-in-africa.com	X					X	X	Yes	
Content claim standard CCS http://textileexchange.org/CCS	X						X	No	Any raw material
Organic content standard (OCS) http://textileexchange.org/OCS	X			X			X	Yes	Organic materials
Recycled claim standard (RCS) http://textileexchange.org/RCS	X			X			X	Yes	Recycled materials

Continued

Table 15.2 Continued

Labelling scheme	Life cycle assessment			Emphasis on			3rd Party verification	Product labelling	Input material
	Single-attribute	Multi-attributes	Life cycle oriented	Ecological criteria	Social criteria	Both			
Global recycled standard, v3 (GRS) http://textileexchange.org/GRS						X		Yes	Recycled material
Intertek recycled PET management certification scheme http://www.intertek.com/packaging/test/pet-rpet/	X				X		Verified by standard setting organization through testing, first-party		Recycled PET
SMaRT consensus sustainable product standards http://mts.sustainableproducts.com/SMaRT_product_standard.html			X			X	X		
Cradle to cradle http://www.c2ccertified.org/			X			X	X	Yes	

[a]The certification criteria are concentrated on examining the final product on its influence on the user, however, the certification is possible at each stage of the production process and if the products are certified in an earlier stage (e.g. raw materials), this is taken into consideration.

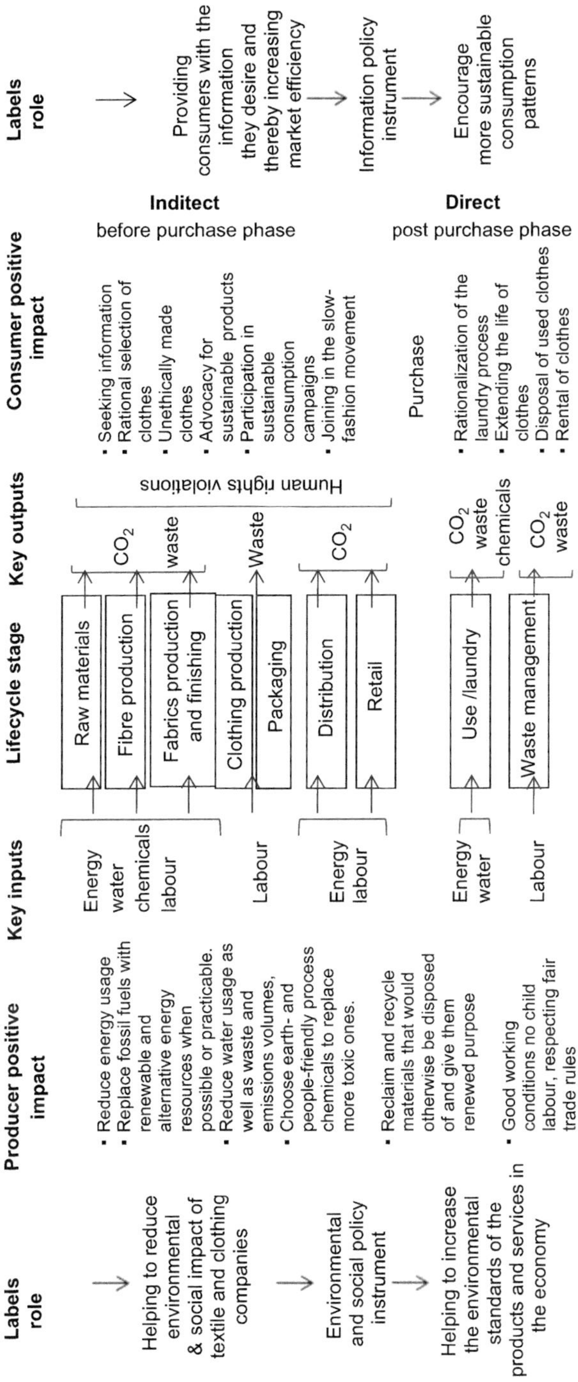

Figure 15.3 The role of environmental and social labels in the development of sustainable textile and clothing markets by life cycle stage.

and therefore less interested in them. The line between the amount of information consumers want and need to feel well informed and an overwhelming amount of overly specific information is very thin. Generally, the weaknesses of labels that somewhat impede further development of labelling systems and have a negative effect on their effectiveness include:

1. Criteria differences across various labelling schemes
2. Market competition between various schemes
3. Technological lock-ins for business that limit innovation (OECD, 1997)
4. International trade concerns – a risk of labels being used to discriminate against imported or even domestic products (OECD, 2007)
5. Greenwashing – environmental labels are frequently abused and counterfeited. The growing popularity of 'false' eco-labels prevents the 'real' labels from becoming more credible. Many products are given labels with flowers, bushes, trees, shrubs, hares and birds to suggest that they are environmentally friendly while in fact they are not (Koszewska, 2011b).
6. Certification costs – the costs of certification fees and compliance
7. Complicated certification process, difficult to understand for all members of the supply chain, following from the global and complex structure of the textile industry.

All these weaknesses notwithstanding, environmental and social labels still belong to the most popular means of attracting attention to the ecological or ethical properties of products (UNEP, 2005; Koszewska, 2011b; MORI, 2000).

Labels cannot be really successful and effective unless consumers perceive them as reliable and clear. For this to be so, transparent standardisation and certification systems are necessary in the first place, as well as the harmonisation and unification of the existing systems, so that the average consumer does not feel confused and can decide on their own what product they want.

Both issues are a huge challenge for all parties interested in continued development of social and ecological labelling systems. To tackle it, many activities and initiatives have recently been undertaken to integrate and standardise the present social and ecological labelling systems (e.g. the activities conducted by GEN and Textile Exchange, or EU's efforts for "Building the Single Market for Green Products": Facilitating better and credible information on the environmental performance of products and organisations).

15.6 Future trends and concluding remarks

The rising interest in ecological and social labelling schemes that has recently been observed in Western Europe is largely driven by the expanding group of socially conscious and wealthy consumers who in addition to using the technological parameters, price, quality, speed of delivery and eco-friendliness to choose products also consider their social and ethical dimensions (Foltynowicz, 2005; Koszewska, 2011a; BSR, 2008). In trying to predict how social and eco-labelling may develop in the future, the following trends should be considered:

1. The clear-cut line between ecological and social labels is gradually blurring. One reason is that the addressing of social issues such as safe working conditions frequently allows mitigating the negative environmental impacts of production processes, a case in point being

the ban on the use of harmful chemicals. At the same time, many labels that are categorised as ecological are also sensitive to some social aspects. This explains why labels giving equal stress to social and environmental issues and thereby contributing to the production of truly sustainable textile and clothing products are becoming increasingly appreciated. In the last decade, labelling instruments spread to earmark social and/or ethical goods and services, reflecting a shift in public opinion from purely environmental concerns to a more holistic approach to sustainability. The trend is likely to grow stronger in coming years, contributing to a wider use of the combination of environmental and social labels.

2. Closely related to the first trend is the harmonisation trend, heralded by efforts to broaden the scope of the existing schemes so that they complement, overlap or even merge with each other.
3. The fastest to develop will apparently be labelling systems that will certify the entire product life cycle as well as take account of the social and ecological aspects of production, and those that ensure high integrity of labels across the product life cycle, for instance through transactional or scope certificates.

   There are two issues that need to be addressed in the context of these three trends. One is that it is practically impossible to fully unify labels based on a diverse range of criteria and the other is that even if this were possible, their use would be limited for two reasons. Firstly, some consumers may find it hard to engage with schemes that bring together many different, often complex issues; secondly, a unified label may distract attention from critical, specific issues associated with particular products (BSR, 2008).

   It seems, however, that the real problem lies not in the scope of information a label gives but rather in how it is communicated. It is essential that complex information on sustainability is communicated in a simple way and is transparent, so that consumers can make informed choices (OECD, 2008).
4. The EU's most recent initiative, 'Building the Single Market for Green Products – Facilitating Better Information on the Environmental Performance of Products and Organisations', may significantly contribute to further development of eco-labelling systems in the community. The initiative has been adopted to remove the present barriers for both producers and consumers to supply and purchase green products, many of which stem from the ambiguity of what constitutes a truly 'green' product and a 'green' organisation. Consumers are targeted today with several types of 'green' marketing information, including labels and environmental declarations. Many of these are based on some forms of life cycle methodologies, but often they are not done in consistent ways, are not verifiable and/or only focus on a single issue. There is no one-to-one relationship between methodologies and labels. In the opinion of the EU's representatives, without specific legislation aimed at avoiding the proliferation of labels, even the existence of a single methodology will not imply per se a reduction in the number of environmental labels that coexist today on the market. This means that the effectiveness of the principle of mutual recognition of different national methods employed to communicate the quantified environmental performance of products will be limited, as it will further augment the diversity of environmental labels and the confusion of consumers.

The Single Market for Green Products initiative consequently proposes a set of actions to overcome the existing problems:

- to select the product environmental footprint and the organisation environmental footprint as the methods for measuring environmental performance across the product life cycle
- to recommend the methods to member states, companies, private organisations and the financial community through a commission recommendation

- to announce a 3-year testing period for developing product- and sector-specific rules through a multi-stakeholder process
- to provide principles for communicating environmental performance, such as transparency, reliability, completeness, comparability and clarity
- to support international efforts toward more coordination in methodological development and data availability.

The last of the important trends in environmental and social labelling is driven by major changes in lifestyles and consumption patterns. More and more consumers reach for the state-of-the-art technologies. Because of the Internet, information flows faster and faster. On top of that, time is in short supply. Consumers expect solutions that will make the buying process simple and quick, and they increasingly use smartphone technology to browse online resources for more detailed and real-time product information (McDougall, 2012). This implies that the buyers of textile products, too, will use the technology more frequently, as has already been observed in the case of consumers of cosmetics and food products. With mobile devices, consumers can get a wealth of information on products from their QR (smart) barcodes. Those who seek sustainable products can learn details about the product's environmental, social and even economic footprint whilst shopping. Some brands allow consumers to use the barcodes to 'track and tell' the product origins (Organic Monitor, 2012). The current environmental and social labelling systems, too, should reach for new technologies.

References

Altun, S., 2012. Life cycle assessment of clothing process. Research Journal of Chemical Sciences 2 (2), 87–89.

BSR, 2008. Eco-promising: Communicating the Environmental Credentials of Your Products and Services. Forum for the Future. Available from: http://www.bsr.org/reports/BSR_Eco-Promising_April_2008.pdf (accessed 14.08.14.).

Blackburn, R.S., 2009. Sustainable Textiles. Life Cycle and Environmental Impact. Woodhead Publishing in Textiles Number 89.

Chen, Y., Chang, C., 2013. Greenwash and green trust: the mediation effects of green consumer confusion and green perceived risk. Journal of Business Ethics 114 (3), 489–500.

Cobut, A., Beauregard, R., Blanchet, P., 2013. Using life cycle thinking to analyze environmental labeling: the case of appearance wood products. International Journal of Life Cycle Assessment 18 (3), 722–742.

Delmas, M.A., Burbano, V., 2011. The drivers of greenwashing. California Management Review 54 (1), 64–87.

Dickson, M.A., 2001. Utility of no sweat labels for apparel consumers: profiling label users and predicting their purchases. Journal of Consumer Affairs 35 (1), 96.

EN ISO 14044 Environmental management, 2006. Life Cycle Assessment – Principles and Framework.

Fliess, B., Lee, H., Dubreuil, O.L., Agatiello, O., 2007. CSR and Trade: Informing Consumers about Social and Environmental Conditions of Globalised Production: Part I. OECD Trade Policy Papers No. 47. OECD Publishing. http://dx.doi.org/10.1787/.

Foltynowicz, Z., 2005. Ekoznakowanie a zintegrowana polityka produktowa. Towaroznawcze Problemy Jakości 2 (3), 2005.

Frieder, R., Frankl, P., 2005. Future of Eco-labelling: Making Environmental Product Information Systems Effective. GBR: Greenleaf Publishing, Sheffield, South Yorkshire.

Gallastegui, I., 2002. The use of eco-labels: a review of the literature. European Environment: The Journal of European Environmental Policy (Wiley) 12 (6), 316–331. http://dx.doi.org/10.1002/eet.304.

GEN Global Ecolabelling Network, 2004. Information Paper: Introduction to Ecolabelling. http://www.globalecolabelling.net/docs/documents/intro_to_ecolabelling.pdf (accessed 03.07.14.).

Genç, E., 2013. An analytical approach to greenwashing: certification versus noncertification. Journal of Management & Economics 20 (2), 151–175.

Gill, A., 2012. Understanding Integrity. An Overview of the System of Standards and Strategies Necessary to Suport Sustainability Claims for Fibres. Textile Exchange Webinar.

Hartlieb, S., Jones, B., 2009. Humanising business through ethical labelling: progress and paradoxes in the UK. Journal of Business Ethics 88 (3), 583–600.

Hilowitz, J., 1997. Social labelling to combat child labour: some considerations. International Labour Review 136 (2), 215.

Koszewska, M., 2011a. The Ecological and Ethical Consumption Development Prospects in Poland Compared with the Western European Countries. In: Comparative Economic Research, vol.14. Number 2, Versita, Warsaw, pp. 101–123 – 9p. http://versita.metapress.com/content/p7w4q3j117rq7826/fulltext.pdf.

Koszewska, M., 2011b. Social and eco-labelling of textile and clothing goods as a means of communication and product differentiation. FIBRES & TEXTILES in Eastern Europe 4 (87), 20–26. http://www.fibtex.lodz.pl/article536.html.

Mintel, 2006. Attitudes towards Ethical Foods – UK, Market Intelligence, August 2006. Mintel Information Services, London.

Moore, S.B., Wentz, M., 2009. Eco-labeling for textiles and apparel. In: Blackburn, R.S. (Ed.), Sustainable Textiles. Life Cycle and Environmental Impact. Woodhead Publishing.

MORI (Market and Opinion Research International), on behalf of CSR Europe 2000, November 2000. The First Ever European Survey of Consumers Attitudes towards Corporate Social Responsibility. CSR Europe, Brussels, Belgium. http://www.ipsos-mori.com (accessed 16.1.10.).

OECD, 1997. Eco-labelling: Actual Effects of Selected Programmes.

OECD, 2007. CSR and Trade: Informing Consumers about Social and Environmental Conditions of Globalised Production.

OECD, 2008. Promoting Sustainable Consumption. Good Practices in OECD Countries. France.

Organic Monitor, 2011. The French Market for Ethical Textiles. http://www.organicmonitor.com/130180.htm (accessed 03.08.14.).

Organic Monitor, 2012. Consumers Getting Smart by Mobile Technology. http://www.organicmonitor.com/r0501.htm.

Parguel, B., Benoît-Moreau, F., Larceneux, F., 2011. How sustainability ratings might deter 'greenwashing: a closer look at ethical corporate communication. Journal of Business Ethics 102 (1), 15–28.

Regulation (EC) No 1980/2000 of the European Parliament and of the Council of 17 July 2000 on a revised Community eco-label award scheme.

Regulation (EC) No 1980/2000 of the European Parliament and of the Council of 17 July 2000 on a revised Community eco-label award scheme, Official Journal of the European Communities.

McDougall, A., 2012. Smartphone Technology Use to Increase as Mobile Platforms Become Key Info Source for Cosmetics. William Reed Business Media. Available from: http://www.cosmeticsdesign-europe.com/Market-Trends/Smartphone-technology-use-to-increase-as-mobile-platforms-become-key-info-source-for-cosmetics (accessed 03.08.14.).

Spillemaeckers, S., 2007. The Belgian Social Label: A Governmental Application of Social LCA. Available from: http://userpage.fu-berlin.de/ffu/calcas/Spillemaeckers_Belgian.pdf (accessed 14.09.14.).

Steinrücken, T., Jaenichen, S., 2007. The fair trade idea: towards an economics of social labels. Journal of Consumer Policy 30 (3), 201–217.

TerraChoice, 2009. Environmental Claims in Consumer Markets Summary Report: North America. Available from: http://sinsofgreenwashing.org/index3c24.pdf (accessed 14.09 14.).

TerraChoice, 2010. The Sins of Greenwashing Home and Family Edition, a Report on Environmental Claims Made in the North American Consumer Market. Available from: http://sinsofgreenwashing.org/index35c6.pdf (accessed 14.09.14.).

UNEP, 2005. The Trade and Environment Effects of Ecolabels: Assessment and Response. Available from: http://www.unep.ch/etb/publications/Ecolabelpap141005f.pdf (accessed 16.01.13.).

UNOPS, 2009. A Guide to Environmental Labels - for Procurement Practitioners of the United Nations System. UNEP.

Vitalis, V., 2002. Round Table on Sustainable Development. Private Voluntary Eco-labels: Trade Distorting, Discriminatory and Environmentally Disappointing OECD. Available from: http://www.oecd.org/sd-roundtable/papersandpublications/39362947.pdf (accessed 16.01.13.).

Zadek, S., Lingayah, S., Forstater, M., 1998. Social Labels: Tools for Ethical Trade. New Economics Foundation for the European Commission.

Sources for further information

Below are some helpful Internet links:
http://www.c2ccertified.org/.
http://www.ecolabelindex.com.
http://www.ecotextilelabels.com.
http://ec.europa.eu/environment/ecolabel/.
http://www.global-standard.org.
http://www.globalecolabelling.net.
https://www.oeko-tex.com.
http://textileexchange.org.
http://sinsofgreenwashing.org.

Clothing disposal habits and consequences for life cycle assessment (LCA)

16

K. Laitala, I.G. Klepp
National Institute for Consumer Research (SIFO), Oslo, Norway

16.1 Introduction

Consumers' decisions in the disposal phase of clothing are crucial from an environmental point of view, as they affect the lifespan of clothing, as well as the potential for reuse and recycling. This chapter discusses the effects of different end-of-life scenarios to the life cycle assessment (LCA) calculations. In doing this, examples of Norwegian consumers' clothing use and disposal practices are used, including the channels they select, how frequently they dispose of clothing and whether there are any differences between consumer groups. Based on these results, the weaknesses of many existing LCA studies related to the use phase are discussed and improvements on gaining more realistic data are suggested.

The first section of this study provides an overview of the environmental benefits of clothing recycling and reuse. This is followed by a discussion that defines *disposal* and how the term is used in this chapter. The statistics for the current situation in Norway for the end-of-life textiles and clothing are presented, followed by a short overview of the literature on clothing disposal practices. Then the two methods that are employed, a wardrobe study and a survey, are presented, followed by results on disposal methods and frequencies. Finally, the implications of the present study are discussed, and suggestions for policy measures and future research directions are suggested.

16.1.1 End-of-life counts

Waste prevention and management are politically highly prioritised areas within the European Union (EU). Waste minimisation strategies are given in the Waste Framework Directive in order of their environmental impact (European Parliament and the Council, 2008). The highest impact is gained by preventing the waste from occurring, for example using fewer materials or keeping products for longer. The second highest impact strategy is to take measures to prepare products for reuse. This may include, for example, cleaning, sorting and repairing so that the product can be used again. The third option is material recycling, that is turning waste into a new products. The fourth strategy includes other forms of recovery, for example incineration with energy recovery. The final and least preferred option is disposal, such as landfilling or incineration without energy recovery.

Handbook of Life Cycle Assessment (LCA) of Textiles and Clothing. http://dx.doi.org/10.1016/B978-0-08-100169-1.00016-2

This has also been proven in several studies that compare the environmental impacts of different end-of-life scenarios of waste treatment; for most of the cases, they show that landfilling is the least preferred option (e.g. Finnveden et al., 2005; Glew et al., 2012; Lanoë et al., 2013).

The Department for Environment, Food and Rural Affairs (Defra) in the United Kingdom has recognised sustainable consumption and production (SCP) as a key priority and identified clothing as one of the priority product groups (Madsen et al., 2007). Textiles and clothing recycling and reuse are also focused on by the Nordic Council of Ministers, which recently initiated several research projects within the topic (Tojo et al., 2012; Palm et al., 2014a,b; Watson et al., 2014; Klepp et al., 2014). This strong political focus is supported by data from LCA studies on textiles. Figure 16.1 presents results from two LCA studies on cotton T-shirts that compare different end-of-life scenarios and document the environmental impacts across the textile life cycle (Fisher et al., 2011; Farrant, 2008 as in Morley et al., 2009). The selected environmental indicator in both of the studies is carbon dioxide equivalents.

Reuse through organisation includes collection, sorting and redistribution of the products, while direct reuse omits these steps and therefore gives higher environmental gains, more comparable to longer use with the first owner. In LCA calculations it is important to take into account the replacement rate, for example the number of new clothing items replaced by the second-hand clothing. Farrant et al. (2010) have estimated that the purchase of 100 second-hand garments would save between 60 and 85 new garments depending on the place of reuse. They estimated that replacement rate is higher in Africa (85%) than in the Northern Europe (60% in Sweden and 75% in Estonia). Replacement rate is higher when consumers are wearing mostly

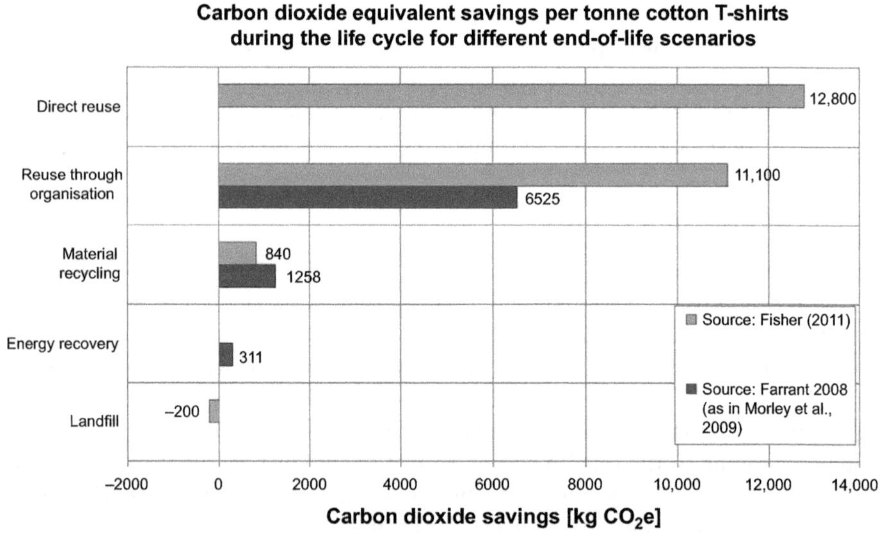

Figure 16.1 Comparison of different end-of-life scenarios of cotton T-shirts and the potential saving of kg CO_2 equivalent per tonne textiles during the life cycle (Fisher et al., 2011; Farrant, 2008; Morley et al., 2009).

second-hand clothing and it truly replaces a new clothing item, especially at lower income areas. Lowest replacement rates occur when the second-hand clothing is bought in addition to other items and does not replace other products. An example of this could be acquiring clothing for a fancy dress costume party, where the consumer would not buy anything new even if no suitable second-hand item was found.

Both studies that are referred to in Figure 16.1 use a material recycling example where T-shirts are cut and used as industrial wipers, replacing paper towels. This is a commonly used textile recycling method today in Europe (Morley et al., 2006). Figure 16.1 clearly demonstrates the benefits of reuse over recycling and that material recycling is better than energy recovery. However, methods where recycled textiles would replace new textile products instead of paper are likely to give even higher environmental gains. The environmental savings vary for different fibre types. It is estimated that replacing 1 kg of virgin cotton fibres through use of second-hand clothing saves 65 kWh; the figure is even higher for polyester clothing, 90 kWh per kilogram (Woolridge et al., 2006). In addition to energy consumption, production of several fibre types requires large amounts of chemicals, water and land area (Sandin et al., 2013; Shen et al., 2010), which can be at least partially avoided through recycling and reuse.

Chang et al. (1999) have evaluated the market applications for recycled post-consumer fibres and concluded that there are several textile sectors with potential increase in amounts of recycled fibres replacing virgin fibres, such as carpet cushions, home insulation, fiberfill stuffing, clean-up products, mattress pads/futons, geotextiles, landscaping and concrete reinforcements. These product areas were evaluated most suitable, as they generate articles with appropriate quality for end use with competitive prices as the only main reprocessing needed is fibre shredding. Other potential areas for the future are within disposable clothing, filtering materials, stationery, harvest mesh bags and shades for poultry houses and greenhouses. In addition to existing recycling solutions, a lot of research on new technologies is being carried out. For example, a project in Sweden is researching possibilities for making new regenerated cellulose fibres from used clothing materials such as cotton (Mistra Future Fashion, 2013) and the International Solid Waste Association has a project on designing jeans for recycling (Van Bommel, 2013). These types of recycling possibilities would benefit from having the environmental requirements integrated early in the design process, such as using homogenous materials that can be recycled in the same way (Bovea and Pérez-Belis, 2012).

16.1.2 What is disposal?

When preparing national statistics of textile waste, only textiles that have reached their end of life are included. This means that such statistics only include things that are thrown away as trash, and exclude items that are given to reuse or recycling either to family, friends or charity. This definition of disposal is based on the way textiles are disposed of. Another extreme would be to say that all the clothes a user stops using are disposed of regardless of what happens to them. A third alternative is that the disposal is linked to the owner of the garment and not to the user and that disposal happens when an owner disposes of the garment. Here is an example that clarifies this difference. We have a family with two children. The oldest has grown out of a sweater, and the younger sibling takes it over. When both kids have outgrown the sweater, the mother gives it to a charity. According to the first definition, the sweater has not been disposed of, as it has not become waste. According to the second option, where the user is in the centre, the garment has been disposed of twice, that is each time a child has outgrown it. The third way of thinking is that the sweater can be regarded as the family's (or the mother's) property and disposed of only when it changes the owner as it is given to charity and thus is no longer the family's property. This happens once in our example.

From here on, we will use an understanding of disposal where we emphasise owning and using, and not solely the method of disposal. This means that when we say that something is disposed, it still may get a new usage period with a new user. This is sometimes referred to as disposition (Jacoby et al., 1977), but we prefer to use the term disposal, as disposition might easily be misunderstood.[1] Our example with the children's sweater was not chosen randomly. The largest portion of reuse in Norway occurs through inheritance within families and between close relatives. This is more common and of a much larger scale than buying used clothing at a second-hand markets (Laitala and Klepp, 2014). These forms of disposal and reuse are not included in the official statistics. Even though interest in reuse from an environmental perspective is growing, the informal channels are often forgotten or left outside the scope of such work (Watson et al., 2014; Palm et al., 2014b).

16.1.3 Relationship between lifespan and disposal

One of the reasons that disposal has substantial environmental impacts is the fact that it affects the lifespan of a garment. When something is given to reuse, the disposition is not the end of the garment's lifespan but also the beginning of a new use period. It is the overall life of the garment that has the most to say environmentally. To put it in another way, if a garment technically tolerates about 300 times of use, it can be either done by one user that uses the garment 300 times or by three users that use it 100 times each. Therefore, reuse is not a goal in itself but only one way to solve the problem that loads of clothes are disposed of before they are worn out.

In Norway, as in many Western countries, the declining prices and consumers' increased purchasing power contributed to a growing amount of textiles (European Environment Agency, 2014). With more clothes, every single garment can easily have a longer lifespan in years because it is used less often. This longer ownership period also increases the possibility that the owner stops using the garment before it is worn out. There can be several reasons for this, such as the owner gains weight, their clothing needs change perhaps because they start a different type of job or they change their style preferences. A smaller portion of the clothes' potential use period is thus utilised by the first owner, and the environmental implications of the way the article of clothing is disposed of increase correspondingly.

[1] Disposition means also a habit, a preparation, a state of readiness or a tendency to act in a specified way.

16.1.4 Norwegian textile waste and reuse statistics

Similar to many other Western European countries, only small amounts of textiles and clothing are produced in Norway, and therefore the majority of these products are imported (European Environment Agency, 2014). Total imports in 2011 were equivalent to 43 kg textiles per capita annually, out of which 14 kg were clothing (Statistics Norway, 2012). Exportation statistics for textiles show that the largest category of textiles that are exported from Norway are used clothing and other worn textile articles, 20,745 tons in 2011, which is almost half of the total textile exportation. This is equivalent to about 4.2 kg per capita annually (Laitala et al., 2012). This trade in used clothing is a large global industry, where the largest clothing streams often originate from the 'Global North' and end up in the 'Global South' (Norris, 2012).

About 113,000 tons of textile waste was generated in Norway in 2011, which equals 23 kg per capita (Statistics Norway, 2013). This is close to the average amount of textile waste generated in the EU-27, which is 26 kg per capita (Laitala et al., 2012). The largest portion (42%) of the Norwegian textile waste originated from households, 9.7 kg/capita, followed by service industries (31%). Clothing and footwear constitutes principal parts of this, about 57%, followed by upholstery textiles and household products (30%). The amount of textile waste from private households has increased from 29,000 tons in 1995 to 52,000 tons in 2010, which is almost an 80% increase in 15 years.

Based on composition analyses of household waste, it has been estimated that about 28% of binned textiles could have been reused (Laitala et al., 2012). The majority of the textile waste (66%) is incinerated with energy recovery together with the rest of the household refuge (Statistics Norway, 2013). The proportion of incinerated materials has increased greatly during the last few years since landfilling of biodegradable waste was banned in 2009 (SFT, 2008). The difference between textile import and waste statistics indicates that potentially large amounts of clothing are kept in storage or disposed of through different methods that are not captured by the statistics.

In Norway, municipalities are responsible for household waste collection and management. The collection systems vary greatly between the different areas, but often products such as glass, metal and paper are collected separately and delivered to recycling. Textile collection is mostly organised by voluntary organisations such as the Salvation Army and Humana People to People. These organisations have to consider profitability when choosing the areas where to collect clothing. Therefore, the possibility of delivering textiles to recycling is often worse for consumers living in sparsely populated areas.

The Norwegian system could be improved so that more of the textiles' potential life is used. Today, all the largest clothes' collectors inform that they only wish to get clothing that is reusable as it is, most likely because sales from clothing for reuse are more profitable than sales from clothing that is going to recycling (Laitala et al., 2012). However, from an environmental point of view, this strategy is not the best possible for two reasons. Firstly, it decreases the collection rates of textiles that could be recycled. Secondly, it also decreases the collection rates of textiles that could be reused, as consumers are very uncertain of where to draw a limit on what is reusable (Laitala and Klepp, unpublished). Some consumers assume that only clothing that they would be prepared to use themselves is donated, thus excluding many items that may have been useable for others.

16.1.5 Clothing disposal practices

Most research on sustainable clothing consumption is related to the acquisition phase, while less attention has been given to the use and disposal phases (Hetherington, 2004; de Coverly et al., 2008). Even though clothing disposal is a less researched area than acquisition, some studies on the topic are currently being conducted. A recent comprehensive review on clothing disposal literature summarises the main findings of 36 research projects that were published during the past 30 years (Laitala, 2014). Most of these studies concentrate on disposal methods, behavioural motivations, disposal reasons and demographics of consumers with specific behaviours. Despite the fact that some of the results present inconsistencies, several main tendencies can be observed. Many of these studies point out that most consumers prefer to deliver clothing for reuse rather than to dispose of them, but convenience is paramount for this behaviour (Domina and Koch, 2002; Joung and Park-Poaps, 2013; Ha-Brookshire and Hodges, 2009; Bianchi and Birtwistle, 2010).

Common disposal reasons for apparel were wear and tear, poor fit and fashion or boredom, in addition to lack of storage space, but the order of importance between these reasons varied between studies (Chun, 1987; Klepp, 2001; Laitala and Boks, 2012; Ungerth and Carlsson, 2011; Fisher et al., 2008). Studied samples consisted mainly of young students with an overrepresentation of females (Koch and Domina, 1997; Morgan and Birtwistle, 2009; Shim, 1995; Lee et al., 2013).

This indicates the need for studies on more heterogeneous samples and comparisons of disposal habits of different demographic groups. The review also points out that survey methods were most common, which indicates the need for other research designs that preferably also include the clothing items (Laitala, 2014). Our current research aimed to cover some of these less-studied areas and employ mixed methods in doing so. This way, clothing disposal parameters were studied, including disposal frequencies, reasons for selecting the specific channels and variations in behaviour between different consumer groups.

16.2 Methods

The aim of the study has been to gain more information on consumer habits during the clothing disposal phase. The analysis is based on mixed methods with both qualitative and quantitative parts. The qualitative method was based on a wardrobe study of clothing to be disposed of, and the quantitative method was based on a consumer survey.

16.2.1 Interviews and a wardrobe study

A qualitative method was chosen for gaining empirical data on clothing disposal practices that is based on concrete garments instead of general opinions of disposal.

The qualitative part of the project had disposed clothing as the starting point. For receiving this detailed information, a strategic sample of 16 households was selected. These households stored all clothes that were to be taken out of use during a period of six months – a total of 620 garments. During the in-depth interview, the households were asked questions related to the disposal of each clothing item:

- Type of garment and which household member used it
- Reasons for why it is going out of use
- How old it is and when it was used the last time
- When and how much it has been used
- Where would it have been disposed to (if it had not been collected to the project)
- Other aspects that may be relevant, such as the acquisition method, especially if the garment was inherited.

This method is called wardrobe study, which consists of an inventory of all or some selection of clothes in a wardrobe. This inventory involves cataloguing the garments combined with a tape recording of what the owner, user or caretaker has to say about the individual garment and its history of use. The material frames of practices, such as the clothes, the wardrobes and laundry baskets are studied. The goal is to look at the relationship between the individual item of clothing and the larger material totalities. This method has been described and discussed in more detail by Klepp and Bjerck (2012). The benefit of using the wardrobe study method is that the material objects (clothes) are used in the interviews to get more specific information of behaviour related to the specific garments.

The 16 households that participated in the study included 35 persons (8 children, 2 teenagers, 16 adult women and 9 adult men). They disposed of 620 garments that were registered during the six-month study period. Participants stopped using on average 18 garments per person, but the figures varied from 0 to 71 items per person. Each household had one adult as main informant. Demographic information of the main informants is given in Table 16.1. Their ages varied between 24 and 67 years, with an average age of 38 years. All participants were given pseudonyms in a way that family members have their pseudonyms starting with same letter.

The intention was to interview individuals with different life situations, ages, genders, civil status, family size and so on. This way of sampling is often referred to as *strategic sampling*, but also *judgement* or *purposeful* sampling. The distribution of informants is not representative of the population, but the informants were actively hand-picked in such a way that it maximises the chance that many different and even conflicting sides of the phenomenon in question are encountered (Eneroth, 1984; Marshall, 1996). It gives examples of different consumers combined with large number of clothing items that are suitable to discuss the article's research questions.

The interviews were recorded and transcribed. Qualitative data from the interviews were analysed with ATLAS.ti software that allows coding and structuring interview material for easier analysis. Quantitative information related to the 620 disposed clothing items was coded in Excel and finally analysed with SPSS software.

Table 16.1 Background variables of the 16 main informants

Property	Variables	Main informants
Sex	Women	13
	Men	3
Age	20–34	8
	35–49	6
	50+	2
Family	No children	7
	Parents with small children	7
	Parents (adult children)	2
Relationship status	Single/living alone	6
	Living with partner	10
Area of living	West Oslo	5
	East Oslo	3
	Neighbour counties to Oslo	5
	Trondheim area	3
Nationality	All Norwegian	12
	Foreign household member(s)	4
Education	Vocational	1
	Bachelor level	6
	Graduate level	9
Employment situation	Working[a]	12
	Student[b]	3
	Retired	1

[a]Three of them only work part time.
[b]All three students had part-time jobs.

16.2.2 *Quantitative survey*

As the representativeness of the qualitative study is limited, additional data were collected through quantitative surveys that are more suitable to answering questions where magnitude of the studied phenomenon is of interest. Here, they included consumers' experiences, practices and opinions concerning clothing use and disposal

Table 16.2 **Respondents divided by background variables and compared to Norwegian population in 2010 (15 years and older) (Agerskov, 2010)**

Background variables		Sample	Norwegian population
Number of respondents (N)		546	-
Gender	Male	23%	50%
	Female	77%	50%
Average age		39.2	45.9
Age groups	15–24 years	9%	16%
	25–39 years	48%	25%
	40–59 years	33%	33%
	60 + years	9%	26%

in Norway. These include consumers' motivations and behaviour related to clothing disposal. Respondents were recruited through different channels. The majority of respondents were found through questionnaires sent by mail to 1200 randomly selected households from the telephone catalogue. The different counties were presented in proportion to the number of inhabitants. Sixty per cent of the questionnaires were sent to men, as earlier experience has shown that men have a lower response rate. Due to low response rate (10.3%), additional respondents were recruited with a second round of contacting 1000 households that were selected randomly from the phone catalogue, in addition to recruitment through personal and work-related networks, and publicity in media. A total of 546 answers were received during the year the survey was available online (Table 16.2).[2]

The received data are not fully representative of the Norwegian population, which has been taken into account when drawing conclusions. Even though the sample is quite randomly selected, the size is too small and it may be biased because the respondents who chose to respond may be different from those who chose not to respond. In addition, the constitution with overrepresentation of women and non-representative age distribution does not correspond to the division in the Norwegian adult population, and therefore is not sufficient to generalise research findings. However, the preponderance of women in the sample also has some benefits because women often take responsibility for a larger portion of tasks related to the family's clothing consumption such as acquisition, laundering and repair (Laitala et al., 2009). It is also argued that young women purchase more clothing than other consumer

[2] The total number of respondents is lower (N = 480) for answer distribution presented in Figure 16.4, because this question was replaced in a later version of the survey questionnaire online.

groups, and therefore focusing on them may generate information on certain volumes of clothing (Birtwistle and Moore, 2007; McGrath, 2012; Joung and Park-Poaps, 2013). These cases are not weighted. The received data were analysed with the help of SPSS software. As this is a convenience sample where the assumptions about probability sampling are violated, it is not appropriate to use statistical significance tests. However, the sample contains a large number of different respondents that can be compared with one another, and can be used as an example of consumers in Norway. They also give a wider context for the results from qualitative study of the 16 households, and therefore these two methods complement each other.

16.3 Results

In our discussion, the results related to general clothing disposal methods based on survey results are presented first. This is followed by disposal methods of specific clothing items based on wardrobe study, and finally the disposal frequency based on survey data. These themes include a discussion of demographic characteristics of consumers that select specific disposal methods. Interviews are used throughout the results section in order to interpret and explain the quantitative data. Participants of the qualitative interviews are referred to as informants, while the participants to the quantitative survey are referred to as respondents.

16.3.1 Disposal methods

There are several ways clothing can be disposed of and end up either for reuse, recycling, storage or as waste. This was researched in the survey through asking the use frequency of the most common methods of clothing disposal.

One of the worst options consumers can choose for clothing disposal is to burn it in private settings. It is estimated that about 45,000 tonnes of waste are burned yearly in small-scale furnaces or open bonfires in Norway (SFT, 2001). Even though the amount of waste burned illegally in this way only represents a small part of what is burned in large incineration plants, the total emissions from small stoves for some pollutants is significantly higher than the total emissions from all large combustion plants combined because the large plants burn the waste more effectively and have better cleaning systems for the exhaust fumes (SFT, 2001). It is difficult to study this phenomenon because respondents may be reluctant to talk about their habits that are illegal.

In the survey the respondents were asked, ‘What do you usually do with clothes that are not worn out, but that you don’t use anymore?’ and ‘What do you usually do with clothes that are worn out or damaged?’ The results show that consumers selected different disposal methods for clothing that was worn out or damaged and for clothing that was intact and usable (Figure 16.2). A majority (59%) said they often binned damaged or worn-out items, while intact clothing was most often donated to charity (69%). Selling was the least commonly used of the given disposal channels. When it comes to undamaged clothing disposal behaviour, these results are similar to

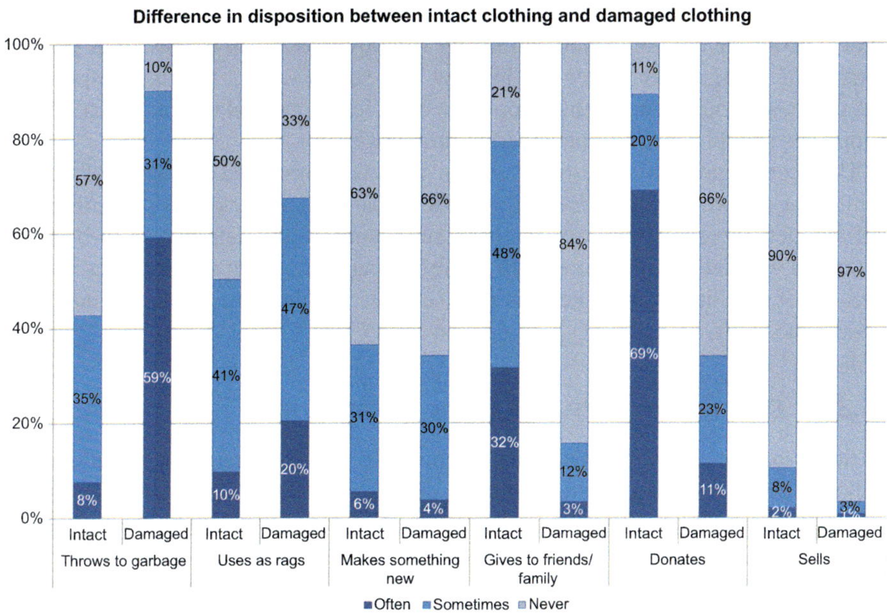

Figure 16.2 Disposal methods of intact and damaged clothing (N = 546).

reported behaviour in the neighbouring country of Sweden, but 53% of the Swedish respondents reported that they donated some of their damaged clothing as well (Ekström et al., 2012).

Men described using clothing more often until it is worn out, but they also were more likely to throw clothing into the garbage. Women were more likely to give intact clothing to charity and to friends and family. They were also more likely to repair and make something new of clothing than men were.

Some disposal behaviours varied between respondents of different ages. Younger respondents were more likely to give clothing to friends and family, while older respondents more often said that they used clothing until it was worn out and used it as rags. No substantial differences between age groups related to the other disposal channels were observed.

Respondents that said they were interested in environmental questions were more likely to give intact clothing to charity and make something new of it, but there were no significant differences in their other behaviours related to selection of clothing disposal channels.

In addition to the disposal methods listed in the survey, interviews revealed at least one alternative method. Here, 59-year-old Pia explains how they dispose of waste, including some textiles, about twice a year: *We've for example cleaned up some of the old stuff from the farm, and that has been burned in the evening (laughs a little). It is not really allowed to burn like that…* This behaviour has not been researched in detail, but it is an old practice and it is still reported to occur in the countryside, even though it is against the regulations (SFT, 2001). This might be one of the reasons that explains the difference between clothing import and waste statistics. Informing consumers about the environmental and health consequences of emissions from private burning of waste could reduce the burning that seems to be occurring despite the fact that it is not allowed and that people are fined for doing it (Hanssen, 2012). These campaigns should especially be aimed at rural areas.

This section gave survey results based on general self-reported behaviour of a large number of Norwegian consumers. Some factors are not that easy to study through self-reported surveys. Therefore, we have supplemented the data related to burning of textiles with qualitative material. In the following section, the selection of disposal methods of specified clothing items is analysed based on the same qualitative material, a wardrobe study.

16.3.2 Differences between types of clothing

In the wardrobe study, the participating households told where they would have disposed the collected clothing items if they had not been given to the study. Figure 16.3 gives the distribution of planned destinations of the clothing. In this figure, 'consumable' clothing items such as socks, stockings and underwear are separated from other types of clothing items, as the material showed that these types of garments were treated differently when they were to be disposed of. Consumables were in most cases to be thrown away with waste, with a few expectations of giving some to charity, family or friends. Other types of clothing items were mostly planned to be donated to charity, followed by throwing away with waste and giving to family or friends.

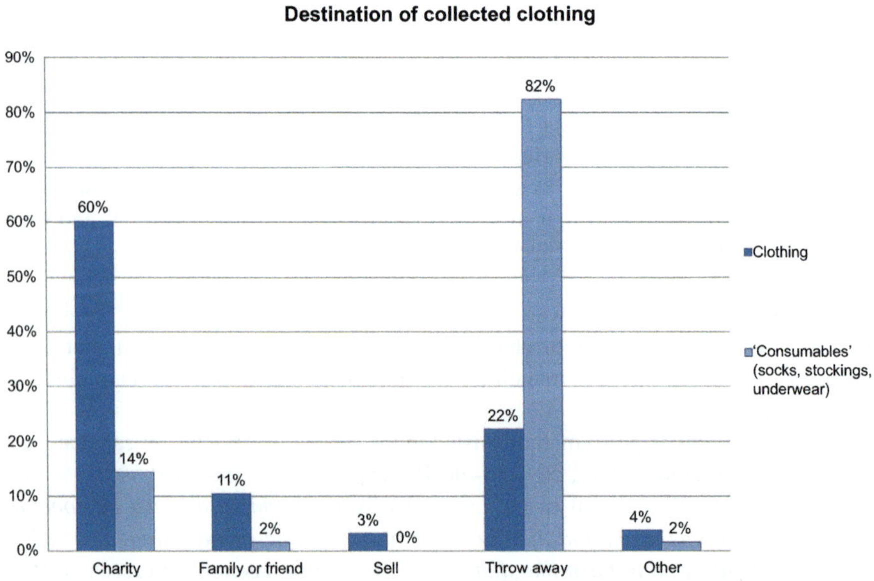

Figure 16.3 Distribution of destinations of clothing. Separately for 'consumables' such as socks, stockings, and underwear, and for other (larger) garments (N = 620).

Very few items were planned to be sold or used for purposes such as storing the item as a memento or intended heirloom, or planning to repurpose it either as clothing or something else. The results indicate that three-quarters of garments were intended for reuse (excluding consumables). This was most common for all clothing, except for the clothing that had changed during use (e.g. got holes, pilling, colour changes or other changes in appearance) and was then more likely to be thrown away as waste. There was also a big difference in this matter between children's and adults' clothing, as 30% of adults' clothing were to be disposed compared to only 7% of children's and teenagers' clothing (excluding consumables). This may be explained by the shorter use periods of children's and teens' clothing, as they outgrow clothing usually within a season.

Consumables and other types of garments were treated differently when disposed of. This difference in behaviour might also occur during the other consumption stages, such as acquisition. In general, research on clothing and fashion consumption is focused on these larger garments, and even some of the informants in the wardrobe study had not collected the socks and underwear they were going to dispose of because they did not consider it to be clothing.

Both of the two largest used clothing collectors in Norway, Fretex and UFF, say that they only want undamaged clothing that can be reused as it is (Fretex, 2013; UFF Norge, 2013). Therefore, many of the informants were uncertain about where to draw the line on whether clothing could be delivered to reuse or not. Some informants chose to give almost anything to charity. Erik (30 years old) describes his jeans with a 15-cm wide tear in the crotch: 'I had given to the charity, because some people like pants with holes in, that are worn. Fretex can choose whether they want it or not'. Some informants also often assumed that if there were only small errors in garments, they would be repaired before the clothes would be given or resold, such as small holes or missing buttons. Camilla (29 years old) felt that the requirements for used clothing collectors were high: *It might be a bit like, at least in Norway, it should be so high standard to be sold as used as well… If I first do not want it for some reason, then maybe often others do not want it either. That it may have become outdated or that it is a bit washed out and not as nice […] I feel that maybe some charity shops only want the very nice clothes, or the very fun clothes for dressing up*. She did not know that a large portion of the clothing is reused or recycled in other countries. Most of the informants did not like to throw clothing into the trash but did not know of other options if they felt it was not directly suited for reuse.

As previous studies also indicated, convenience with easy access and savings in time consumption and effort were important in disposal decisions. These decisions were seldom made due to environmental considerations, here exemplified again by Camilla: *I guess I always have ambitions to give away clothes that are not damaged and like that. But it is more because I think that it is stupid to throw away clothes that someone can use, not necessarily because of the environment […] But often it's a bit cumbersome to deliver clothes to a specific place. Then I don't get it done anyway, and then it just lies there, until I eventually throw it away*. This shows one of the problems related to the disposal process. The consumers may have good intentions, but the possibilities and practical hindrances may stop the intended behaviour.

16.3.3 Disposal frequency

Clothing disposal frequency was studied with both of the applied methods. In the survey, two questions concerning the most common disposal frequency and disposal occasions were stated, the first being 'How often do you give or throw away your own clothing?' and the second 'When or at which kind of occasions do you usually give or throw away your own clothing?' The results show that it was most common to dispose of clothing once or twice a year (48%). Every third respondent (32%) disposed of clothing more often than this, and every fifth (20%) more seldom. Women were more likely to dispose of clothing more often than male respondents (Figure 16.4). It was most common to go through the wardrobes during spring and autumn, when some of the winter and summer clothing are circulated between different storage areas (30%). At this stage, clothing that was not too damaged or worn out to be used could be taken out of use. Damaged clothing was more often disposed of directly. The largest portion of respondents that disposed of clothing once or twice a year were in the group over age 60. This practice is explained here by 67-year-old Olivia: *I put summer clothes away in another closet or up in the attic, in plastic boxes or such, and take down the woollen clothes that I have not used during the summer. So I have this autumn and spring cleaning up….* This is also common advice given for taking care of clothing, where the focus lies on cleaning and airing clothing before storing it.

Every fourth respondent (27%) also said that they disposed of clothing when the wardrobe got too full. This was normal practice for 31-year-old Fiona: *I collect and collect and collect, and then finally there is no more room, and then you just have*

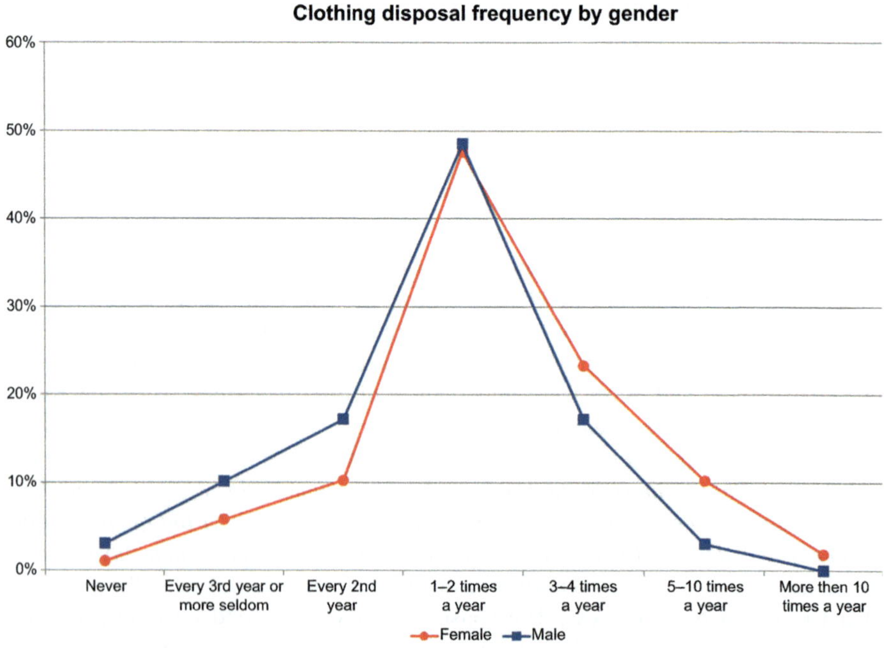

Figure 16.4 Clothing disposal frequency (N = 480).

to take action. It happens very rarely, maybe every third year. The disposal frequency based on lack of space could vary, however, as informants that shopped more frequently would need to dispose of clothing more frequently as well. These results are important when collection systems for used clothing are planned.

Thirteen per cent said that they disposed of clothing right away after they stopped wearing them. A few informants also reported to give away clothing when somebody asked for it, for example when there was a charitable collection or friends who have younger children came to visit. In addition, moving was one of the occasions when it was common to dispose of clothing. Barbara, 25 years old, describes that even though she thinks it is difficult to dispose of clothing, she still does it sometimes, for example when moving: *Then I give maybe three or four black plastic bags away, either to charity or my little sister. And then I throw away maybe a couple of boxes or garbage bags, because it is so worn that I do not think it's nice to give it away.* This implies that she gives almost twice as much clothing to reuse than to waste disposal.

Even though many garments were collected during the study, informants still described that they had kept a lot of clothing that was not in active use. Thirty-five-year-old Anita explains, *I have lots of clothes in the wardrobe that I never use, but I'm really bad at clearing them out. When they are in good condition and could be used, I can't get myself to get rid of them.* This shows that for her and many other informants, it was easier to dispose of clothing that was worn out or damaged than clothing that was in good condition but was not used for other reasons.

16.4 Discussion

The ways consumers dispose of clothing have significant consequences for the environment as well as LCA calculations. The parameters that need to be considered include the disposal methods, type and condition of the clothing and the disposal frequency. The selection of disposal method is affected by several factors such as available alternatives in the country where the garments are used and disposed of, type of clothing, condition of the clothing as well as consumers' values and personal interests. This can be included as a distribution of how common the various ways of disposing are in the specific area. When giving clothing to reuse, the environmental benefits of using formal channels are likely to be smaller due to the higher need of energy consumption in transportation, sorting and retail operations.

Results indicate that types of clothing can be divided roughly in two, related to consumers disposal behaviour. The majority of consumable items such as underwear, socks and stockings were binned, while larger garments were mainly intended for reuse and given to charity, friends or family. Another relevant factor for some consumers was the garment's closeness to body, as outerwear was more likely to be delivered for reuse than underwear or other next-to-skin products. However, further division can be made between children's and adults' clothing, as it was a lot more common to give children's clothing for reuse within informal channels such as family, friends, neighbours and so on.

The condition of clothing also played significant role, as damaged or worn-out clothing is more likely to be binned and intact clothing to be delivered to reuse. This may however be related to the total length of use period of garments, as worn-out items are likely to have been used more than the intact items, which is also essential in LCA calculations that use functional units such as 'day of wearing the garment' instead of just number or weight of produced garments as a measuring unit.

The results on disposal frequency showed that some factors contribute to more frequent disposal, such as age and gender. Therefore, the intended user group can be taken into account when estimating the length of use period before disposal.

There are clear benefits of following the waste hierarchy also within textiles, as prolonged textile lifespans give environmental savings compared to reuse, which is again usually more beneficial than recycling. Material recycling where the materials are transferred to new products is usually more beneficial than energy recovery, where the materials are incinerated. However, all these solutions are better than landfilling or other ways of disposal that have negative environment impacts. When making LCA calculations on reuse, it is important to take into account the replacement rate, that is, how many new garments the second-hand garment substitutes for, as that is a crucial point for the outcome of the study (Fisher et al., 2011).

Clothing for reuse is not defined as waste because it is not intended to be disposed but reused. Therefore, the Norwegian municipalities are not obligated to collect this fraction. In order to guarantee recycling access for all consumers, municipalities could also take responsibility for collecting clothing for reuse. In many cases, it would still be most feasible to delegate this task to other organisations, but the collections should be agreed on even in sparsely populated areas and to include all clean and dry textiles independent of the usability. Even though there are many clothes collection points in Norway, consumers living in the less-populated areas may have to drive long distances to deliver clothing for reuse and recycling. Better availability of collection systems, including take-back schemes in retail stores, could increase clothes collection rates in these areas. As consumers disposed of clothing usually once or twice a year, separate biannual collections during spring and autumn could be an economically viable solution. Textiles that get wet or soiled by other waste loose value and are less, if at all, suitable for reuse or recycling, and therefore the collection systems should be separated from the wet waste fractions (Morley et al., 2009).

The weakness of many existing LCA studies is that they only assume the lifespans and disposal methods for clothing. Methodologically, it would be beneficial to study actual consumer behaviour instead of only using surveys based on questionnaires. There are several reasons for this. First, disposal is a morally loaded issue, and consumers may be reluctant to report behaviour that is considered to be 'wrong', such as wastefulness, or even illegal such as private burning of waste. Second, many consumers are not aware of all the details of their daily practices such as amounts of clothing that are disposed of in various ways. Third, one does not get very objective information about the condition of clothing through surveys, as consumers have varying opinions on what clothing is suitable for reuse and what is not, for example, based on whether the person himself could use clothing with holes, stains or pills. In addition, details such as the clothing brand, original price, personal use experiences and how close to the body the garment is used play a role in these decisions. The wardrobe studies showed to be an interesting and suitable method for studying clothing consumption, as detailed empirical data that combined the material and behavioural aspects were received.

The limitation of this and many other studies on clothing disposal is that the material on some consumer groups' clothing disposal practices is less strong, including men and the elderly. Therefore, these consumer groups should be given special focus in future studies.

Another way of getting data for LCA calculations is an opposite method based on top-down approach and use the national statistics on import, export and waste, as presented in the introduction. This method can be more useful when LCA is calculated for large product groups rather than single, specific clothing items.

16.5 Conclusions

We need to incorporate knowledge about clothing use period, length of lifespans as well as the disposal methods for the LCA results to become more accurate. The disposal is a complicated interaction between the clothing technical and aesthetic durability and adaptability, and the consumers' knowledge and access to simple and reliable solutions for further use. In the current situation, not only the knowledge of these different factors is lacking but also a good organisation that could safeguard these resources. This applies not only in Norway but also in many Western countries.

This study has implications both for the academics and practitioners. Even though not statistically significant, it contributed in understanding the underlying dynamics of the Norwegian consumers' clothing disposal practices. The practical implications of this study are relevant for several actors within the clothing chain, such as producers, designers, retailers, consumers, collectors and recyclers, as well as educators and public policymakers, who wish to diminish the environmental impacts of clothing consumption through longer lifespans and increased reuse and recycling rates.

Instead of assuming that all clothes are disposed of equally regardless of type of garment, person and place, the LCA analyses should be nuanced in relation to knowledge of disposal practices. These vary based on the parameters given in this chapter, but many LCA studies on clothing use assumed end-of-life scenarios based on general distribution of used textiles (Collins and Aumônier, 2002; Steinberger et al., 2009). As Muthu et al. (2011) point out, 'as far as the end-of-life phase is concerned, it lies mostly in the hands of the public. Also it is vitally important to use real values rather than assumptions, which can be obtained only from the actual users'. An easy starting point is to distinguish between consumable products and garments. The LCA methods should also be updated when the systems are improved.

An important goal for the environmental work within textiles is to exploit their potential use life. To reach this goal, both the clothes and the practices associated with them should be improved. Knowledge about these practices, such as disposal practices that we have discussed here, are important to facilitate change.

Acknowledgments

We would like to thank Tone Skårdal Tobiasson and Casper Boks for valuable comments, Madeline Buck for clothing registrations, as well as Orkla ASA and the Norwegian Research Council for financial support on the project entitled 'From textile waste to material resources in a grave to cradle perspective', which this article is part of. The funding bodies have not been involved in any parts of the research work.

References

Agerskov, U., 2010. Nordic Statistical Yearbook 2010, vol. 48. Nordic Council of Ministers, Copenhagen.

Bianchi, C., Birtwistle, G., 2010. Sell, give away, or donate: an exploratory study of fashion clothing disposal behaviour in two countries. The International Review of Retail, Distribution and Consumer Research 20, 353–368.

Birtwistle, G., Moore, C.M., 2007. Fashion clothing – where does it all end up? International Journal of Retail & Distribution Management 35, 210–216.

Bovea, M.D., Pérez-Belis, V., 2012. A taxonomy of ecodesign tools for integrating environmental requirements into the product design process. Journal of Cleaner Production 20, 61–71.

Chang, Y., Chen, H.-L., Francis, S., 1999. Market applications for recycled postconsumer fibers. Family and Consumer Sciences Research Journal 27, 320–340.

Chun, H.-K., 1987. Differences between Fashion Innovators and Non-fashion Innovators in Their Clothing Disposal Practices. Apparel, Interiors and Merchandising. Oregon State University, Corvallis.

Collins, M., Aumônier, S., 2002. Streamlined Life Cycle Assessment of Two Marks & Spencers plc Apparel Products. Environmental Resources Management, Oxford.

de Coverly, E., Mcdonagh, P., O'malley, L., Patterson, M., 2008. Hidden mountain: the social avoidance of waste. Journal of Macromarketing 28, 289–303.

Domina, T., Koch, K., 2002. Convenience and frequency of recycling: implications for including textiles in curbside recycling programs. Environment and Behavior 34, 216–238.

Ekström, K.M., Gustafsson, E., Hjelmgren, D., Salomonson, N., 2012. Mot en mer hållbar konsumtion: En studie om konsumenters anskaffing och avyttring av kläder (Towards a More Sustainable Consumption: A Study of Consumers' Acquisition and Disposal of Clothing). Högskolan i Borås, Borås. Vetenskap för profession 20:2012.

Eneroth, B., 1984. Hur mäter man 'vackert'? Grundbok i kvalitativ metod (How to Measure 'Beautiful'? Basics of Qualitative Method). Stockholm Akademilitteratur, Stockholm.

European Environment Agency, 2014. Environmental Indicator Report 2014. Environmental Impacts of Production-consumption Systems in Europe. Luxembourg.

European Parliament and the Council, 2008. Directive 2008/98/EC on Waste and Repealing Certain Directives. Official Journal of the European Union, Brussels.

Farrant, L., 2008. Environmental Benefits from Reusing Clothes. Department of Management Engineering. Kgs. Lyngby Technical University of Denmark.

Farrant, L., Olsen, S., Wangel, A., 2010. Environmental benefits from reusing clothes. The International Journal of Life Cycle Assessment 15, 726–736.

Finnveden, G., Johansson, J., Lind, P., Moberg, Å., 2005. Life cycle assessment of energy from solid waste - part 1: general methodology and results. Journal of Cleaner Production 13, 213–229.

Fisher, K., James, K., Maddox, P., 2011. Benefits of Reuse Case Study: Clothing. WRAP, Banbury.

Fisher, T., Cooper, T., Woodward, S., Hiller, A., Goworek, H., 2008. Public Understanding of Sustainable Clothing: A Report to the Department for Environment Food and Rural Affairs. London, Defra.

Fretex, 2013. Hva kan jeg gi (What Can I Donate?).

Glew, D., Stringer, L.C., Acquaye, A.A., Mcqueen-Mason, S., 2012. How do end of life scenarios influence the environmental impact of product supply chains? Comparing biomaterial and petrochemical products. Journal of Cleaner Production 29-30, 122–131.

Ha-Brookshire, J.E., Hodges, N.N., 2009. Socially responsible consumer behavior? Exploring used clothing donation behavior. Clothing and Textiles Research Journal 27, 179–196.

Hanssen, A.-K., 2012. Gigant-bot etter søppelbrenning (Gigantic Fine for Burning Garbage). Fremover, Narvik.

Hetherington, K., 2004. Secondhandedness: consumption, disposal, and absent presence. Environment and Planning D: Society and Space 22, 157–173.

Jacoby, J., Berning, C.K., Dietvorst, T.F., 1977. What about disposition? Journal of Marketing 41, 22–28.

Joung, H.-M., Park-Poaps, H., 2013. Factors motivating and influencing clothing disposal behaviours. International Journal of Consumer Studies 37, 105–111.

Klepp, I.G., 2001. Hvorfor går klær ut av bruk? Avhending sett i forhold til kvinners klesvaner (Why Are Clothes No Longer Used? Clothes Disposal in Relationship to Women's Clothing Habits). SIFO, Oslo. Report No. 3-2001.

Klepp, I.G., Bjerck, M., 2012. A methodological approach to the materiality of clothing: wardrobe studies. International Journal of Social Research Methodology 1–14.

Klepp, I.G., Laitala, K., Schragger, M., Follér, A., Paulander, E., Tobiasson, T.S., Eder-Hansen, J., Palm, D., Elander, M., Rydberg, T., Watson, D., Kiørboe, N., 2014. Sustainable Textile Initiatives and Suggestions for a Nordic Roadmap. Nordiske Arbejdspapirer. Nordisk Ministerråd, Copenhagen.

Koch, K., Domina, T., 1997. The effects of environmental attitude and fashion opinion leadership on textile recycling in the US. Journal of Consumer Studies & Home Economics 21, 1–17.

Laitala, K., 2014. Consumers' clothing disposal behaviour – a synthesis of research results. International Journal of Consumer Studies 38, 444–457.

Laitala, K., Boks, C., 2012. Sustainable clothing design: use matters. Journal of Design Research 10, 121–139.

Laitala, K., Hauge, B., Klepp, I.G., 2009. Large? Clothing Size and Size Labeling. TemaNord 2009:503. Nordic Council of Ministers, Copenhagen.

Laitala, K., Klepp, I.G., 2014. Arv og deling av klær, sko og sportsutstyr. In: Lavik, R., Borgeraas, E. (Eds.), Forbrukstrender 2014: SIFO Survey. SIFO, Oslo.

Laitala, K., Klepp, I.G. (Unpublished). Clothing Reuse: Used clothes as gifts? *Manuscript submitted for publication.*

Laitala, K., Klepp, I.G., Morley, N., Meistad, T., Chapman, A., Chen, W., Hebrok, M., Daae, J., Austgulen, M.H., 2012. Potensiale for økt materialgjenvinning av tekstilavfall og andre avfallstyper (papir/papp, metall og glass) (Potential for Increased Material Recycling of Textile Waste and Other Waste Types). SIFO for Klima- og forurensningsdirektoratet, Oslo.

Lanoë, T., Simões, C.L., Simoes, R., 2013. Improving the environmental performance of bedding products by using life cycle assessment at the design stage. Journal of Cleaner Production 52, 155–164.

Lee, J.Y., Halter, H., Johnson, K.K.P., Ju, H., 2013. Investigating fashion disposition with young consumers. Young Consumers: Insight and Ideas for Responsible Marketers 14, 67–78.

Madsen, J., Hartlin, B., Perumalpillai, S., Selby, S., Aumônier, S., 2007. Mapping of Evidence on Sustainable Development Impacts that Occur in Life Cycles of Clothing: A Report to Defra. Environmental Resources Management Ltd, London.

Marshall, M.N., 1996. Sampling for qualitative research. Family Practice 13, 522–525.

McGrath, A.S., 2012. Fashioning Sustainability: How the Clothes We Wear Can Support Environmental and Human Well-being. Environmental Sciences Senior Thesis Symposium. University of California, Berkeley.

Mistra Future Fashion, 2013. Reuse, Recycling and End of Life Issues. Borås.

Morgan, L.R., Birtwistle, G., 2009. An investigation of young fashion consumers' disposal habits. International Journal of Consumer Studies 33, 190–198.

Morley, N., Bartlett, C., Mcgill, I., 2009. Maximising Reuse and Recycling of UK Clothing and Textiles. Incl. Appendix 1-Technical Report. A Research Report Completed for Defra by Oakdene Hollins Ltd, London.

Morley, N., Slater, S., Russell, S., Tipper, M., Ward, G.D., 2006. Recycling of Low Grade Clothing Waste. Prepared for Defra Aylesbury, Uk, Oakdene Hollins Ltd Salvation Army Trading Company Ltd and Nonwovens Innovation & Research Institute Ltd.

Muthu, S.S., Li, Y., Hu, J.Y., Mok, P.Y., 2011. Carbon footprint of shopping (grocery) bags in China, Hong Kong and India. Atmospheric Environment 45, 469–475.

Norris, L., 2012. Trade and transformations of secondhand clothing: introduction. Textile: The Journal of Cloth and Culture 10, 128–143.

Palm, D., Elander, M., Watson, D., Kiørboe, N., Lyng, K.-A., Gíslason, S., 2014a. Towards a New Nordic Textile Commitment: Collection, Sorting, Reuse and Recycling. TemaNord 2014:540.

Palm, D., Elander, M., Watson, D., Kiørboe, N., Salmenperä, H., Dahlbo, H., Moliis, K., Lyng, K.-A., Valente, C., Gíslason, S., Tekie, H., Rydberg, T., 2014b. Towards a Nordic Textile Strategy – Collection, Sorting, Reuse and Recycling of Textiles. TemaNord 2014:538.

Sandin, G., Peters, G.M., Svanström, M., 2013. Moving down the cause-effect chain of water and land use impacts: an LCA case study of textile fibres. Resources, Conservation and Recycling 73, 104–113.

SFT, 2001. Åpen brenning og brenning av avfall i småovner: Veiledning for kommuner (Open Burning and Waste Incineration in Small Incinerators). Climate and Pollution Agency, Oslo.

SFT, 2008. Mer avfall skal gjenvinnes (More Waste Is Going to Be Recycled). Climate and Pollution Agency, Oslo.

Shen, L., Worrell, E., Patel, M.K., 2010. Environmental impact assessment of man-made cellulose fibres. Resources, Conservation and Recycling 55, 260–274.

Shim, S., 1995. Environmentalism and consumers' clothing disposal patterns: an exploratory study. Clothing and Textiles Research Journal 13, 38–48.

Statistics Norway, 2012. Table 08812: External Trade in Goods, by Commodity Group (Two-Digit SITC), Country and Mode of Transport (Tonnes) (1989–2011). SSB.

Statistics Norway, 2013. Større avfallsmengder og mer gjenvinning (Increased Amounts of Waste and More Recycling). Oslo.

Steinberger, J., Friot, D., Jolliet, O., Erkman, S., 2009. A spatially explicit life cycle inventory of the global textile chain. The International Journal of Life Cycle Assessment 14, 443–455.

Tojo, N., Kogg, B., Kiørboe, N., Kjær, B., Aalto, K., 2012. Prevention of Textile Waste: Material Flows of Textiles in Three Nordic Countries and Suggestions on Policy Instruments. TemaNord 2012:545. Nordic Council of Ministers.

Uff Norge, 2013. Innsamling av klær og sko (Collection of Clothing and Shoes).

Ungerth, L., Carlsson, A., 2011. Vad händer sen med våra kläder? Enkätundersökning, Stockholm, Konsumentföreningen.

Van Bommel, H., 2013. ISWA product chain management project – design jeans for recycling. In: International Research Workshop on Sustainable Fashion. Copenhagen Business School, Copenhagen.

Watson, D., Kiørboe, N., Palm, D., Tekie, H., Harris, S., Ekvall, T., Lindhqvist, T., Lyng, K.-A., 2014. EPR Systems and New Business Models – Reuse and Recycling of Textiles in the Nordic Region. TemaNord 2014:539.

Woolridge, A.C., Ward, G.D., Phillips, P.S., Collins, M., Gandy, S., 2006. Life cycle assessment for reuse/recycling of donated waste textiles compared to use of virgin material: an UK energy saving perspective. Resources, Conservation and Recycling 46, 94–103.

Index

Note: Page numbers followed by "f" and "t" indicate figures and tables, respectively.

A

ABS. *See* Acrylonitrile-butadiene-styrene (ABS)
Acidification potential (AP), 309
Acrylonitrile-butadiene-styrene (ABS), 315
AFNOR. *See* Association française de normalisation (AFNOR)
Agricultural land occupation (ALO), 259–260
AHP. *See* Analytic hierarchy process (AHP)
Air handling unit (AHU), 178
ALCA. *See* Attributional life cycle assessment (ALCA)
Allocation
 of coproducts, 263–264
 sensitivity of results to allocation method, 242
ALO. *See* Agricultural land occupation (ALO)
Analogue models, 157
Analytic hierarchy process (AHP), 156–157
Animal fibres, 128
Anthropocene, 301
AP. *See* Acidification potential (AP)
Apparel, 97–98, 103–106
 cotton fabrics for, 204–207
 industry, 104
  characteristics of textiles, 104–105
  design-led approaches for effective recycling, 119
  future of recycled textiles, 119–120
  innovations in stakeholder engagement, 118–119
 processing, 248
Artificial fibre products, 128–129
Assessment matrix, 155
Association française de normalisation (AFNOR), 140
ASTM environmental standards, 135
Attributional approach, 6
Attributional life cycle assessment (ALCA), 224
 CLCA *vs.*, 224–225
Automotive components LCA, 312–315

B

Baseline study, 152
Best available techniques (BAT), 53–54, 169
Bio-composites. *See* Eco-composites
Biocapacity, 63
Biodegradable textiles, 116
 manufacture and application of, 114
Biophysical allocation, 228, 242, 249–250
Bivoltines (BV), 257
 silks, 256
Blue Angel label, 139
Blue water footprint (BWF), 259–260
Bluesign® standard, 141
BMP certified cotton, 143
Bombyx mori (*B. mori*), 256
Bottle to fibre, 108
British study, 293
Brown Lung, 129
Building sector, LCA in, 315–318
BV. *See* Bivoltines (BV)
BWF. *See* Blue water footprint (BWF)
Byssinosis, 129

C

C2C methodology. *See* Cradle-to-cradle methodology (C2C methodology)
Capital goods, 263
Carbon, 303
Carbon Care® Asia, 21
Carbon dioxide (CO_2), 5, 94–95, 97–98
 measured label, 21
Carbon dioxide equivalents (CO_2e), 221

Carbon emission (CE), 43
Carbon emissions signature (CES), 43
Carbon footprint/footprinting, 3, 4f, 10t–11t. *See also* Ecological footprint (EF); Energy footprints
  calculation
    of textile manufacturing processes, 12–16
    of textile products, 16–17, 18t–20t
  grocery shopping bags, 296–297
  key issues in calculating, 5–9
  labelling of textile products, 21–22
  methodological issues
    goal and scope, 6
    holistic assessments, 9
    impact assessment, 8
    inventory analysis, 7–8
  shopping bags, 295–296
  standards, 9–12
    for organisational assessment, 12
    for product assessment, 11–12
Carbon Reduction Label, 21, 44
Carbon Trust Standard, 43–44
Carbonfund.org, 22
Carrying capacity, 288–289
Cause-and-effect diagrams, 155
CB. *See* Crossbreeds (CB)
CBA approach. *See* Cost-benefit analysis approach (CBA approach)
CE. *See* Carbon emission (CE)
CED. *See* Cumulative energy demand (CED)
Certification, 334
CES. *See* Carbon emissions signature (CES)
Characterisations factors, 167
Checklists, 154
Chemical oxygen demand (COD), 309
China Environmental Labelling, 143
Chinese silk, 267
CLCA. *See* Consequential life cycle assessment (CLCA)
Closed-loop recycling (CLR), 106. *See also* Open-loop recycling (OLR)
  barriers to effective, 117–118
  examples of apparel companies, 115–116
  implications, 116
  LCA findings, 115
  manufacture and application of biodegradable textiles, 114
  methods for chemical recycling of polymer-based fibres, 114
  textiles and apparel context, 111–114
Closed-loop reuse, 114
Clothing disposal habits
  clothing
    disposal practices, 350
    for reuse, 360
  data for LCA calculations, 361
  differences between type of clothing, 356–357
  disposal, 347–348, 360–361
    frequency, 358–359
    methods, 354–356, 359
  end of life counts, 345–347
  interviews, 350–351
  Norwegian textile waste and reuse statistics, 349–350
  quantitative survey, 352–354
  relationship between lifespan and disposal, 348
  wardrobe study, 350–351
Clothing life cycle
  use phase, 93
    consumer use phase in LCA studies, 98–100
    quantification, 93–98
Clothing production stages, 37
CLR. *See* Closed-loop recycling (CLR)
CmiA cotton, 202–203, 211
Cocoon drying
  and cooking, 262
  and reeling, 257, 262
COD. *See* Chemical oxygen demand (COD)
Company-based carbon footprint analysis, 51
Component method, 65–66, 79
Conditioning system, 178
Consequential approach. *See* Attributional approach
Consequential life cycle assessment (CLCA), 224–225
  ALCA *vs.*, 224–225
Consideration of alternatives, 152
Consultation, 153
Consumer behaviour, 93–94, 100
Continental clothing company, 53
Conventional cotton, 202
Coop Naturaline label, 141

Corporate Social Responsibility (CSR), 159–160
Cost-benefit analysis approach (CBA approach), 158
Cotton, 128
 denim, 95–96
 polyester *vs.*, 277–278
Cotton shopping bags
 environmental concerning, 285
 LCA studies, 287–288, 287t
  carbon footprint of grocery shopping bags, 296–297
  ExcelPlas Australia, 289–292, 291t
  Finnish study, 295–296
  Nolan ITU study–plastic shopping bags–analysis, 288–289
  supermarket carrier bags, 293–295
  Zero Waste South Australia–UK study, 292–293
 manufacturing, 285, 286f
 merits, 285–287
Cotton textiles, 94–96, 195. *See also* Linen textiles; Silk; Wool textiles and clothing; Woolen textiles
 application areas, 198
 Cotton denim, 95–96
 current production scenario, 195–198
 factors influencing sustainability, 212–214
 LCA, 199, 200f, 201
  cotton fabrics for apparel, 204–207
  fibre, 201–204
  methodology, 199–201
  results discussion, 211–212
  technical cotton textiles, 207–211
  yarn, 201–204
 sustainability aspects, 198–199
 T-shirts, 94–95
 types comparison, 202f
Country-based sector analysis, 54
Cradle-to-cradle methodology (C2C methodology), 111–114
Cradle-to-gate approach, 13, 201–202
Crossbreeds (CB), 257
CSR. *See* Corporate Social Responsibility (CSR)
Cumulative energy demand (CED), 259–260

D
Decision-making, 153
Defra. *See* Department for Environment, Food and Rural Affairs (Defra)
Delphi survey, 162
Department for Environment, Food and Rural Affairs (Defra), 346
Department of Energy (DOE), 239–240
Descriptive checklists, 154
Digraphs, 155
Direct energy, 41
Directed graphs. *See* Digraphs
Disposable product, 83, 85
Disposal, 347–348
Disposition, 348
DOE. *See* Department of Energy (DOE)
Doretex spinning macro-process, 173
 STCS for, 177t
 TCIM for, 174t–176t
Drying, 93–95, 98
Dyeing process, 37–38

E
ECIs. *See* Environmental condition indicators (ECIs)
Eco-certification rules, 133, 328–329
Eco-composites, 304–306
Eco-consumption, 331
Eco-design, strategies for, 303
Eco-functional assessment, 86, 284
eco-INSTITUT, 139–140
Eco-label, 136–137
 eco-labelling certifications, 132
 in Europe, 139–142
 in World, 142–143
Eco-text standards, 138–139
Ecoindicator (EI), 66
Ecological footprint (EF), 63. *See also* Carbon footprint/footprinting; Energy footprints
 calculation of textile manufacturing plant, 67–74
  description of production process, 67–68
  dressmaking factory process flowsheet, 67f
  environmental evaluation, 68–74
 calculation of textile products, 74–77
  description of case study, 74–76
  EF results, 76–77

Ecological footprint (EF) (*Continued*)
  future trends and challenges, 78–79
  indicator, 63
  methodologies and standards, 65–66
Ecological labels, 326–327, 329
Ecological sustainability index (ESI), 213–214
Ecoproof, 140
EES. *See* Environmental evaluation system (EES)
EF. *See* Ecological footprint (EF)
Egg production, 262
EI. *See* Ecoindicator (EI); Environmental impact index (EI)
EIA. *See* Environmental impact assessment (EIA)
EIPRO. *See* Environmental Impact of Products (EIPRO)
EIS. *See* Environmental Impact Statement (EIS)
Electric energy, 37
ELV. *See* Emission limit values (ELV); End of Life Vehicles (ELV)
Emission limit values (ELV), 169
EMP. *See* Environmental management plan (EMP)
EMS. *See* Environmental management system (EMS)
End of life
  counts, 345–347
  scenario, 249, 294
End of Life Vehicles (ELV), 311–312
Energy, 31
  analysis
    in textile industry, 53–54
    from view of environmental approaches, 54
  balance approach, 41
  consumption
    contribution, 179t
    of world textile industry, 33–35
  conversion factors, 32t
  energy-based trends, 55
  units, 32t
Energy and utility management support system (EUMSS), 53–54
Energy footprints, 44–45. *See also* Carbon footprint/footprinting; Ecological footprint (EF)
  calculation of textile products, 44–47
  energy analysis
    from view of environmental approaches, 54
    in textile industry, 53–54
  energy consumption of World textile industry, 33–35
  energy types, 37–41
  future trends, 55–56
  key issues in calculating, 31–41
  methodologies and standards, 41–44
  reduction in textile and clothing products, 47–51
    supply chain, 52–53
  textile production stages, 36–37
Environment, 200–201
Environmental and social labels, 325
  classifications, 327–330, 328f
  comparison
    for textile and clothing products, 336t–338t
    in terms of LCA, 333–335
  environmental and ecological labels, 326–327
  ISO environmental labels types comparison, 329t–330t
  social labels, 327
  in textile and clothing development, 339f
  utility for consumers, 335–340
Environmental condition indicators (ECIs), 43
Environmental evaluation system (EES), 156, 162
Environmental impact assessment (EIA), 150, 152f
  illustrative example, 173–186
  review process and methods, 151–159
  SustainTex system, 159–173, 161f
  tools, 152f
Environmental impact index (EI), 213–214
Environmental Impact of Products (EIPRO), 149, 165–167
Environmental Impact Statement (EIS), 153
  review of, 153
Environmental impacts, textiles and clothing, 126–132
Environmental labels, 326–327
Environmental Management Life Cycle Assessment, 308

Environmental management plan (EMP), 153
Environmental management system (EMS), 43, 133
  ASTM environmental standards, 135
  GOTS, 135–136
  ISO 14000 family, 133–135
  NATURTEXTIL BEST, 136
  UNEP, 133
Environmental performance evaluation method (EPE method), 43
Environmental performance indicators (EPIs), 43
Environmental profile of cotton and polyester-cotton fabrics–bed sheets, 277–278
Environmentally conscious design. *See* Environmentally sustainable design
Environmentally sustainable design, 125, 128
EP. *See* Eutrophication potential (EP)
EPA. *See* U.S. Environmental Protection Agency (EPA)
EPE method. *See* Environmental performance evaluation method (EPE method)
EPIs. *See* Environmental performance indicators (EPIs)
ESI. *See* Ecological sustainability index (ESI)
EU. *See* European Union (EU)
EU Ecolabel, 137–138
EU-27, 149, 150f
EUMSS. *See* Energy and utility management support system (EUMSS)
European Union (EU), 235, 311, 326, 345
Eutrophication potential (EP), 309
Evaluation and assessment of significance, 152
ExcelPlas Australia, 289–292, 291t
Exergy analysis, 42
Expert systems, 156–157

F

Fabric wet processing, 130
Fair trade, 329–330
Farm production, 238
Farmyard manure (FYM), 256
FE. *See* Freshwater eutrophication (FE)
Fibre, 104, 201–204
Fibre-reinforced composites
  bio-derived resins, 307t
  eco-composites, 304–306
  green composites, 306
  LCA
    composites, 311–318
    environmental impact assessment methods, 308–309
    environmental impact classification factors, 309–311
    method description, 306–308
    percentage of GW impact, 313f–314f
    ReCipe Endpoint method, 310f
  material groups, 302t
  sustainable development, 301
  traditional, 303–304
Field and laboratory experimental methods, 157
Finnish study, 295–296
Firewood, 263
Flock, 233–234
Flocking. *See* Flock
Freshwater consumption, 227
Freshwater eutrophication (FE), 259–260
FU. *See* Functional unit (FU)
Functional unit (FU), 85, 94–95, 201–202, 290, 293–297
FYM. *See* Farmyard manure (FYM)

G

GDP. *See* Gross domestic product (GDP)
GEN. *See* Global Ecolabelling Network (GEN)
Genetically modified cotton, 202
Geographical information system (GIS), 156
GHG. *See* Greenhouse gas (GHG)
GIS. *See* Geographical information system (GIS)
Global Ecolabelling Network (GEN), 143, 326–327
Global Footprint Network, 66
Global Organic Textile Standard (GOTS), 135–136, 325
Global Recycled Standard (GRS), 142, 325
Global warming (GW), 312

Global warming potential (GWP), 5, 203–204, 221, 237, 241t, 243t, 293, 309
  fibres types, 203f
Good Environmental Choice Australia, 143
GOTS. *See* Global Organic Textile Standard (GOTS)
Green claim, 331
Green composites, 306
  SWOT analysis, 307f
Green consumption, 331
Green Shape, 140
Greenhouse gas (GHG), 134–135, 221
  Protocol, 9–11
Greenwashing, 331–332
Grocery shopping bags carbon footprint, 296–297
Gross domestic product (GDP), 195
GRS. *See* Global Recycled Standard (GRS)
GW. *See* Global warming (GW)
GWP. *See* Global warming potential (GWP)

H

H&M. *See* Hennes & Mauritz (H&M)
Heat energy, 41
Hennes & Mauritz (H&M), 116
Higher heating value (HHV), 263
Holistic approaches, 330
Hotspot, 93–94, 96, 98

I

Identification of main impacts, 152
ILO. *See* International Labour Organization (ILO)
ILUC. *See* Indirect land use change (ILUC)
IMO. *See* Institute for Marketecology (IMO)
Impact
  assessment, 126, 294–295
  identification methods, 154–157
  matrices, 155
  prediction methods, 152, 157
  significance evaluation methods, 158
Incineration, 293
India Eco-Mark, 143
Indian silk LCI, 259
  allocation of coproducts, 263–264
  capital goods, 263
  cocoon drying, cooking and reeling, 262
  firewood, 263
  goal and scope, 259–260
  mulberry production, 260–262
  rearing silkworms, 262
  sensitivity analysis, 264
  system boundaries, 260f
Indirect energy, 41
Indirect land use change (ILUC), 6
Informants, 354
Information, 153
Institute for Marketecology (IMO), 142
Integrated Pollution Prevention and Control (IPPC), 169
Internal assurance, 334
Internal rate of return (IRR), 172
International Labour Organization (ILO), 141
International Organization for Standardization (ISO), 199, 283, 308, 326–328
  ISO 14000 family, 133–135
  ISO 14020 family, 329
  ISO 14031 environmental standard, 43
  ISO 14044 standard, 333
Internationale Verband der Naturtextilwirtschaft (IVN), 136
Interviews, 350–351
Inventory
  analysis, 4, 4f, 293–294
  data, 7
    challenges, 228–229
IPPC. *See* Integrated Pollution Prevention and Control (IPPC)
IRR. *See* Internal rate of return (IRR)
ISO. *See* International Organization for Standardization (ISO)
Itex finishing macro-process
  STCS for, 184t
  TCIM for, 182t–183t
IVN. *See* Internationale Verband der Naturtextilwirtschaft (IVN)

J

Japan EcoMark, 142
Japan Environment Association (JEA), 142
Joint Research Centre (JRC), 149, 167
Judgement sampling, 351

K
Key performance indicators (KPIs), 162
King of fibres. *See* Cotton
Knitting, 129–130
KPIs. *See* Key performance indicators (KPIs)
KRAV label, 141

L
Labels, 21
Land occupation. *See* Land use
Land use, 227
LCA. *See* Life cycle assessment (LCA)
LCI. *See* Life cycle inventory (LCI)
LCIA. *See* Life cycle impact assessment (LCIA)
LCT. *See* Life cycle thinking (LCT)
Leopold Matrix, 155
Life cycle assessment (LCA), 3, 47, 85, 93, 103–104, 126, 132, 151, 195, 217, 255, 275, 283, 301, 330, 345. *See also* Clothing disposal habits; Fibre-reinforced composites
  cotton textiles, 199, 200f, 201
    cotton fabrics for apparel, 204–207
    fibre, 201–204
    methodology, 199–201
    results discussion, 211–212
    technical cotton textiles, 207–211
    yarn, 201–204
  environmental and social labels comparison in, 333–335
  indicators, 292
  reusability importance in, 84
  textiles and clothing, 126–132, 331–332
Life cycle impact assessment (LCIA), 5–6, 213–214, 235, 308–309
Life cycle inventory (LCI), 213–214, 235, 238, 258, 277, 308
  background data, 240
  distribution, 238–239
  end-of-life phase, 240
  production, 238
  use, 239–240
Life cycle thinking (LCT), 103–104
Life cycle–oriented labels, 333
Lifestyles of Health and Sustainability (LOHAS), 234–235
Linen textiles, 96–97. *See also* Cotton textiles; Woolen textiles
LOHAS. *See* Lifestyles of Health and Sustainability (LOHAS)
Lyocell, 114

M
Machinery, 37
MADE-BY Track&Trace system, 334
Man-made fibres, 276
  processing, 36
Manufactured fibres, 105
Material flow analysis (MFA), 151
Materials, 68
Mathematical models, 157
Matrix method, 154–155
Measurement of energy, 31
MEDCs. *See* More economically developed countries (MEDCs)
Merino, 217–218
  sheep farming, 248
Method selection in EIA, 158–159
MFA. *See* Material flow analysis (MFA)
Mitigation, 153
  solution evaluation method, 169–173
Mobile irrigation systems, 213
Monitoring and environmental auditing, 153
More economically developed countries (MEDCs), 229
Moriculture, 256–257, 264
MSW. *See* Municipal solid waste (MSW)
Mulberry leaf production, 256–257
Mulberry production, 260–262
Multi-attribute labels, 333
Multi-attribute methods, 158
Multi-criteria methods, 158
Multifunctional processes and coproducts, 228
Multivoltine (MV), 256
Municipal solid waste (MSW), 106
MV. *See* Multivoltine (MV)

N
National Environmental Policy Act (1969), 154–155
Natural and man-made textiles, 276
  LCA studies, 276
    benchmarking study on textiles, 279–280

Natural and man-made textiles (*Continued*)
cotton *vs*. polyester, 277–278
textile fibres biodiversity study, 278–279
Natural cellulosic fibres, 128
Natural fibres, 36, 105, 304, 305t
Natural gas consumption, Contribution to, 185t
NATURTEXTIL BEST, 136
Nearly Zero Energy Buildings (NZEB), 311
Negotiation, 153
Net present value (NPV), 172
Network diagram, 155
NF Environnement label, 140
Nitrogen oxides (NO_x), 94–95, 97–98
Nolan ITU study–plastic shopping bags–analysis, 288–289
Nonwoven fabric production, 233–234
Nonwoven production, 129–130
Nordic Ecolabel, 141
Norwegian textile waste and reuse statistics, 349–350
Notification, 153
NO_x. *See* Nitrogen oxides (NO_x)
NPV. *See* Net present value (NPV)
NZEB. *See* Nearly Zero Energy Buildings (NZEB)

O

OCS. *See* Organic Content Standard (OCS)
OEKO-TEX® Standard 100 plus, 139
OEKO-TEX® Standard 100, 138–139, 334–335
Öko-Tex standard, 132
OLR. *See* Open-loop recycling (OLR)
Open-loop recycling (OLR), 106–111. *See also* Closed-loop recycling (CLR)
apparel companies, 110
applications for recycled fibres, 108–109
barriers to effective, 117–118
discussion of methods for mechanically recycling textiles, 107–108
implications, 111
LCA findings, 109–110
Organic agriculture, 329
Organic Content Standard (OCS), 333
Organic cotton, 202
Overlay techniques, 156
OVERSEER model, 238

P

Participation, 153
Particulates (PMs), 94–95, 97–98
PAS2050. *See* Publically Available Standard 2050 (PAS2050)
PBS. *See* Poly(butylene succinate) (PBS)
PCRs. *See* Product Category Rules (PCRs)
PDCA cycle. *See* Plan-Do-Check-Act cycle (PDCA cycle)
PED. *See* Primary energy demand (PED)
PEF. *See* Product environmental footprint (PEF)
Pesticide application, 256
Pests and disease, 212–213
PET. *See* Polyethylene terephthalate (PET)
PHA. *See* Polyhydroxyalkanoates (PHA)
PLA. *See* Polylactic acid (PLA)
Plan-Do-Check-Act cycle (PDCA cycle), 134
PMCs. *See* Polymeric matrix composites (PMCs)
PMs. *See* Particulates (PMs)
Poly(butylene succinate) (PBS), 306
Polyester
cotton *vs*., 277–278
fibre, 128–129
garments, 97–98
Polyethylene terephthalate (PET), 105
Polyhydroxyalkanoates (PHA), 306
Polylactic acid (PLA), 116, 306
Polymer composites, 312
Polymer-based fibres, methods for chemical recycling of, 114
Polymeric matrix composites (PMCs), 303
Post-consumer
waste, 116
wool-rich knitwear, 233
Pre-consumer waste, 115–116
Prediction of impacts, 152
Primary energy demand (PED), 237
Primary recycling, 106
Process analysis, 7
Product carbon footprint, 221
Product Category Rules (PCRs), 9–11
Product environmental footprint (PEF), 3
Product/material recovery, 131
PTA. *See* Purified terephthalic acid (PTA)
Public consultation and communication, 153

Public involvement, 153, 153t
Publically Available Standard 2050 (PAS2050), 9–11
Purified terephthalic acid (PTA), 277
Purposeful sampling, 351

Q

Qualitative method, 350–351
Quantitative methods, 156
Questionnaire checklist, 154

R

Raw material
  extraction, 293
  preparation, 128–129
Raw silk, 257
Rearing silkworms, 262
Rearing waste, 263
ReCipe Endpoint method, 310f
Recycled polyethylene terephthalate (RPET), 108
Recycling, 105–106, 131–132
  textiles, 103–106, 119–120
Reduce, reuse and recycle (3R's), 83
Reducing CO_2 label, 21
Reducing technique, 131–132
Reeling industry, 267
Respondents, 354
Reusable/reusability, 83
  implications in textiles and clothing products, 84–85
  importance in LCA, 84–86
  measurement of textile products, 83
  of product, 83
  quantification of textiles and clothing products, 87–90
Reuse/reusing, 105–106, 131–132
  and recycling phases, 230
    collection of waste textiles, 232f
    data, 230
    pathways for post-consumer textiles, 231f
    for wool textiles, 231–235
RPET. *See* Recycled polyethylene terephthalate (RPET)

S

Sankey diagrams, 42, 71, 72f
Scaling checklists, 154
SCAP. *See* Sustainable Clothing Action Plan (SCAP)
Schadstoffgeprüft-Zeichen Label (SG Label), 140
Scope certificates, 325–326
Scoping, 152
SCP. *See* Sustainable consumption and production (SCP)
Screening, 152
SEC. *See* Singapore Environment Council (SEC); Specific energy consumption (SEC)
Secondary recycling, 106
SEEA. *See* System of economic and environmental accounting (SEEA)
Sensitivity analysis, 264, 267–269
Sericulture, 255–256, 259, 264
  Chinese, 267
  Indian, 263, 269–270
SG Label. *See* Schadstoffgeprüft-Zeichen Label (SG Label)
SGLS. *See* Singapore Green Label Scheme (SGLS)
Sheep (*Ovis aries*), 217, 223, 225
Shopping bags, 283–284. *See also* Cotton shopping bags
  environmental impacts, 284–285
Silk. *See also* Cotton textiles; Wool textiles and clothing
  cocoon drying and reeling, 257
  comparison with Chinese silk, 267
  constraints for improvement, 269–270
  coproduct using, 267
  environmental impact, 264–267, 265t
  Indian silk LCI, 259–264
  literature review, 257–259
  mulberry leaf production, 256–257
  opportunities for improvement, 269
  production, 255–256, 261t
  regions, 256
  sensitivity analysis, 267–269
  silkworm rearing, 257
Silkworm rearing, 257
Simple checklists, 154
Simple matrix, 155
Singapore Environment Council (SEC), 142–143
Singapore Green Label Scheme (SGLS), 142–143

Single-attribute labels, 333
SIPs. *See* Structural insulated panels (SIPs)
Skal label, 140
Soft systems methodology (SSM), 53–54
Soil fertility, 212
Specific energy consumption (SEC), 42
Spinning
  company, 173–178
  processes, 36
SSM. *See* Soft systems methodology (SSM)
Standards
  and associations for environmental management systems, 133–136
  relationship with LCA and eco-labelling certifications, 132
STCS. *See* Sustainable Textile&Clothing Scorecard (STCS)
STeP. *See* Sustainable Textile Production (STeP)
Stichting Milieukeur label, 141
Stock units (SU), 238
Strategic sampling, 351
Strengths, weaknesses, opportunities and threats analysis (SWOT analysis), 306
Structural insulated panels (SIPs), 316
SU. *See* Stock units (SU)
Sulfur dioxide (SO_2), 94–95, 97–98
supermarket carrier bags, 293–295
Sustainability, 63–64, 79
  cotton textiles, 212–214
Sustainable Clothing Action Plan (SCAP), 48
Sustainable consumption, 331
Sustainable consumption and production (SCP), 346
Sustainable development, 198
Sustainable Textile Production (STeP), 139, 325
Sustainable Textile&Clothing Scorecard (STCS), 160, 162–164, 165t
SustainTex system, 159–173, 161f
  mitigation solution evaluation method, 169–173
  Textile&Clothing Assessment Matrix, 160–169
  Textile&Clothing Mitigation Solution Database, 169, 170t–171t
SWOT analysis. *See* Strengths, weaknesses, opportunities and threats analysis (SWOT analysis)
Synthetic clothing, 97–98
Synthetic fibres, 105, 128–129
System of economic and environmental accounting (SEEA), 151

T

T-shirts, 94–95
  case study, 277
TCA. *See* Total cost assessment (TCA)
TCAM. *See* Textile&Clothing Assessment Matrix (TCAM)
TCIM. *See* Textile&Clothing Inventory Matrix (TCIM)
Tertiary recycling, 106
Textile
  carbon footprints calculation of textile manufacturing processes, 12–16
  and clothing industry, 149
  fibres, 107–108
    biodiversity study, 278–279
    types, 36
  production stages, 36–37
    energy types, 37–41
  recycling method, 347
  surface structuring processes, 37
  waste management, 131–132
Textile-based trends, 55–56
Textile&Clothing Assessment Matrix (TCAM), 160–169
  STCS, 162–164, 165t
  TCIM, 160–162, 163t
Textile&Clothing Inventory Matrix (TCIM), 160–162, 163t
Textile&Clothing Mitigation Solution Database, 169, 170t–171t
Textiles and clothing. *See also* Wool textiles and clothing
  eco-labelling and certification, 136–143
  environmental and social labels, 336t–338t, 339f
  greenwashing, 331–332
  LCA, 331–332
  LCA and environmental impacts of, 126–132
  production stage, 129–131
Thai Green label, 143

3R's. *See* Reduce, reuse and recycle (3R's)
Tier 2' model, 228–229
Tiers of costs, 172t
Total cost assessment (TCA), 172
Traditional fibre-reinforced composites, 303–304
Transactional certificates, 325–326, 334–335
Transportation scenario, 294
Tree diagrams, 155

U
U.S. Environmental Protection Agency (EPA), 83, 199
Umberto® 5.5 software, 68
UNEP. *See* United Nations Environment Programme (UNEP)
Unit of energy footprint, 31–33
United Nations Environment Programme (UNEP), 133, 328–329
United States (US), 235
Use phase impacts, 93–95

V
VAUDE company, 140
Vertical plant, 12–13
Volatile organic compounds (VOCs), 309

W
Wardrobe study, 350–351
Warehousing and distribution, 248
Waste Electrical and Electronic Equipment (WEEE), 311
Water, 213–214, 227
Weaving, 129–130
  and finishing company, 178–186
WEEE. *See* Waste Electrical and Electronic Equipment (WEEE)
Weighted matrices, 155
Wet processing, 37
White Label, 327
Winding process, 178
Wool textiles and clothing. *See also* Cotton textiles; Silk
  LCA studies, 221–224, 222t
    ALCA *vs*. CLCA, 224–225
    case study results, 240–242
    data requirements and limitations, 236–237
    environmental impact categories, 226–227
    goal, scope and system boundary, 225–226, 235–236
    GWP, 241t, 243t
    interpretation and conclusions, 248–249
    inventory data challenges, 228–229
    LCI, 238–240
    methodological and data challenges, 224
    multifunctional processes and coproducts, 228
    primary energy demand, 244t–245t
    resource use, 226–227
    reuse and recycling phases, 230–235
    sheep population, 226f
    water consumption, 246t–247t
  life cycle
    of textile producing from, 220f
    of wool, 219f
  wool textile value chain, 217–221
Wool-rich knitwear, 233
Woolen textiles, 96. *See also* Cotton textiles; Linen textiles
World Health Organisation, 212
World textile industry, energy consumption of, 33–35
World Trade Organisation (WTO), 33–34, 328–329

Y
Yarn, 201–204
Yarn spinning, 129

Z
Zero Waste South Australia–UK study, 292–293

9780081001691